MATRICES AND VECTOR SPACES

PURE AND APPLIED MATHEMATICS

A Program of Monographs, Textbooks, and Lecture Notes

MONOGRAPHS AND TEXTBOOKS IN
PURE AND APPLIED MATHEMATICS

1. *K. Yano*, Integral Formulas in Riemannian Geometry (1970)*(out of print)*
2. *S. Kobayashi*, Hyperbolic Manifolds and Holomorphic Mappings (1970) *(out of print)*
3. *V. S. Vladimirov*, Equations of Mathematical Physics (A. Jeffrey, editor; A. Littlewood, translator) (1970) *(out of print)*
4. *B. N. Pshenichnyi*, Necessary Conditions for an Extremum (L. Neustadt, translation editor; K. Makowski, translator) (1971)
5. *L. Narici, E. Beckenstein, and G. Bachman*, Functional Analysis and Valuation Theory (1971)
6. *D. S. Passman*, Infinite Group Rings (1971)
7. *L. Dornhoff*, Group Representation Theory (in two parts). Part A: Ordinary Representation Theory. Part B: Modular Representation Theory (1971, 1972)
8. *W. Boothby and G. L. Weiss (eds.)*, Symmetric Spaces: Short Courses Presented at Washington University (1972)
9. *Y. Matsushima*, Differentiable Manifolds (E. T. Kobayashi, translator) (1972)
10. *L. E. Ward, Jr.*, Topology: An Outline for a First Course (1972) *(out of print)*
11. *A. Babakhanian*, Cohomological Methods in Group Theory (1972)
12. *R. Gilmer*, Multiplicative Ideal Theory (1972)
13. *J. Yeh*, Stochastic Processes and the Wiener Integral (1973) *(out of print)*
14. *J. Barros-Neto*, Introduction to the Theory of Distributions (1973) *(out of print)*
15. *R. Larsen*, Functional Analysis: An Introduction (1973) *(out of print)*
16. *K. Yano and S. Ishihara*, Tangent and Cotangent Bundles: Differential Geometry (1973) *(out of print)*
17. *C. Procesi*, Rings with Polynomial Identities (1973)
18. *R. Hermann*, Geometry, Physics, and Systems (1973)
19. *N. R. Wallach*, Harmonic Analysis on Homogeneous Spaces (1973) *(out of print)*
20. *J. Dieudonné*, Introduction to the Theory of Formal Groups (1973)
21. *I. Vaisman*, Cohomology and Differential Forms (1973)
22. *B. -Y. Chen*, Geometry of Submanifolds (1973)
23. *M. Marcus*, Finite Dimensional Multilinear Algebra (in two parts) (1973, 1975)
24. *R. Larsen*, Banach Algebras: An Introduction (1973)
25. *R. O. Kujala and A. L. Vitter (eds.)*, Value Distribution Theory: Part A; Part B: Deficit and Bezout Estimates by Wilhelm Stoll (1973)
26. *K. B. Stolarsky*, Algebraic Numbers and Diophantine Approximation (1974)
27. *A. R. Magid*, The Separable Galois Theory of Commutative Rings (1974)
28. *B. R. McDonald*, Finite Rings with Identity (1974)
29. *J. Satake*, Linear Algebra (S. Koh, T. A. Akiba, and S. Ihara, translators) (1975)

30. *J. S. Golan*, Localization of Noncommutative Rings (1975)
31. *G. Klambauer*, Mathematical Analysis (1975)
32. *M. K. Agoston*, Algebraic Topology: A First Course (1976)
33. *K. R. Goodearl*, Ring Theory: Nonsingular Rings and Modules (1976)
34. *L. E. Mansfield*, Linear Algebra with Geometric Applications: Selected Topics (1976)
35. *N. J. Pullman*, Matrix Theory and Its Applications (1976)
36. *B. R. McDonald*, Geometric Algebra Over Local Rings (1976)
37. *C. W. Groetsch*, Generalized Inverses of Linear Operators: Representation and Approximation (1977)
38. *J. E. Kuczkowski and J. L. Gersting*, Abstract Algebra: A First Look (1977)
39. *C. O. Christenson and W. L. Voxman*, Aspects of Topology (1977)
40. *M. Nagata*, Field Theory (1977)
41. *R. L. Long*, Algebraic Number Theory (1977)
42. *W. F. Pfeffer*, Integrals and Measures (1977)
43. *R. L. Wheeden and A. Zygmund*, Measure and Integral: An Introduction to Real Analysis (1977)
44. *J. H. Curtiss*, Introduction to Functions of a Complex Variable (1978)
45. *K. Hrbacek and T. Jech*, Introduction to Set Theory (1978)
46. *W. S. Massey*, Homology and Cohomology Theory (1978)
47. *M. Marcus*, Introduction to Modern Algebra (1978)
48. *E. C. Young*, Vector and Tensor Analysis (1978)
49. *S. B. Nadler, Jr.*, Hyperspaces of Sets (1978)
50. *S. K. Segal*, Topics in Group Rings (1978)
51. *A. C. M. van Rooij*, Non-Archimedean Functional Analysis (1978)
54. *L. Corwin and R. Szczarba*, Calculus in Vector Spaces (1979)
53. *C. Sadosky*, Interpolation of Operators and Singular Integrals: An Introduction to Harmonic Analysis (1979)
54. *J. Cronin*, Differential Equations: Introduction and Quantitative Theory (1980)
55. *C. W. Groetsch*, Elements of Applicable Functional Analysis (1980)
56. *I. Vaisman*, Foundations of Three-Dimensional Euclidean Geometry (1980)
57. *H. I. Freedman*, Deterministic Mathematical Models in Population Ecology (1980)
58. *S. B. Chae*, Lebesgue Integration (1980)
59. *C. S. Rees, S. M. Shah, and C. V. Stanojević*, Theory and Applications of Fourier Analysis (1981)
60. *L. Nachbin*, Introduction to Functional Analysis: Banach Spaces and Differential Calculus (R. M. Aron, translator) (1981)
61. *G. Orzech and M. Orzech*, Plane Algebraic Curves: An Introduction Via Valuations (1981)
62. *R. Johnsonbaugh and W. E. Pfaffenberger*, Foundations of Mathematical Analysis (1981)
63. *W. L. Voxman and R. H. Goetschel*, Advanced Calculus: An Introduction to Modern Analysis (1981)
64. *L. J. Corwin and R. H. Szcarba*, Multivariable Calculus (1982)
65. *V. I. Istrătescu*, Introduction to Linear Operator Theory (1981)
66. *R. D. Järvinen*, Finite and Infinite Dimensional Linear Spaces: A Comparative Study in Algebraic and Analytic Settings (1981)

67. *J. K. Beem and P. E. Ehrlich*, Global Lorentzian Geometry (1981)
68. *D. L. Armacost*, The Structure of Locally Compact Abelian Groups (1981)
69. *J. W. Brewer and M. K. Smith, eds.*, Emmy Noether: A Tribute to Her Life and Work (1981)
70. *K. H. Kim*, Boolean Matrix Theory and Applications (1982)
71. *T. W. Wieting*, The Mathematical Theory of Chromatic Plane Ornaments (1982)
72. *D. B. Gauld*, Differential Topology: An Introduction (1982)
73. *R. L. Faber*, Foundations of Euclidean and Non-Euclidean Geometry (1983)
74. *M. Carmeli*, Statistical Theory and Random Matrices (1983)
75. *J. H. Carruth, J. A. Hildebrant, and R. J. Koch*, The Theory of Topological Semigroups (1983)
76. *R. L. Faber*, Differential Geometry and Relativity Theory: An Introduction (1983)
77. *S. Barnett*, Polynomials and Linear Control Systems (1983)
78. *G. Karpilovsky*, Commutative Group Algebras (1983)
79. *F. Van Oystaeyen and A. Verschoren*, Relative Invariants of Rings: The Commutative Theory (1983)
80. *I. Vaisman*, A First Course in Differential Geometry (1984)
81. *G. W. Swan*, Applications of Optimal Control Theory in Biomedicine (1984)
82. *T. Petrie and J. D. Randall*, Transformation Groups on Manifolds (1984)
83. *K. Goebel and S. Reich*, Uniform Convexity, Hyperbolic Geometry, and Nonexpansive Mappings (1984)
84. *T. Albu and C. Năstăsescu*, Relative Finiteness in Module Theory (1984)
85. *K. Hrbacek and T. Jech*, Introduction to Set Theory, Second Edition, Revised and Expanded (1984)
86. *F. Van Oystaeyen and A. Verschoren*, Relative Invariants of Rings: The Noncommutative Theory (1984)
87. *B. R. McDonald*, Linear Algebra Over Commutative Rings (1984)
88. *M. Namba*, Geometry of Projective Algebraic Curves (1984)
89. *G. F. Webb*, Theory of Nonlinear Age-Dependent Population Dynamics (1985)
90. *M. R. Bremner, R. V. Moody, and J. Patera*, Tables of Dominant Weight Multiplicities for Representations of Simple Lie Algebras (1985)
91. *A. E. Fekete*, Real Linear Algebra (1985)
92. *S. B. Chae*, Holomorphy and Calculus in Normed Spaces (1985)
93. *A. J. Jerri*, Introduction to Integral Equations with Applications (1985)
94. *G. Karpilovsky*, Projective Representations of Finite Groups (1985)
95. *L. Narici and E. Beckenstein*, Topological Vector Spaces (1985)
96. *J. Weeks*, The Shape of Space: How to Visualize Surfaces and Three-Dimensional Manifolds (1985)
97. *P. R. Gribik and K. O. Kortanek*, Extremal Methods of Operations Research (1985)
98. *J.-A. Chao and W. A. Woyczynski, eds.*, Probability Theory and Harmonic Analysis (1986)
99. *G. D. Crown, M. H. Fenrick, and R. J. Valenza*, Abstract Algebra (1986)
100. *J. H. Carruth, J. A. Hildebrant, and R. J. Koch*, The Theory of Topological Semigroups, Volume 2 (1986)

101. *R. S. Doran and V. A. Belfi,* Characterizations of C*-Algebras: The Gelfand-Naimark Theorems (1986)

102. *M. W. Jeter,* Mathematical Programming: An Introduction to Optimization (1986)

103. *M. Altman,* A Unified Theory of Nonlinear Operator and Evolution Equations with Applications: A New Approach to Nonlinear Partial Differential Equations (1986)

104. *A. Verschoren,* Relative Invariants of Sheaves (1987)

105. *R. A. Usmani,* Applied Linear Algebra (1987)

106. *P. Blass and J. Lang,* Zariski Surfaces and Differential Equations in Characteristic p > 0 (1987)

107. *J. A. Reneke, R. E. Fennell, and R. B. Minton.* Structured Hereditary Systems (1987)

108. *H. Busemann and B. B. Phadke,* Spaces with Distinguished Geodesics (1987)

109. *R. Harte,* Invertibility and Singularity for Bounded Linear Operators (1988).

110. *G. S. Ladde, V. Lakshmikantham, and B. G. Zhang,* Oscillation Theory of Differential Equations with Deviating Arguments (1987)

111. *L. Dudkin, I. Rabinovich, and I. Vakhutinsky,* Iterative Aggregation Theory: Mathematical Methods of Coordinating Detailed and Aggregate Problems in Large Control Systems (1987)

112. *T. Okubo,* Differential Geometry (1987)

113. *D. L. Stancl and M. L. Stancl,* Real Analysis with Point-Set Topology (1987)

114. *T. C. Gard,* Introduction to Stochastic Differential Equations (1988)

115. *S. S. Abhyankar,* Enumerative Combinatorics of Young Tableaux (1988)

116. *H. Strade and R. Farnsteiner,* Modular Lie Algebras and Their Representations (1988)

117. *J. A. Huckaba,* Commutative Rings with Zero Divisors (1988)

118. *W. D. Wallis,* Combinatorial Designs (1988)

119. *W. Więsław,* Topological Fields (1988)

120. *G. Karpilovsky,* Field Theory: Classical Foundations and Multiplicative Groups (1988)

121. *S. Caenepeel and F. Van Oystaeyen,* Brauer Groups and the Cohomology of Graded Rings (1989)

122. *W. Kozlowski,* Modular Function Spaces (1988)

123. *E. Lowen-Colebunders,* Function Classes of Cauchy Continuous Maps (1989)

124. *M. Pavel,* Fundamentals of Pattern Recognition (1989)

125. *V. Lakshmikantham, S. Leela, and A. A. Martynyuk,* Stability Analysis of Nonlinear Systems (1989)

126. *R. Sivaramakrishnan,* The Classical Theory of Arithmetic Functions (1989)

127. *N. A. Watson,* Parabolic Equations on an Infinite Strip (1989)

128. *K. J. Hastings,* Introduction to the Mathematics of Operations Research (1989)

129. *B. Fine,* Algebraic Theory of the Bianchi Groups (1989)

130. *D. N. Dikranjan, I. R. Prodanov, and L. N. Stoyanov,* Topological Groups: Characters, Dualities, and Minimal Group Topologies (1989)

Other Volumes in Preparation

MATRICES AND VECTOR SPACES

William C. Brown
Michigan State University
East Lansing, Michigan

Marcel Dekker, Inc.　　　New York • Basel • Hong Kong

Library of Congress Cataloging--in--Publication Data

Brown, William C. (William Clough)
    Matrices and vector spaces/William C. Brown.
        p. cm. -- -- (Monographs and textbooks in pure and applied
    mathematics; 145)
    Includes bibliographical references and index.
    ISBN 0-8247-8419-7 (acid-free paper)
    1. Algebras, Linear. I. Title. II. Series.
QA 184.B764 1991
512'.5-- --dc20   90-28562
                                                               CIP

This book is printed on acid-free paper.

MARCEL DEKKER, INC.
270 Madison Avenue, New York, New York 10016

Current printing (last digit):
10 9 8 7 6 5 4 3 2 1

PRINTED IN THE UNITED STATES OF AMERICA

TO KATY, JEANNE, and MELISSA

Preface

I have designed *Matrices and Vector Spaces* to be used as a textbook for a one-semester or two-quarter course in linear algebra. I have written this text for those students who have a serious interest in mathematics, and for those college instructors who want a challenging book for their students. My primary aim was to create a suitable text for an honors course at the junior level in linear algebra. This book can also be used by seniors and graduate students who find they need a more thorough grounding in the fundamentals of the subject.

The text in *Matrices and Vector Spaces* is deliberately formal in style. The idea is to present the subject matter in the same way most students will see it in advanced courses and applications. I have taken care to present clear and concise statements of all definitions and theorems in the book. Students who plan to take advanced courses in linear algebra must learn how the basic theorems in the subject are proven. I have made an extra effort to present clear, understandable proofs of all major results in the text. In particular, whenever an argument is given that uses an algorithm or a reduction process, I have tried to present several steps in the procedure rather than resort to induction. In this way, I hope to make the algorithm or reduction process clear for future use.

Most of the material in *Matrices and Vector Spaces* is presented from a matrix point of view when appropriate. The corresponding statements for linear transformations, if not given in the text, can be found in several advanced texts on the same topic. A list of references is provided for this purpose at the end of this book. All definitions, theorems, and important formulas are clearly displayed in the text with an accompanying number. Notation for specific ideas, for example, $\mathbb{R}$ for the field of real numbers, is introduced in the main body of the text as needed. A glossary of notation has been provided at the back of the book for the reader's convenience. Answers and suggestions for selected problems have also been included at the end of the text.

There is no section of this book that I would recommend the reader skip on a first reading of this text. All the material in the book is important and fundamental to advanced courses and applications. There are no esoteric topics or examples here. My advice to the reader is to begin on page 1 and read the text straight through to the end.

William C. Brown

Contents

MATRICES AND
VECTOR SPACES

I
Matrices and Linear Systems of Equations

1. MATRICES

In this book, the symbol $\mathbb{R}$ will denote the set of real numbers, $\mathbb{C}$ will denote the set of complex numbers. Of course, $\mathbb{R} \subseteq \mathbb{C}$. An $m \times n$ matrix A with entries in $\mathbb{R}$ is defined as follows.

Definition 1.1 Let m and n be positive integers. An $m \times n$ matrix A with real entries is a rectangular array of m rows and n columns of real numbers.

We will use the following notation to represent an $m \times n$ matrix A:

1.2
$$A = \begin{bmatrix} a_{11} & \cdots & a_{1n} \\ \vdots & & \vdots \\ a_{m1} & \cdots & a_{mn} \end{bmatrix}$$

For those readers who prefer a more precise definition, the $m \times n$ matrix A given in equation (1.2) is nothing but a function whose domain is the set of ordered pairs $\{(i,j) \mid 1 \leqslant i \leqslant m; \ 1 \leqslant j \leqslant n\}$ and whose range is the subset

1

$\{a_{ij}|1 \leqslant i \leqslant m;\ 1 \leqslant j \leqslant n\}$ of $\mathbb{R}$. The value of A at (i,j) is a_{ij}, and the function A has simply been identified with its range.

If a matrix A has m rows and n columns as in equation (1.2), then A is called an $m \times n$ (read m by n) matrix. If A is an $m \times n$ matrix, we will refer to $m \times n$ as the size of A and write size$(A) = m \times n$. Thus, the size of a matrix A is a pair of positive integers, the first integer giving the number of rows of A and the second integer giving the number of columns of A. When the size of A is understood from the context or is not important, we will shorten our notation and write $A = (a_{ij})$, or just A. If $A = (a_{ij})$, then a_{ij} is the entry of A appearing in the ith row and jth column of A.

An $m \times n$ matrix with complex entries has precisely the same definition as in Definition 1.1 except the word "real" is replaced with the word "complex." We need names for these two sets of matrices.

Definition 1.3 The set of all $m \times n$ matrices with real entries will be denoted by $M_{m \times n}(\mathbb{R})$. The set of all $m \times n$ matrices with complex entries will be denoted by $M_{m \times n}(\mathbb{C})$.

Since $\mathbb{R} \subseteq \mathbb{C}$, $M_{m \times n}(\mathbb{R}) \subseteq M_{m \times n}(\mathbb{C})$. In most of this chapter, it will make little difference if A has real or complex entries. Thus, we will usually drop any reference to $\mathbb{R}$ or $\mathbb{C}$ and simply write $A \in M_{m \times n}$ to indicate A is an $m \times n$ matrix.

Two matrices $A = (a_{ij})$ and $B = (b_{pq})$ are equal if A and B have the same size (say $m \times n$), and $a_{ij} = b_{ij}$ for all $i = 1, \ldots, m$ and $j = 1, \ldots, n$. If A and B are equal, we will write $A = B$. Notice that two matrices can be equal only if they have the same size. In particular, two matrices of different sizes cannot be equal no matter what their entries.

Many classes of matrices have special names in this subject matter. An $m \times 1$ matrix A is called a column vector of size m. Thus, $M_{m \times 1}$ is the set of all column vectors of size m. If A is a column vector in $M_{m \times 1}$, we will usually dispense with our double index notation and simply write A as follows.

1.4
$$A = \begin{bmatrix} x_1 \\ \vdots \\ x_m \end{bmatrix}$$

Column vectors are important enough to warrant simplifying our notation one step further. The set of all column vectors of size m and having real entries will henceforth be denoted by $\mathbb{R}^m$. Thus, $\mathbb{R}^m = M_{m \times 1}(\mathbb{R})$. The set of all column vectors of size m and having complex entries will be denoted by $\mathbb{C}^m$. Clearly, $\mathbb{R}^m \subseteq \mathbb{C}^m$.

A $1 \times n$ matrix B is called a row vector (of size n). Our notation for a typical row vector will be as follows.

1.5 $$B = [y_1 \quad \cdots \quad y_n]$$

Because of the way in which this text has been written, column vectors play a more important role than row vectors. Consequently, we will adopt no special notation other than $M_{1 \times n}$ for the set of row vectors of size n.

Now suppose $A = (a_{ij})$ is an $m \times n$ matrix. Let $\text{Row}_i(A)$ denote the ith row of A. Thus, $\text{Row}_i(A) = [a_{i1}, \ldots, a_{in}]$ is a row vector of size n. We can think of A as a set of m row vectors $\text{Row}_1(A), ,\ldots, \text{Row}_m(A)$. Let $\text{Col}_j(A)$ denote the jth column of A. Thus,

1.6
$$\text{Col}_j(A) = \begin{bmatrix} a_{1j} \\ a_{2j} \\ \vdots \\ a_{mj} \end{bmatrix}, \quad j = 1, \ldots, n$$

$\text{Col}_j(A)$ is a column vector of size m, and we can think of A as a set of n column vectors $\text{Col}_1(A), \ldots, \text{Col}_n(A)$.

The column and row vectors of a matrix A are special cases of submatrices of A. A submatrix of A is a matrix obtained from A by deleting certain rows and/or columns of A. Consider the following example.

Example 1.7 Let
$$A = \begin{bmatrix} 1 & 2 \\ 3 & 4 \end{bmatrix}$$

Then
$$[1], [2], [3], [4], \begin{bmatrix} 1 \\ 3 \end{bmatrix}, \begin{bmatrix} 2 \\ 4 \end{bmatrix} [1, 2], [3, 4], \begin{bmatrix} 1 & 2 \\ 3 & 4 \end{bmatrix}$$

is a complete list of all submatrices of A.

A partition of A is a series of horizontal and vertical lines drawn in A which divide A into various submatrices. Thus, a partition of A does not affect the entries of A. The lines only serve as a useful tool in thinking about various submatrices of A.

Example 1.8   Let

$$A = \begin{bmatrix} 1 & 2 & 3 & 4 \\ 0 & 1 & 2 & 1 \\ 0 & 1 & 3 & 2 \\ 1 & 0 & 0 & 6 \end{bmatrix}$$

Then

$$A = \left[\begin{array}{cc|cc} 1 & 2 & 3 & 4 \\ 0 & 1 & 2 & 1 \\ \hline 0 & 1 & 3 & 2 \\ 1 & 0 & 0 & 6 \end{array}\right] \quad \text{and} \quad A = \left[\begin{array}{ccc|c} 1 & 2 & 3 & 4 \\ 0 & 1 & 2 & 1 \\ 0 & 1 & 3 & 2 \\ \hline 1 & 0 & 0 & 6 \end{array}\right]$$

are two different partitions of A. In the partition on the left, A is divided into four 2×2 submatrices

$$A = \left[\begin{array}{c|c} A_1 & A_2 \\ \hline A_3 & A_4 \end{array}\right]$$

where

$$A_1 = \begin{bmatrix} 1 & 2 \\ 0 & 1 \end{bmatrix}$$

etc. In the partition on the right, A is again divided into four submatrices

$$A = \left[\begin{array}{c|c} B_1 & B_2 \\ \hline B_3 & B_4 \end{array}\right].$$

Here

$$B_1 = \begin{bmatrix} 1 & 2 & 3 \\ 0 & 1 & 2 \\ 0 & 1 & 3 \end{bmatrix}, \quad B_2 = \begin{bmatrix} 4 \\ 1 \\ 2 \end{bmatrix}$$

etc.

The two most important partitions of A are its column partition and its row partition. The column partition of A is as follows:

1.9 $A = [\mathrm{Col}_1(A) \,|\, \mathrm{Col}_2(A) \,|\, \cdots \,|\, \mathrm{Col}_n(A)]$

The row partition of A has the following form.

1.10

$$A = \begin{bmatrix} \text{Row}_1(A) \\ \hline \text{Row}_2(A) \\ \hline \vdots \\ \hline \text{Row}_m(A) \end{bmatrix}$$

We will use both of these partitions of A frequently in subsequent sections in this book.

Let $A = (a_{ij}) \in M_{m \times n}$. The matrix A is square if $m = n$. A is upper triangular if $a_{ij} = 0$ whenever $i > j$. A is lower triangular if $a_{ij} = 0$ whenever $i < j$. A is diagonal if $a_{ij} = 0$ whenever $i \neq j$. These types of matrices have the following forms when $m \leqslant n$.

1.11

$$A = \begin{bmatrix} a_{11} & \cdots & a_{1n} \\ \vdots & & \vdots \\ a_{n1} & \cdots & a_{nn} \end{bmatrix}, \quad \begin{array}{l} \text{a square matrix of} \\ \text{size } n \times n \end{array}$$

$$A = \begin{bmatrix} a_{11} & a_{12} & a_{13} & \cdots & & a_{1n} \\ 0 & a_{22} & a_{23} & \cdots & & a_{2n} \\ 0 & 0 & a_{33} & \cdots & & a_{3n} \\ \vdots & \vdots & & & & \vdots \\ 0 & \cdots & \cdots & 0 & a_{mm} & \cdots & a_{mn} \end{bmatrix}, \quad \begin{array}{l} \text{an upper triangular} \\ m \times n \text{ matrix} \end{array}$$

$$A = \begin{bmatrix} a_{11} & 0 & \cdots & \cdots & \cdots & \cdots & 0 & \cdots & 0 \\ a_{21} & a_{22} & 0 & \cdots & & \cdots & 0 & \cdots & 0 \\ a_{31} & a_{32} & a_{33} & 0 & & \cdots & 0 & \cdots & 0 \\ \vdots & \vdots & & & & & & & \vdots \\ a_{m1} & a_{m2} & \cdots & \cdots & 0 & a_{mm} & \cdots & 0 & \cdots & 0 \end{bmatrix}, \quad \begin{array}{l} \text{a lower} \\ \text{triangular} \\ m \times n \text{ matrix} \end{array}$$

$$A = \begin{bmatrix} a_{11} & 0 & \cdots & \cdots & \cdots & 0 & \cdots & 0 \\ 0 & a_{22} & 0 & \cdots & & 0 & \cdots & 0 \\ \vdots & \vdots & & & \vdots & & \vdots \\ 0 & 0 & \cdots & \cdots & a_{mm} & 0 & \cdots & 0 \end{bmatrix}, \quad \begin{array}{l} \text{a diagonal } m \times n \\ \text{matrix} \end{array}$$

In any $m \times n$ matrix $A = (a_{ij})$, the entries $a_{11}, a_{22}, \ldots, a_{rr}$ where $r = \min\{m, n\}$ will be called the diagonal entries of A. The $n \times n$ diagonal matrix which has every diagonal entry equal to 1 is called the identity matrix of size n. We will let I_n denote the identity matrix of size n. Thus

1.12

$$I_1 = [1], \quad I_2 = \begin{bmatrix} 1 & 0 \\ 0 & 1 \end{bmatrix}, \quad I_3 = \begin{bmatrix} 1 & 0 & 0 \\ 0 & 1 & 0 \\ 0 & 0 & 1 \end{bmatrix}$$

etc. When the size of I_n is clear, we will write I for I_n.

The $m \times n$ matrix all of whose entries are zero is called the zero matrix and written O. Notice that there are infinitely many zero matrices, one for each size. We use the same notation O for all of them. This will cause no real confusion in the sequel. The reader will always be able to tell the size of O from the context. Here is an example of how our notation works.

Example 1.13 Let

$$A = \left[\begin{array}{ccc|c} 1 & 2 & 3 & 0 \\ 0 & 1 & 2 & 0 \\ 0 & 1 & 3 & 0 \\ \hline 0 & 0 & 0 & 6 \end{array} \right]$$

This partition can be written as

$$A = \left[\begin{array}{c|c} B & O \\ \hline O & C \end{array} \right]$$

where

$$B = \begin{bmatrix} 1 & 2 & 3 \\ 0 & 1 & 2 \\ 0 & 1 & 3 \end{bmatrix}, \quad C = [6] \quad \text{and} \quad O = \begin{bmatrix} 0 \\ 0 \\ 0 \end{bmatrix} \quad \text{or} \quad [0 \ 0 \ 0].$$

Before closing this section, let us point out the connection between systems of linear equations and matrices. A linear system of m equations in n unknowns $x_1, \ldots, x_n$, is a system of equations of the following form.

1.14

$$
\begin{aligned}
a_{11}x_1 + a_{12}x_2 + \cdots + a_{1n}x_n &= b_1 \\
a_{21}x_1 + a_{22}x_2 + \cdots + a_{2n}x_n &= b_2 \\
\vdots \qquad\quad \vdots \qquad\qquad\quad \vdots \quad\ &\ \ \vdots \\
a_{m1}x_1 + a_{m2}x_2 + \cdots + a_{mn}x_n &= b_m
\end{aligned}
$$

In equation (1.14), the a_{ij}, $i = 1, \ldots, m$ and $j = 1, \ldots, n$ are constants either real or complex as the case may be. The b_k are also constants. If $b_1 = \cdots = b_m = 0$, then the equations in (1.14) are called a homogeneous system of (linear) equations.

The system of equations in (1.14) has two matrices associated with it which will be of interest. Namely,

1.15

$$A = \begin{bmatrix} a_{11} & \cdots & a_{1n} \\ \vdots & & \vdots \\ a_{m1} & \cdots & a_{mn} \end{bmatrix} \quad \text{and} \quad \begin{bmatrix} a_{11} & \cdots & a_{1n} & b_1 \\ \vdots & & \vdots & \vdots \\ a_{m1} & \cdots & a_{mn} & b_m \end{bmatrix}$$

The matrix A in (1.15) is called the matrix of coefficients of the linear system. The second matrix in (1.15) is called the augmented matrix of the system.

When studying a linear system such as that in (1.14), there are always three fundamental questions involved. Does the system have a solution? If the system has a solution, how many different solutions are there? How do we find all possible solutions to the system? The answers to these questions are all embedded in the two matrices listed in (1.15). We will see how to answer these questions in theorems presented later in this book.

Most readers have already had some experience with the third question posed above. The algorithm used to compute the solutions to the equations in (1.14) reduces the system of equations to an equivalent system whose augmented matrix is upper triangular. The solutions are then computed by what is called back substitution. The reduction procedure is called Gaussian elimination. We will discuss this procedure in detail in Section 4 of this chapter. For now, consider the following simple example.

Example 1.16 Find the solutions (if any) of the following system:

(a)
$$\begin{aligned} -6x - y + 13z &= 7 \\ 2x + y - 4z &= -1 \\ 8x + 6y - 12z &= 6 \end{aligned}$$

First interchange equations one and two. Then add 3 times the first equation to the second equation, and -4 times the first equation to the third equation. We get the following equivalent system of equations:

(b)
$$\begin{aligned} 2x + y - 4z &= -1 \\ 2y + z &= 4 \\ 2y + 4z &= 10 \end{aligned}$$

Now subtract the second equation from the third in (b), and obtain the following equivalent system of equations:

(c)
$$2x + y - 4z = -1$$
$$2y + z = 4$$
$$3z = 6$$

The augmented matrix of the system in (c) is the upper triangular matrix

$$\begin{bmatrix} 2 & 1 & -4 & -1 \\ 0 & 2 & 1 & 4 \\ 0 & 0 & 3 & 6 \end{bmatrix}$$

The equations in (c) are easily solved by back substitution. The third equation implies $z = 2$. Substituting $z = 2$ in the second equation gives $y = 1$. Substituting $z = 2$ and $y = 1$ in the first equation gives $x = 3$.

The important point to notice here is that the variables x, y, and z play no real role in the solution of the problem. The reduction to upper triangular form is carried out on the entries of the augmented matrix of the system. If we eliminate the variables from the example above, then the solution to (a) is computed as follows. Write down the augmented matrix of the system (a). Reduce this augmented matrix to an upper triangular matrix by suitable operations on its rows. Solve the system represented by the upper triangular matrix by back substitution. Thus, the solution to 1.16 looks like

$$\begin{bmatrix} -6 & -1 & 13 & 7 \\ 2 & 1 & -4 & -1 \\ 8 & 6 & -12 & 6 \end{bmatrix} \mapsto \begin{bmatrix} 2 & 1 & -4 & -1 \\ -6 & -1 & 13 & 7 \\ 8 & 6 & -12 & 6 \end{bmatrix}$$

$$\mapsto \begin{bmatrix} 2 & 1 & -4 & -1 \\ 0 & 2 & 1 & 4 \\ 0 & 2 & 4 & 10 \end{bmatrix}$$

$$\mapsto \begin{bmatrix} 2 & 1 & -4 & -1 \\ 0 & 2 & 1 & 4 \\ 0 & 0 & 3 & 6 \end{bmatrix}$$

Then back substitution from the last matrix gives $z = 2$, $y = 1$, $x = 3$.

We can give complete answers to all three of our questions about linear systems by studying the matrices involved. The answers require a certain amount of matrix arithmetic which is developed in the next section. We will return to systems of linear equations in Section 4.

Exercises for Section 1

1.1   Give an example of an $m \times n$ matrix A in $M_{m \times n}(\mathbb{C})$ but not in $M_{m \times n}(\mathbb{R})$.

1.2   Let $A \in M_{m \times n}$. Suppose we delete rows $i_1 < \cdots < i_r$ and columns $j_1 \cdots j_s$ from A. Here $1 \leqslant r \leqslant m$ and $1 \leqslant s \leqslant n$. Exactly what function does the resulting submatrix represent?

1.3   List all possible partitions and submatrices of

$$A = \begin{bmatrix} 1 & 2 & 3 \\ 0 & 1 & 2 \end{bmatrix}$$

1.4   Write down the 3×4 matrix A whose (i,j)th entry is given by $a_{ij} = (-1)^{i+j}/(i+j)$.

1.5   Write out what the matrices look like in (1.11) when $m \geqslant n$.

1.6   Consider the linear system

$$ax + by = c$$
$$dx + ey = f$$

Here $a, b, c, d, e, f \in \mathbb{R}$. Considering this system as the equations of two straight lines in the plane, discuss the possibilities for solutions to this system.

1.7   Contrast the situation in Exercise 1.6 with the case of two nonlinear equations in x and y, e.g., the equations of a circle and a parabola.

1.8   What system of linear equations does the augmented matrix

$$\begin{bmatrix} 1 & 2 & 0 & 1 \\ 0 & 4 & 3 & 1 \\ 0 & 0 & 0 & 0 \end{bmatrix}$$

represent? Discuss the solutions to this system.

1.9   Let C be the augmented matrix of a system of linear equations as in (1.14). The row operations we perform on C to reduce C to an upper triangular matrix are as follows:
(a) Interchange two rows of C.

(b) Add a multiple of one row of C to another row of C.
It is also convenient to include
(c) Multiply a row of C by a nonzero constant.
Show that each of these operations when applied to C produces a new matrix B which is the augmented matrix of a linear system of equations having precisely the same solutions as our original system.

1.10 Estimate how many arithmetic operations, i.e., additions, multiplications, etc., it takes to row reduce an $m \times n$ matrix A to an upper triangular matrix.

1.11 Give an example of a linear system of equations which has no solution.

1.12 Give an example of a linear system of equations which has infinitely many solutions.

1.13 In each of the linear systems listed below, write down the augmented matrix of the system, row reduce it to an upper triangular form, and compute the solutions to the system by back substitution.

(a) $2x_1 - 3x_2 = -4$ (b) $3x + y = 6$ (c) $3x - y = 5$
     $4x_1 - 2x_2 = 0$ $2x + 3y = 6$ $6x - 2y = 4$

(d) $4x - y + z = 4$ (e) $x + y - z = 1$
     $x + 2y + 3z = 5$ $2x + 3y + 2z = 2$
     $-5x + 2y - z = -3$ $4x + 5y + z = 3$

(f) $x + y - z - 2w + r = -3$
         $2x + z - 3r = 1$
         $-2x + 3y + z - 2w + 3r = 0$
         $y + z + w + r = 6$
         $x - z - r = 0$

1.14 Consider the following linear system of equations:

$$x_1 + x_2 = 6$$
$$2x_1 + ax_2 = 5$$

For what values of a does this system have a solution?

1.15 Consider the linear system

$$x_1 + x_2 + x_3 - x_4 = 1$$
$$3x_1 - x_2 + x_3 + x_4 = 2$$

(a) Find a solution to this system in which x_3 and x_4 are arbitrary.
(b) Find a solution to this system in which x_1 and x_2 are arbitrary.

(c) Is there a solution which is common to both solution sets in (a) and (b)?

1.16 Consider the system

$$ax + by = c_1$$
$$cx + dy = c_2 \qquad (*)$$

Set $\Delta = ad - bc$. Suppose $\Delta \neq 0$. Show $(*)$ has a unique solution. Find a formula for this solution.

1.17 Exhibit two 2×2 matrices A and B such that A cannot be obtained from B by any number of row operations as in Exercise 1.9.

2. MATRIX ARITHMETIC

In this section, we will develop the arithmetic of matrices. In order to facilitate our computations, we introduce the following notation. If $A \in M_{m \times n}$, then the (i, j)th entry of A will be denoted by $[A]_{ij}$. Thus, if $A = (a_{ij})$, then $[A]_{ij} = a_{ij}$ for all i and j.

Matrix arithmetic begins with the definition of addition.

Definition 2.1 Let A and B be $m \times n$ matrices. The sum $A + B$ of A and B is the $m \times n$ matrix whose (i,j)th entry is given by $[A + B]_{ij} = [A]_{ij} + [B]_{ij}$.

Thus, for all $i = 1, \ldots, m$ and $j = 1, \ldots, n$, the (i, j)th entry of $A + B$ is the sum of the (i,j)th entry of A and the (i,j)th entry of B. Notice that the sum of two matrices is defined only when both matrices have the same size. We cannot, for instance, add a 1×3 matrix to a 2×4 matrix.

Example 2.2

$$\begin{bmatrix} 1 & 2 & 3 & 4 \\ 0 & 3 & 1 & 2 \end{bmatrix} + \begin{bmatrix} 1 & -1 & 0 & 2 \\ 0 & 1 & 1 & 3 \end{bmatrix} = \begin{bmatrix} 2 & 1 & 3 & 6 \\ 0 & 4 & 2 & 5 \end{bmatrix}$$

If $A = (a_{ij})$ and $B = (b_{ij})$ are $m \times n$ matrices, then $A + B = (a_{ij} + b_{ij})$. In particular, $A, B \in M_{m \times n}(\mathbb{R})$ imply $A + B \in M_{m \times n}(\mathbb{R})$. Similarly, $A, B \in M_{m \times n}(\mathbb{C})$ imply $A + B \in M_{m \times n}(\mathbb{C})$.

Example 2.3

$$\begin{bmatrix} 1 + i & -i \\ 2 & 3i \end{bmatrix} + \begin{bmatrix} 2 & 4 + i \\ i & 2 - i \end{bmatrix} = \begin{bmatrix} 3 + i & 4 \\ 2 + i & 2 + 2i \end{bmatrix}$$

In Example 2.3, i is the imaginary number $i = \sqrt{-1}$.

Scalar multiplication or multiplying a matrix A by a constant c is defined as follows.

Definition 2.4   Let $A \in M_{m \times n}(\mathbb{R})$. If $c \in \mathbb{R}$, then cA is the $m \times n$ matrix whose (i,j)th entry is given by $[cA]_{ij} = c[A]_{ij}$.

Thus, the (i,j)th entry of cA is c times the (i,j)th entry of A. Scalar multiplication for complex matrices is defined in the same way.

Example 2.5

$$2 \begin{bmatrix} 1 & 2 & 0 & 1 \\ 4 & 1 & 2 & 3 \end{bmatrix} = \begin{bmatrix} 2 & 4 & 0 & 2 \\ 8 & 2 & 4 & 6 \end{bmatrix}, \quad (1+i) \begin{bmatrix} 2 \\ 3-i \\ i \end{bmatrix} = \begin{bmatrix} 2+2i \\ 4+2i \\ -1+i \end{bmatrix}$$

Since matrix addition and scalar multiplication are done entry by entry, the usual rules for addition and multiplication of real (or complex) numbers imply similar results for matrices.

Theorem 2.6   Let A, B, $C \in M_{m \times n}(\mathbb{R})$. Let x, $y \in \mathbb{R}$. Then

(a) $A + B = B + A$,
(b) $(A + B) + C = A + (B + C)$,
(c) $A + O = A$,
(d) $A + (-1)A = 0$,
(e) $1A = A$,
(f) $(xy)A = x(yA)$,
(g) $x(A + B) = xA + xB$,
(h) $(x + y)A = xA + yA$.

Proof.   The proofs of these eight assertions are all very easy. In each case, compute the (i,j)th entry of each side of the equation and compare the results. We illustrate this procedure by proving (g). The proofs of the rest of these statements are left as exercises.

Let $A = (a_{ij})$ and $B = (b_{ij})$. Using Definitions 2.1 and 2.4, we have $[xA + xB]_{ij} = xa_{ij} + xb_{ij}$ for all i and j. Similarly, the (i,j)th entry of $x(A + B)$ is given by $[x(A + B)]_{ij} = x(a_{ij} + b_{ij})$. Now the usual distributive law in $\mathbb{R}$ implies $xa_{ij} + xb_{ij} = x(a_{ij} + b_{ij})$. Therefore, $[xA + xB]_{ij} = [x(A + B)]_{ij}$ for all $i = 1, \ldots, m$ and $j = 1, \ldots, n$. We conclude that $xA + xB = x(A + B)$. $\qquad \Box$

Replacing $\mathbb{R}$ with $\mathbb{C}$ in Theorem 2.6, gives the corresponding theorem for complex matrices. In Chapter II, we will introduce the definition of an abstract vector space. The reader might glance at the eight conditions of Definition 1.1 in Chapter II. Theorem 2.6 says $M_{m \times n}(\mathbb{R})$ is a vector space over $\mathbb{R}$.

Before discussing matrix multiplication, let us point out some of the implications of Theorem 2.6. The statement in Theorem 2.6(a) is that matrix addition is commutative. Thus, in forming a sum of two matrices, it does not matter which comes first. The statement in Theorem 2.6(b) is that matrix addition is associative. This property allows us to drop all parentheses when dealing with sums of matrices. Addition is a binary operation, i.e. matrices are added two at a time. Thus, when adding three matrices A, B, and C, we can form $A + B$ and then $(A + B) + C$, or $B + C$ and then $A + (B + C)$. Theorem 2.6(b) implies that these two processes result in the same answer. In particular, we can drop the parentheses from the sum and write $A + B + C$. No matter which way the reader interprets this sum the answer is the same. By repeated use of Theorem 2.6(b), we can argue that any parentheses put in a sum of p matrices $A_1, \ldots, A_p$ results in the same answer. Consider the following example.

Example 2.7 Show that

$$(A_1 + A_2) + (A_3 + A_4) = A_1 + (A_2 + (A_3 + A_4)).$$

To see that this is true, set $A_1 = A$, $A_2 = B$, and $C = A_3 + A_4$. Now use Theorem 2.6(b). We have

$$(A_1 + A_2) + (A_3 + A_4) = (A + B) + C = A + (B + C)$$
$$= A_1 + (A_2 + (A_3 + A_4)).$$

The general argument is done using induction on p. See Exercise 2.19. Since any parentheses put in a sum of p matrices $A_1, \ldots, A_p$ result in the same answer, we will leave all parentheses out of sums and simply write $A_1 + \cdots + A_p$.

Example 2.8

$$\begin{bmatrix} 1 \\ 2 \\ 3 \end{bmatrix} + \begin{bmatrix} 0 \\ 1 \\ 1 \end{bmatrix} + \begin{bmatrix} 3 \\ 4 \\ 2 \end{bmatrix} + \begin{bmatrix} -6 \\ 0 \\ -1 \end{bmatrix} + \begin{bmatrix} 0 \\ 1 \\ 2 \end{bmatrix} = \begin{bmatrix} -2 \\ 8 \\ 7 \end{bmatrix}$$

The statements in Theorem 2.6(g) and (h) are called distributive laws.

Combining addition of matrices with scalar multiplication produces expressions which are called linear combinations of matrices. Thus, if $A_1, \ldots, A_p \in M_{m \times n}$ and $x_1, \ldots, x_p$ are constants from $\mathbb{R}$ (or $\mathbb{C}$), then $x_1 A_1 + \cdots + x_p A_p$ is called a linear combination of $A_1, \ldots, A_p$. The constants $x_1 \ldots, x_p$ in this linear combination are usually called the coefficients or scalars of the expression.

Example 2.9 Let

$$A_1 = \begin{bmatrix} 1 & 2 \\ 0 & 1 \end{bmatrix}, \quad A_2 = \begin{bmatrix} 1 & 0 \\ 1 & 0 \end{bmatrix}, \quad A_3 = \begin{bmatrix} -3 & 4 \\ 1 & 5 \end{bmatrix}.$$

Then

$$3A_1 + 2A_2 - 4A_3 = \begin{bmatrix} 17 & -10 \\ -2 & -17 \end{bmatrix}$$

Notice a common piece of notation in Example 2.9. We will write $-A$ for $(-1)A$. More generally, if $x \in \mathbb{R}$ is negative, we will write xA as $-|x|A$. Thus, we write

$$-4 \begin{bmatrix} -3 & 4 \\ 1 & 5 \end{bmatrix}$$

instead of

$$(-4) \begin{bmatrix} -3 & 4 \\ 1 & 5 \end{bmatrix}$$

The peculiar form that matrix multiplication takes comes directly from linear substitutions. Consider the following system of m equations in n unknowns $x_1, \ldots, x_n$.

2.10
$$a_{11}x_1 + a_{12}x_2 + \cdots + a_{1n}x_n = c_1$$
$$a_{21}x_1 + a_{22}x_2 + \cdots + a_{2n}x_n = c_2$$
$$\vdots \qquad \vdots \qquad \vdots \qquad \vdots$$
$$a_{m1}x_1 + a_{m2}x_2 + \cdots + a_{mn}x_n = c_m$$

The matrix of coefficients in (2.10) is the following $m \times n$ matrix.

2.11
$$A = \begin{bmatrix} a_{11} & \cdots & a_{1n} \\ \vdots & & \vdots \\ a_{m1} & \cdots & a_{mn} \end{bmatrix}$$

Suppose we introduce a new system of variables $y_1, \ldots, y_p$ which are related to $x_1, \ldots, x_n$ by a series of linear equations of the following form.

2.12
$$x_1 = b_{11}y_1 + b_{12}y_2 + \cdots + b_{1p}y_p$$
$$x_2 = b_{21}y_1 + b_{22}y_2 + \cdots + b_{2p}y_p$$
$$\vdots \qquad \vdots \qquad \vdots \qquad \qquad \vdots$$
$$x_n = b_{n1}y_1 + b_{n2}y_2 + \cdots + b_{np}y_p$$

The linear change of variables represented in equation (2.12) is completely determined by the $n \times p$ matrix B.

2.13
$$B = \begin{bmatrix} b_{11} & \cdots & b_{1p} \\ \vdots & & \vdots \\ b_{n1} & \cdots & b_{np} \end{bmatrix}$$

Now substitute (2.12) into (2.10). We obtain a linear system of m equations in p unknowns $y_1, \ldots, y_p$. Notice the constants $c_1, \ldots, c_m$ remain unchanged. A natural question arises here. What is the coefficient matrix of our new system of equations?

This question is easy to answer and motivates the definition of matrix multiplication. Suppose we look at the ith equation of 2.10 after the substitution. We have

2.14
$$c_i = a_{i1}x_1 + a_{i2}x_2 + \cdots + a_{in}x_n$$
$$= a_{i1}(b_{11}y_1 + b_{12}y_2 + \cdots + b_{1p}y_p)$$
$$+ a_{i2}(b_{21}y_1 + b_{22}y_2 + \cdots + b_{2p}y_p)$$
$$\vdots$$
$$+ a_{in}(b_{n1}y_1 + b_{n2}y_2 + \cdots + b_{np}y_p)$$

Now gather like terms together in (2.14). Switching to summation notation, equation (2.14) becomes

2.15
$$\left(\sum_{k=1}^{n} a_{ik}b_{k1} \right) y_1 + \left(\sum_{k=1}^{n} a_{ik}b_{k2} \right) y_2 + \cdots + \left(\sum_{k=1}^{n} a_{ik}b_{kp} \right) y_p = c_i$$

The sums in parentheses in equation (2.15) are the entries in the ith row of the new coefficient matrix. We can summarize these observations in the following theorem.

Theorem 2.16 Consider a linear system of m equations in n unknowns $x_1, \ldots, x_n$

$$a_{11}x_1 + a_{12}x_2 + \cdots + a_{1n}x_n = c_1$$
$$\vdots \qquad \vdots \qquad \qquad \vdots \qquad \vdots$$
$$a_{m1}x_1 + a_{m2}x_2 + \cdots + a_{mn}x_n = c_m \qquad (*)$$

Let

$$x_1 = b_{11}y_1 + b_{12}y_2 + \cdots + b_{1p}y_p$$
$$\vdots \qquad \vdots \qquad \vdots \qquad \qquad \vdots \qquad \qquad (**)$$
$$x_n = b_{n1}y_1 + b_{n2}y_2 + \cdots + b_{np}y_p$$

be a linear change of variables. Substituting $(**)$ into $(*)$, produces a linear system of m equations in p unknowns $y_1, \ldots, y_p$. The coefficient matrix of this new system is the following $m \times p$ matrix.

2.17
$$\begin{bmatrix} \sum\limits_{k=1}^{n} a_{1k}b_{k1} & \sum\limits_{k=1}^{n} a_{1k}b_{k2} & \cdots & \sum\limits_{k=1}^{n} a_{1k}b_{kp} \\ \vdots & \vdots & & \vdots \\ \sum\limits_{k=1}^{n} a_{mk}b_{k1} & \sum\limits_{k=1}^{n} a_{mk}b_{k2} & \cdots & \sum\limits_{k=1}^{n} a_{mk}b_{kp} \end{bmatrix}$$

Let us introduce some notation which makes the matrix listed in 2.17 a little easier to digest.

Definition 2.18 Let $\alpha = [x_1 \cdots x_n]$ be a row vector of size n. Let

$$\beta = \begin{bmatrix} y_1 \\ \vdots \\ y_n \end{bmatrix}$$

be a column vector of size n. The product $\alpha\beta$ of α and β is the number $x_1y_1 + \cdots + x_ny_n$.

If $n = 1$, then $\alpha\beta$ is just the ordinary product of two numbers. If $n = 2$ or 3, and $x_i, y_j \in \mathbb{R}$, then $\alpha\beta$ is the usual dot product of the two vectors α and β.

Let us use this definition to analyze the entries in the matrix in (2.17). Let $A = (a_{rs})$ denote the $m \times n$ coefficient matrix of $(*)$. Let $B = (b_{uv})$ denote the $n \times p$ matrix describing the substitution in $(**)$. Then the (i,j)th entry in the matrix in equation (2.17) is precisely $\text{Row}_i(A)\text{Col}_j(B)$. Hence, the $m \times p$ matrix in (2.17) can be written in the following form.

2.19
$$\begin{bmatrix} \text{Row}_1(A)\text{Col}_1(B) & \text{Row}_1(A)\text{Col}_2(B) & \cdots & \text{Row}_1(A)\text{Col}_p(B) \\ \vdots & \vdots & & \vdots \\ \text{Row}_m(A)\text{Col}_1(B) & \text{Row}_m(A)\text{Col}_2(B) & \cdots & \text{Row}_m(A)\text{Col}_p(B) \end{bmatrix}$$

The matrix product AB of A and B is defined to be the matrix given in equation (2.19). The formal definition is as follows.

Definition 2.20 Let $A \in M_{m \times n}$ and $B \in M_{n \times p}$. The product AB of A and B is the $m \times p$ matrix whose (i,j)th entry is given by $[AB]_{ij} = \text{Row}_i(A)\text{Col}_j(B)$ for all $i = 1, \ldots, m$ and $j = 1, \ldots, p$.

Thus, if

$$A = \begin{bmatrix} a_{11} & \cdots & a_{1n} \\ \vdots & & \vdots \\ a_{m1} & \cdots & a_{mn} \end{bmatrix} \quad \text{and} \quad B = \begin{bmatrix} b_{11} & \cdots & b_{1p} \\ \vdots & & \vdots \\ b_{n1} & \cdots & b_{np} \end{bmatrix}$$

then $[AB]_{ij} = a_{i1}b_{1j} + a_{i2}b_{2j} + \cdots + a_{in}b_{nj}$ for all $i = 1, \ldots, m$ and $j = 1, \ldots, p$. The reader should notice that the product of two matrices A and B is defined only when the number of columns of A is equal to the number of rows of B.

Example 2.21 Let

$$A = \begin{bmatrix} a_{11} & \cdots & a_{1n} \\ \vdots & & \vdots \\ a_{m1} & \cdots & a_{mn} \end{bmatrix}, \quad X = \begin{bmatrix} x_1 \\ \vdots \\ x_n \end{bmatrix}, \quad Z = [z_1 \quad \cdots \quad z_m]$$

$$C = \begin{bmatrix} 1 & 0 & 2 \\ 3 & 0 & 1 \end{bmatrix} \quad \text{and} \quad D = \begin{bmatrix} 1 & 4 & 1 \\ 2 & 1 & 1 \\ 3 & 0 & 1 \end{bmatrix}$$

Then

(a)
$$CD = \begin{bmatrix} 7 & 4 & 3 \\ 6 & 12 & 4 \end{bmatrix}$$

(b) The product DC is not defined since the number of columns of D is not equal to the number of rows of C.

(c)

$$AX = \begin{bmatrix} a_{11}x_1 + a_{12}x_2 + \cdots + a_{1n}x_n \\ \vdots \\ a_{m1}x_1 + a_{m2}x_2 + \cdots + a_{mn}x_n \end{bmatrix}$$

AX is a $m \times 1$ matrix.

(d)

$$ZA = [z_1 a_{11} + z_2 a_{21} + \cdots + z_m a_{m1} \quad \cdots \quad z_1 a_{1n} + z_2 a_{2n} + \cdots + z_m a_{mn}]$$

ZA is a $1 \times n$ matrix.

Using matrix multiplication, substitutions in linear systems of equations become simple. Let us again return to Theorem 2.16. Consider the following matrices.

2.22
$$A = \begin{bmatrix} a_{11} & \cdots & a_{1n} \\ \vdots & & \vdots \\ a_{m1} & \cdots & a_{mn} \end{bmatrix}, \quad B = \begin{bmatrix} b_{11} & \cdots & b_{1p} \\ \vdots & & \vdots \\ b_{n1} & \cdots & b_{np} \end{bmatrix}$$

$$C = \begin{bmatrix} c_1 \\ \vdots \\ c_m \end{bmatrix}, \quad X = \begin{bmatrix} x_1 \\ \vdots \\ x_n \end{bmatrix}, \quad Y = \begin{bmatrix} y_1 \\ \vdots \\ y_p \end{bmatrix}$$

The linear system (*) in Theorem 2.16 can be written succinctly as $AX = C$. The substitution in (**) is $X = BY$. The matrix in equation (2.17) is AB. Hence, Theorem 2.16 can be rewritten as follows.

Theorem 2.23 The linear system of equations $AX = C$ becomes $(AB)Y = C$ under the linear substitution $X = BY$.

In particular, the coefficient matrix A of $AX = C$ becomes AB when substituting BY for X.

Having introduced matrix multiplication, the next order of business is to discuss the basic arithmetic theorems that multiplication satisfies.

Theorem 2.24 Let A, $D \in M_{m \times n}$, $B \in M_{n \times p}$, $C \in M_{p \times q}$, and $E \in M_{r \times m}$. Then the following identities are true:

(a) $(AB)C = A(BC)$,
(b) $(A + D)B = AB + DB$,

(c) $E(A + D) = EA + ED$,
(d) $OA = O$,
(e) $I_m A = AI_n = A$,
(f) $c(AB) = (cA)B = A(cB)$ for any constant c.

Proof. It is instructive to check that the matrices here all have the correct sizes so that the operations indicated in (a)–(f) are well defined. For instance, in (a), the number of columns of A is equal to the number of rows of B. Hence, AB is a well-defined $m \times p$ matrix. Since C has p rows, $(AB)C$ is a well-defined $m \times q$ matrix. Similarly, $A(BC)$ is a well-defined $m \times q$ matrix. Thus, both sides of (a) have the same size.

The proofs of the six assertions in 2.24 are all similar in nature. Carefully compute the (i, j)th entry of each side of the given equation and compare the results. We will prove (a) and (b) and leave the remaining assertions to the exercises.

(a) Let $A = (a_{xy})$, $B = (b_{rs})$, and $C = (c_{tu})$. Then for every $i = 1, \ldots, m$ and $j = 1, \ldots, q$, we have

$$[(AB)C]_{ij} = \sum_{k=1}^{p} [AB]_{ik} c_{kj} = \sum_{k=1}^{p} \left(\sum_{r=1}^{n} a_{ir} b_{rk} \right) c_{kj}$$

$$= \sum_{k=1}^{p} \sum_{r=1}^{n} a_{ir} b_{rk} c_{kj} = \sum_{r=1}^{n} \sum_{k=1}^{p} a_{ir} b_{rk} c_{kj}$$

$$= \sum_{r=1}^{n} a_{ir} \left(\sum_{k=1}^{p} b_{rk} c_{kj} \right) = \sum_{r=1}^{n} a_{ir} [BC]_{rj} = [A(BC)]_{ij}.$$

We conclude $(AB)C = A(BC)$.

(b) Let $A = (a_{xy})$, $D = (a'_{xy})$, and $B = (b_{rs})$. Then for all $i = 1, \ldots, m$ and $j = 1, \ldots, p$, we have

$$[(A + D)B]_{ij} = \sum_{k=1}^{n} [A + D]_{ik} b_{kj} = \sum_{k=1}^{n} (a_{ik} + a'_{ik}) b_{kj}$$

$$= \sum_{k=1}^{n} a_{ik} b_{kj} + \sum_{k=1}^{n} a'_{ik} b_{kj} = [AB]_{ij} + [DB]_{ij} = [AB + DB]_{ij}.$$

We conclude $(A + D)B = AB + DB$. $\square$

The assertion in Theorem 2.24(a) is that matrix multiplication is associative. Hence, we can make the same comments concerning the use of parentheses in products as we made for sums. If $A_1, \ldots, A_p$ are matrices for which $A_i A_{i+1}$ is defined, then we can write $A_1 A_2 \cdots A_p$ for the product of these matrices. Any insertion of parentheses in $A_1 A_2 \cdots A_p$ will result in the same answer.

One special case is worth mentioning here. Suppose A is a square matrix of size n. Then for any positive integer k, the product $AA \cdots A$ (k factors) is well defined. We will abbreviate this product by writing A^k. Thus, $A^1 = A$, $A^2 = AA$, $A^3 = AAA$, etc. It will also be useful to set $A^0 = I_n$. The reader can easily check that the usual laws of exponents $A^p A^q = A^{p+q}$ and $(A^p)^q = A^{pq}$ are valid for all nonnegative integers p and q.

Another comment about Theorem 2.24 worth emphasizing here is that there is no analog of 2.6(a). Matrix multiplication is not commutative. If $A \in M_{m \times n}$ and $B \in M_{n \times p}$, then AB is defined. But in order for BA to be defined, we must have $p = m$ which may or may not be the case. The matrices C and D in Example 2.21 are just such an example. CD is not equal to DC because DC does not exist. Even if $p = m$, so that both AB and BA are well-defined products, it need not happen that $AB = BA$. Consider the following example.

Example 2.25 Let

$$A = \begin{bmatrix} 1 & 2 \\ 0 & 1 \end{bmatrix} \quad \text{and} \quad B = \begin{bmatrix} 1 & 1 \\ 2 & 1 \end{bmatrix}$$

Then

$$AB = \begin{bmatrix} 5 & 3 \\ 2 & 1 \end{bmatrix} \quad \text{and} \quad BA = \begin{bmatrix} 1 & 3 \\ 2 & 5 \end{bmatrix}$$

Thus, $AB \neq BA$.

Extra care must be taken when performing matrix calculations because multiplication is not commutative. One must not interchange adjacent factors in a product unless these factors commute.

Example 2.26 Let P, A, X, and B be $n \times n$ matrices. Suppose $AX = B$. Multiply both sides of this equation on the left with P and obtain $PAX = PB$. Multiply on the right by P and obtain $AXP = BP$. The two equations $PAX = PB$ and $AXP = BP$ are in general different from each other.

Another major difference between ordinary arithmetic in $\mathbb{R}$ (or $\mathbb{C}$) and matrix arithmetic is that the cancellation law does not hold for matrices. If x, y, $z \in \mathbb{R}$, with $x \neq 0$ and $xy = xz$, then we can cancel x and conclude $y = z$. The corresponding statement for matrices is not true in general.

Example 2.27 Let

$$A = \begin{bmatrix} 1 & 0 \\ 0 & 0 \end{bmatrix}, \quad B = \begin{bmatrix} 0 & 0 \\ 1 & 0 \end{bmatrix}, \quad C = \begin{bmatrix} 0 & 0 \\ 0 & 1 \end{bmatrix}$$

Then $A \neq O$. Also, $AB = AC$, but $B \neq C$.

Because the cancellation law does not hold for matrices, special care must be exercised when dealing with matrix equations. Suppose, for example, we want to find all X for which $AX = AC$. Here A, X, and C are matrices for which these products are all defined. Certainly, $X = C$ is one solution to this equation. However, as Example 2.27 indicates, C may not be the only solution to $AX = AC$.

There is one important case when A can be cancelled from the equation $AX = AC$. Namely, when A is invertible.

Definition 2.28 A square matrix $A \in M_{n \times n}$ is invertible or nonsingular if there exists a square matrix $B \in M_{n \times n}$ such that $AB = BA = I_n$.

Clearly, a 1×1 matrix $[x]$ is invertible if and only if $x \neq 0$. Then $[x][1/x] = [1/x][x] = [1]$. If $n \geqslant 2$, then a square matrix A of size $n \times n$ need not be invertible.

Example 2.29 Let

$$A = \begin{bmatrix} 2 & 5 \\ 1 & 3 \end{bmatrix} \quad \text{and} \quad D = \begin{bmatrix} 1 & 0 \\ 0 & 0 \end{bmatrix}$$

A is invertible since

$$\begin{bmatrix} 2 & 5 \\ 1 & 3 \end{bmatrix}\begin{bmatrix} 3 & -5 \\ -1 & 2 \end{bmatrix} = \begin{bmatrix} 3 & -5 \\ -1 & 2 \end{bmatrix}\begin{bmatrix} 2 & 5 \\ 1 & 3 \end{bmatrix} = \begin{bmatrix} 1 & 0 \\ 0 & 1 \end{bmatrix}$$

D is not invertible since for any 2×2 matrix

$$C = \begin{bmatrix} a & b \\ c & d \end{bmatrix} \quad CD = \begin{bmatrix} a & 0 \\ c & 0 \end{bmatrix} \neq I_2$$

If A is invertible, and $AB = BA = I_n$ for some matrix B, then B is the unique $n \times n$ matrix with this property. Suppose $C \in M_{n \times n}$ and $AC = CA = I_n$. Then using Theorem 2.24, we have $B = BI_n = B(AC) = (BA)C = I_nC = C$. When A is invertible, the unique matrix B for which $AB = BA = I_n$ is called the inverse of A. The inverse of A is denoted by A^{-1}. Thus, in Example 2.29,

$$A^{-1} = \begin{bmatrix} 3 & -5 \\ -1 & 2 \end{bmatrix}$$

The matrix

$$\begin{bmatrix} 1 & 0 \\ 0 & 0 \end{bmatrix}$$

is not invertible. Notice that the zero matrix O is never invertible, and the identity matrix I_n is invertible with $(I_n)^{-1} = I_n$.

Two important facts about invertible matrices are given in the next theorem.

Theorem 2.30   Let A, $B \in M_{n \times n}$ and X, $Y \in M_{n \times p}$.

(a)   If A and B are invertible, then AB is also invertible. Furthermore, $(AB)^{-1} = B^{-1}A^{-1}$.
(b)   If A is invertible and $AX = AY$, then $X = Y$.

Proof.   (a) If A and B are invertible, then A^{-1} and B^{-1} both exist. $B^{-1}A^{-1} \in M_{n \times n}$ and $(B^{-1}A^{-1})(AB) = B^{-1}A^{-1}AB = B^{-1}I_nB = B^{-1}B = I_n$. Similarly, $(AB)(B^{-1}A^{-1}) = I_n$. Thus, AB is invertible with inverse $B^{-1}A^{-1}$.

(b) If A is invertible, and $AX = AY$, then multiply both sides of this equation on the left by A^{-1}. We get $X = Y$.   □

Thus, the cancellation law holds if A is invertible. Theorem 2.30 has an interesting corollary for systems of linear equations.

Corollary 2.31   Consider the following system of n equations in n unknowns $x_1 \ldots, x_n$

$$a_{11}x_1 + a_{12}x_2 + \cdots + a_{1n}x_n = c_1$$
$$\vdots \qquad \vdots \qquad\qquad \vdots \quad \vdots \qquad\qquad (*)$$
$$a_{n1}x_1 + a_{n2}x_2 + \cdots + a_{nn}x_n = c_n$$

Set

$$C = \begin{bmatrix} c_1 \\ \vdots \\ c_n \end{bmatrix}$$

If the coefficient matrix $A = (a_{ij})$ of (∗) is invertible, then (∗) has precisely one solution $X = A^{-1}C$.

Proof. Let

$$X = \begin{bmatrix} x_1 \\ \vdots \\ x_n \end{bmatrix}$$

Then (∗) can be written as $AX = C$. If A is invertible, then multiplying on the left by A^{-1}, gives the solution $X = A^{-1}C$. Suppose Y is a second solution to (∗). Then $AY = C$. In particular, $AY = A(A^{-1}C)$. Theorem 2.30(b) then implies $Y = A^{-1}C$. Hence, $A^{-1}C$ is the unique solution to (∗).

□

There are two important conclusions from Corollary 2.31. If A is invertible, then the linear system $AX = C$ always has a solution $A^{-1}C$. Furthermore, $A^{-1}C$ is the only solution to $AX = C$.

Example 2.32 Consider the linear system

$$\begin{aligned} 2x + 5y &= 1 \\ x + 3y &= 2 \end{aligned} \tag{∗}$$

The coefficient matrix of (∗) is

$$A = \begin{bmatrix} 2 & 5 \\ 1 & 3 \end{bmatrix}.$$

We had seen in Example 2.29 that A is invertible with inverse

$$A^{-1} = \begin{bmatrix} 3 & -5 \\ -1 & 2 \end{bmatrix}$$

Thus, the unique solution to (∗) is

$$\begin{bmatrix} x \\ y \end{bmatrix} = \begin{bmatrix} 3 & -5 \\ -1 & 2 \end{bmatrix}\begin{bmatrix} 1 \\ 2 \end{bmatrix} = \begin{bmatrix} -7 \\ 3 \end{bmatrix}$$

Thus, $x = -7$ and $y = 3$.

Let us finish this already long section with one last idea which will be useful in subsequent sections.

Definition 2.33 Let $A \in M_{m \times n}$. The transpose A^t of A is the $n \times m$ matrix whose (i,j)th entry is given by $[A^t]_{ij} = [A]_{ji}$ for all $i = 1, \ldots, n$ and $j = 1, \ldots, m$.

Thus the transpose of A is the matrix whose rows are the columns of A.

Example 2.34 Let

$$A = \begin{bmatrix} 1 & 2 & 0 \\ 6 & 2 & 1 \end{bmatrix}, \quad B = \begin{bmatrix} x_1 \\ \vdots \\ x_n \end{bmatrix}, \quad C = [y_1 \quad \cdots \quad y_m]$$

Then

$$A^t = \begin{bmatrix} 1 & 6 \\ 2 & 2 \\ 0 & 1 \end{bmatrix}, \quad B^t = [x_1 \quad \cdots \quad x_n], \quad C^t = \begin{bmatrix} y_1 \\ \vdots \\ y_m \end{bmatrix}$$

Notice that the transpose of a row vector of size n is a column vector of size n and vice versa. The transpose notation is particularly useful to textbook writers. When we wish to indicate that B is a column vector of size n and save space in writing B, we can write $B = [x_1 \quad \cdots \quad x_n]^t$. We will use this notation whenever it is convenient. The transpose has important mathematical uses which will become apparent later.

The following theorem describes how the transpose relates to some of the other operations we have been discussing.

Theorem 2.35 Let $A, C \in M_{m \times n}$ and $B \in M_{n \times p}$. Then

(a) $(A + C)^t = A^t + C^t$.
(b) $(AB)^t = B^t A^t$.
(c) $(A^t)^t = A$.
(d) If A is invertible, then A^t is also invertible. In this case, we have $(A^t)^{-1} = (A^{-1})^t$.

Proof. (a) and (c) follow directly from the definitions. Let us consider (b).

For all $i = 1, \ldots, p$ and $j = 1, \ldots, m$, we have

$$[(AB)^t]_{ij} = [AB]_{ji} = \sum_{k=1}^{n} [A]_{jk}[B]_{ki}$$

$$= \sum_{k=1}^{n} [B]_{ki}[A]_{jk} = \sum_{k=1}^{n} [B^t]_{ik}[A^t]_{kj} = [B^t A^t]_{ij}.$$

Hence $(AB)^t = B^t A^t$ and the proof of (b) is complete.

(d) Suppose A is invertible. Then necessarily $m = n$. From (b), we have $A^t(A^{-1})^t = (A^{-1}A)^t = I_n^t = I_n$. Similarly, $(A^{-1})^t A^t = I_n$. We conclude that A^t is invertible with inverse $(A^{-1})^t$. ☐

If $A \in M_{m \times n}(\mathbb{C})$, then the conjugate transpose (or Hermitian conjugate) A^* of A is also important. Before introducing A^*, recall the definition of the conjugate of a complex number z. If $z = a + bi$, then the conjugate $\bar{z}$ of z is the complex number $\bar{z} = a - bi$. Here $a, b \in \mathbb{R}$ and $i = \sqrt{-1}$. The definition of the conjugate transpose of A is as follows.

Definition 2.36 Let $A \in M_{m \times n}(\mathbb{C})$. The conjugate transpose A^* of A is the $n \times m$ matrix whose (i, j)th entry is given by $[A^*]_{ij} = \overline{[A]_{ji}}$ for all $i = 1, \ldots, n$ and $j = 1, \ldots, m$.

Thus, to form A^*, transpose A and conjugate the entries of A^t.

Example 2.37 Let

$$A = \begin{bmatrix} 3 & 1 & 1-i \\ i & 2 & 1+i \end{bmatrix}$$

Then

$$A^* = \begin{bmatrix} 3 & -i \\ 1 & 2 \\ 1+i & 1-i \end{bmatrix}$$

Since the conjugate of any real number is itself, clearly $A^* = A^t$ for any $A \in M_{m \times n}(\mathbb{R})$. Theorem 2.35 remains true with t replaced by $*$.

Theorem 2.38 Let $A, C \in M_{m \times n}(\mathbb{C})$ and $B \in M_{n \times p}(\mathbb{C})$. Then

(a) $(A + C)^* = A^* + C^*$.

(b) $(AB)^* = B^* A^*$.

(c) $(A^*)^* = A$.

(d) If A is invertible, then A^* is invertible. Furthermore, $(A^*)^{-1} = (A^{-1})^*$.

The proofs of these assertions are left as exercises at the end of this section.

A square matrix A such that $A = A^t$ is called a symmetric matrix. Clearly, A is symmetric if and only if $[A]_{ij} = [A]_{ji}$ for all i, j. Notice that any (square) diagonal matrix, e.g., O or I_n, is symmetric. A square matrix A for which $A^t = -A$ is called a skew-symmetric matrix. The matrix

$$\begin{bmatrix} 0 & -1 \\ 1 & 0 \end{bmatrix}$$

is skew-symmetric.

Finally, a matrix A with complex entries is said to be Hermitian if $A = A^*$. For example

$$\begin{bmatrix} 2 & i & 1+i \\ -i & 3 & 2+i \\ 1-i & 2-i & 4 \end{bmatrix}$$

is Hermitian.

Exercises for Section 2

2.1 Let

$$A = \begin{bmatrix} 1 & 2 \\ 0 & 1 \end{bmatrix}, \quad B = \begin{bmatrix} 2 & -1 \\ -1 & 3 \end{bmatrix}$$

$$C = \begin{bmatrix} 1 & 4 \\ 6 & 2 \end{bmatrix}, \quad D = \begin{bmatrix} 0 & 1 \\ -1 & 3 \end{bmatrix}$$

Compute the following algebraic expressions:

(a) $2A + 3B - 4C + D$,

(b) $AB - CD$,

(c) $BA - DC$,

(d) $2A^2 - 3BD + D^2$,

(e) $ABCD$,

(f) $2AB^t + D^t B$.

2.2 Prove assertions (a)–(f) and (h) in Theorem 2.6.

2.3 Discuss when $A^2 - B^2 = (A - B)(A + B)$ for two square matrices A and B of the same size.

2.4 Suppose A is invertible. Set $A^{-k} = (A^{-1})^k$ for any positive integer k. Show $A^p A^q = A^{p+q}$ and $(A^p)^q = A^{pq}$ for any integers p, q.

2.5 Show that the column vector $\alpha = [1 \quad 2 \quad 4]^t$ is not any linear combination of $\beta = [1 \quad 0 \quad 1]^t$ and $\delta = [1 \quad -1 \quad 1]^t$.

2.6 Check that all matrices listed in Theorem 2.24 have the correct sizes so that the indicated computations are well-defined.

2.7 Prove the remaining assertions (c)–(f) in Theorem 2.24.

2.8 Let

$$A = \begin{bmatrix} 1 & 0 \\ 0 & 0 \end{bmatrix}$$

Find all $X \in M_{2 \times 2}$ such that $AX = O$.

2.9 Let

$$A = \begin{bmatrix} a & b \\ c & d \end{bmatrix}$$

Set $\Delta = ad - bc$. If $\Delta \neq 0$, show that A is invertible with inverse

$$A^{-1} = \Delta^{-1} \begin{bmatrix} d & -b \\ -c & a \end{bmatrix}$$

2.10 Use the results in Exercise 2.9 to solve the following systems of equations:

(a) $2x - 3y = 1$
    $x + y = 2$

(b) $7x - 5y = 12$
    $4x + 2y = 10$

(c) $x + y - z = 2$
    $2x + 3y + z = 3$
    $x + y + z = 4$

2.11 When is a diagonal matrix invertible and what is its inverse?

2.12 Show that the sum of two nonsingular matrices need not be nonsingular.

2.13 For any z, $w \in \mathbb{C}$, show that the following formulas are true:

(a) $\overline{z + w} = \bar{z} + \bar{w}$,

(b) $\overline{zw} = \bar{z}\bar{w}$,

(c) $\bar{\bar{z}} = z$,

(d) $z\bar{z} = a^2 + b^2$ where $z = a + bi$.

Use these formulas to prove Theorem 2.38.

2.14   Show that any square matrix A is the sum of a symmetric and a skew-symmetric matrix. (Hint: consider $(A + A^t)/2$.)

2.15   Let $A, B \in M_{n \times n}$. Suppose A and B are both upper triangular. Show that AB is also upper triangular. Is the corresponding assertion true for lower triangular matrices? How about for diagonal matrices?

2.16   Let $A \in M_{m \times n}$. Show that both AA^t and A^tA are symmetric.

2.17   A square matrix A is said to be idempotent if $A^2 = A$.

(a) Which of the following four matrices listed below are idempotent?

$$E_{11} = \begin{bmatrix} 1 & 0 \\ 0 & 0 \end{bmatrix}, \quad E_{12} = \begin{bmatrix} 0 & 1 \\ 0 & 0 \end{bmatrix},$$

$$E_{21} = \begin{bmatrix} 0 & 0 \\ 1 & 0 \end{bmatrix}, \quad E_{22} = \begin{bmatrix} 0 & 0 \\ 0 & 1 \end{bmatrix}$$

(b) Use these four matrices to determine all idempotent matrices in $M_{2 \times 2}$.

2.18   A square matrix A is said to be nilpotent if $A^k = 0$ for some $k > 0$. What lower triangular matrices are nilpotent?

2.19   Show that any parentheses placed in the sum $A_1 + \cdots + A_p$ gives the same answer. Here $A_1, \ldots, A_p$ are $m \times n$ matrices. Give a similar statement and proof for products.

3.   THE MULTIPLICATION OF PARTITIONED MATRICES

In this section, we will discuss how to multiply matrices together by blocks. This technique turns up in many diverse matrix computations, and, consequently, is well worth mastering. Let us begin with a simple example.

Example 3.1   Suppose

$$A = \begin{bmatrix} 1 & 2 & 3 \\ 0 & 1 & 2 \end{bmatrix}, \quad B = \begin{bmatrix} 1 & -1 \\ 2 & 2 \\ 0 & 4 \end{bmatrix}$$

A is partitioned into submatrices

$$A_1 = \begin{bmatrix} 1 & 2 \\ 0 & 1 \end{bmatrix} \quad \text{and} \quad A_2 = \begin{bmatrix} 3 \\ 2 \end{bmatrix}$$

and B into submatrices

$$B_1 = \begin{bmatrix} 1 & -1 \\ 2 & 2 \end{bmatrix} \quad \text{and} \quad B_2 = [0 \quad 4].$$

Thus, the product of A and B can be written in the form

$$AB = [A_1 \mid A_2] \begin{bmatrix} B_1 \\ B_2 \end{bmatrix}$$

The sum $A_1 B_1 + A_2 B_2$ is well defined, and the reader can easily check that

$$AB = \begin{bmatrix} 5 & 15 \\ 2 & 10 \end{bmatrix} = \begin{bmatrix} 5 & 3 \\ 2 & 2 \end{bmatrix} + \begin{bmatrix} 0 & 12 \\ 0 & 8 \end{bmatrix} = A_1 B_1 + A_2 B_2$$

Thus, if we treat A as a 1×2 matrix with entries A_1 and A_2 and B as a 2×1 matrix with entries B_1 and B_2, and multiply these matrices together (getting $A_1 B_1 + A_2 B_2$), we get the correct answer. This process is called block multiplication.

In this section, we will explain carefully why block multiplication works. Remember that a product of two matrices A and B is defined only when the number of columns of A is equal to the number of rows of B. This fact allows only certain types of partitions for block multiplication. We will build up to the main theorem by examining some special cases.

We begin with a row vector $\alpha = [x_1 \cdots x_n]$ of size n and a column vector $\beta = [y_1 \cdots y_n]^t$ of size n. Suppose we choose an arbitrary partition of α as follows.

3.2 $$\alpha = [A_1 \mid A_2 \mid \cdots \mid A_k]$$

Each A_i in equation (3.2) is a row vector in its own right. If the size of A_i is n_i, then $n_1 + \cdots + n_k = n$, and $A_1 = [x_1 \cdots x_{n_1}]$, $A_2 = [x_{n_1+1} \cdots x_{n_1+n_2}]$, etc.

Now partition β in exactly the same way.

3.3 $$\beta = \begin{bmatrix} B_1 \\ \vdots \\ B_k \end{bmatrix}$$

Each B_i here is a column vector of size n_i for $i = 1, \ldots, k$. Thus, $B_1 = [y_1 \cdots y_{n_1}]^t$, $B_2 = [y_{n_1+1} \cdots y_{n_1+n_2}]^t$, etc. In particular, each product $A_1 B_1, \ldots, A_k B_k$ is well defined. The product of α and β can be written as $\alpha\beta = \sum_{i=1}^{n} x_i y_i = \sum_{i=1}^{n_1} x_i y_i + \sum_{i=n_1+1}^{n_1+n_2} x_i y_i + \cdots = A_1 B_1 + \cdots + A_k B_k$. Thus, we have established the following identity.

3.4
$$[A_1 | \cdots | A_k] \begin{bmatrix} B_1 \\ \vdots \\ B_k \end{bmatrix} = \sum_{i=1}^{k} A_i B_i.$$

Equation (3.4) holds for row vectors $A_1, \ldots, A_k$ and column vectors $B_1, \ldots, B_k$ for which size $(A_i) =$ size(B_i) for all $i = 1, \ldots, k$. This equation is the simplest example of block multiplication of matrices.

We next discuss a natural generalization of (3.4). Let $A \in M_{m \times n}$ and $B \in M_{n \times p}$. Suppose A and B are partitioned into submatrices as follows.

3.5
$$A = [A_1 | \cdots | A_k] \quad \text{and} \quad B = \begin{bmatrix} B_1 \\ \vdots \\ B_k \end{bmatrix}$$

In equation (3.5), each A_i is an $m \times n_i$ matrix. Each B_i is an $n_i \times p$ matrix. In particular, $n_1 + \cdots + n_k = n$ and each product $A_i B_i$, $i = 1, \ldots, k$, is a well-defined $m \times p$ matrix. Equation (3.4) is still valid in this more general setting, i.e., $AB = A_1 B_1 + \cdots + A_k B_k$. To see this, examine the (i,j)th entry of AB and use equation (3.4). For each $i = 1, \ldots, m$ and $j = 1, \ldots, p$,

$$[AB]_{ij} = \text{Row}_i(A)\text{Col}_j(B)$$

$$= [\text{Row}_i(A_1) | \cdots | \text{Row}_i(A_k)] \begin{bmatrix} \text{Col}_j(B_1) \\ \vdots \\ \text{Col}_j(B_k) \end{bmatrix}$$

$$= \sum_{r=1}^{k} \text{Row}_i(A_r)\text{Col}_j(B_r) = \sum_{r=1}^{k} [A_r B_r]_{ij} \quad \text{(by 3.4)}$$

$$= \left[\sum_{r=1}^{k} A_r B_r \right]_{ij}.$$

We conclude that $AB = \sum_{r=1}^{k} A_r B_r$. Let us summarize what we have proven so far in the following equation.

3.6

$$[A_1|\cdots|A_k]\begin{bmatrix} B_1 \\ \hline \vdots \\ \hline B_k \end{bmatrix} = \sum_{r=1}^{k} A_r B_r$$

Equation (3.6) holds for matrices $A_1, \ldots, A_k$ of sizes $m \times n_1, \ldots, m \times n_k$ respectively, and matrices $B_1, \ldots, B_k$ of sizes $n_1 \times p, \ldots, n_k \times p$, respectively.

We are now ready to discuss the general situation. Suppose $A \in M_{m \times n}$ and $B \in M_{n \times p}$. Suppose these two matrices have been partitioned as follows.

3.7

$$A = \begin{bmatrix} A_{11} & A_{12} & \cdots & A_{1k} \\ A_{21} & A_{22} & \cdots & A_{2k} \\ \vdots & \vdots & & \vdots \\ A_{r1} & A_{r2} & \cdots & A_{rk} \end{bmatrix}, \quad B = \begin{bmatrix} B_{11} & B_{12} & \cdots & B_{1t} \\ B_{21} & B_{22} & \cdots & B_{2t} \\ \vdots & \vdots & & \vdots \\ B_{k1} & B_{k2} & \cdots & B_{kt} \end{bmatrix}$$

In equation (3.7), each A_{ij} is an $m_i \times n_j$ matrix. Thus, $m_1 + \cdots + m_r = m$ and $n_1 + \cdots + n_k = n$. Each B_{jl} is an $n_j \times p_l$ matrix. So, $p_1 + \cdots + p_t = p$. Notice that for each $i = 1, \ldots, r$ and $j = 1, \ldots, t$, the product

3.8

$$[A_{i1}|\cdots|A_{ik}]\begin{bmatrix} B_{1j} \\ \vdots \\ B_{kj} \end{bmatrix}$$

is a well-defined $m_i \times p_j$ matrix. We will call the product in (3.8) the block product of the ith row of the partitioned matrix A with the jth column of the partitioned matrix B. The result in equation (3.6) implies that the value of the block product in (3.8) is precisely the $m_i \times p_j$ matrix $\sum_{q=1}^{k} A_{iq} B_{qj}$.

Now the rows of the submatrix $[A_{i1}|\cdots|A_{1k}]$ of A begin with $M_i = m_1 + \cdots + m_{i-1} + 1$ and end with $M_i' = m_1 + \cdots + m_i$. The columns of the submatrix

$$\begin{bmatrix} B_{1j} \\ \vdots \\ B_{kj} \end{bmatrix}$$

of B begin with $P_j = p_1 + \cdots + p_{j-1} + 1$ and end with $P_j' = p_1 + \cdots + p_j$. Let an index x range from M_i to M_i', and let y range from P_j to P_j'. Then $\text{Row}_x(A)\text{Col}_y(B)$ ranges over the entries of the $m_i \times p_j$ submatrix of AB

given by $\{[AB]_{xy} \mid M_i \leqslant x \leqslant M_i'; P_j \leqslant y \leqslant P_j'\}$. Denote this submatrix of AB by C_{ij}. Then equation (3.6) implies

3.9

$$C_{ij} = (A_{i1} \mid \cdots \mid A_{ik}) \begin{bmatrix} B_{1j} \\ \hline \vdots \\ \hline B_{kj} \end{bmatrix} = \sum_{q=1}^{k} A_{iq} B_{qj}$$

In equation (3.9), $i = 1, \ldots, r$ and $j = 1, \ldots, t$. As i and j vary, the C_{ij} determine a natural partition of the product AB. Each C_{ij} is the block product of the ith row of the partitioned matrix A with the jth column of the partitioned matrix B. We have now proven the following theorem.

Theorem 3.10 Let $A \in M_{m \times n}$ and $B \in M_{n \times p}$. Suppose

$$A = \begin{bmatrix} A_{11} & \cdots & A_{1k} \\ \vdots & & \vdots \\ A_{r1} & \cdots & A_{rk} \end{bmatrix} \quad \text{and} \quad B = \begin{bmatrix} B_{11} & \cdots & B_{1t} \\ \vdots & & \vdots \\ B_{k1} & \cdots & B_{kt} \end{bmatrix}$$

are partitions of A and B such that each A_{ij} is an $m_i \times n_j$ matrix, and each B_{jl} is an $n_j \times p_l$ matrix. Then $m_1 + \cdots + m_r = m$, $n_1 + \cdots + n_k = n$, and $p_1 + \cdots + p_t = p$. For each $i = 1, \ldots, r$ and $j = 1, \ldots, t$, set $C_{ij} = \sum_{q=1}^{k} A_{iq} B_{qj}$. Then

$$AB = \begin{bmatrix} C_{11} & \cdots & C_{1t} \\ \vdots & & \vdots \\ C_{r1} & \cdots & C_{rt} \end{bmatrix}$$

Let us consider another example before exploring the corollaries to Theorem 3.10.

Example 3.11 Let

$$A = \begin{bmatrix} 1 & 2 & 0 & 1 \\ 0 & 1 & -1 & 2 \\ 1 & 0 & -1 & 1 \\ 2 & 1 & 3 & 1 \end{bmatrix} = \begin{bmatrix} A_{11} & A_{12} \\ A_{21} & A_{22} \end{bmatrix}$$

Then $\text{size}(A_{11}) = \text{size}(A_{12}) = 3 \times 2$, and $\text{size}(A_{21}) = \text{size}(A_{22}) = 1 \times 2$.
Let

$$B = \left[\begin{array}{cc|c|cc} 1 & 2 & 0 & 1 & 2 \\ 0 & 1 & 1 & 0 & -1 \\ \hline 1 & -1 & 0 & 0 & 2 \\ -1 & 0 & 1 & 2 & 3 \end{array}\right] = \left[\begin{array}{c|c|c} B_{11} & B_{12} & B_{13} \\ \hline B_{21} & B_{22} & B_{23} \end{array}\right]$$

Then $\text{size}(B_{11}) = \text{size}(B_{13}) = \text{size}(B_{21}) = \text{size}(B_{23}) = 2 \times 2$, and $\text{size}(B_{12}) = \text{size}(B_{22}) = 2 \times 1$. It is easy to check that the block product of A and B is well-defined. Theorem 3.10 implies

$$AB = \left[\begin{array}{c|c} A_{11} & A_{12} \\ \hline A_{21} & A_{22} \end{array}\right]\left[\begin{array}{c|c|c} B_{11} & B_{12} & B_{13} \\ \hline B_{21} & B_{22} & B_{23} \end{array}\right]$$

$$= \left[\begin{array}{c|c|c} A_{11}B_{11} + A_{12}B_{21} & A_{11}B_{12} + A_{12}B_{22} & A_{11}B_{13} + A_{12}B_{23} \\ \hline A_{21}B_{11} + A_{22}B_{21} & A_{21}B_{12} + A_{22}B_{22} & A_{21}B_{13} + A_{22}B_{23} \end{array}\right]$$

Now the reader can easily check that

$$A_{11}B_{11} + A_{12}B_{21} = \begin{bmatrix} 0 & 4 \\ -3 & 2 \\ -1 & 3 \end{bmatrix}, \quad A_{11}B_{12} + A_{12}B_{22} = \begin{bmatrix} 3 \\ 3 \\ 1 \end{bmatrix}$$

$$A_{21}B_{11} + A_{22}B_{21} = [4 \quad 2], \quad A_{21}B_{12} + A_{22}B_{22} = 2$$

$$A_{21}B_{13} + A_{22}B_{23} = [4 \quad 12], \quad A_{11}B_{13} + A_{12}B_{23} = \begin{bmatrix} 3 & 3 \\ 4 & 3 \\ 3 & 3 \end{bmatrix}$$

Hence

$$AB = \left[\begin{array}{cc|c|cc} 0 & 4 & 3 & 3 & 3 \\ -3 & 2 & 3 & 4 & 3 \\ -1 & 3 & 1 & 3 & 3 \\ \hline 4 & 2 & 2 & 4 & 12 \end{array}\right]$$

Directly multiplying A and B together, we see this is indeed the correct answer.

There are several important corollaries to Theorem 3.10 which are worth noting.

Corollary 3.12 Let $A \in M_{m \times n}$. Then for any column vector $\xi = [y_1 \ \cdots \ y_n]^t$, we have $A\xi = y_1 \mathrm{Col}_1(A) + \cdots + y_n \mathrm{Col}_n(A)$.

Proof. Partition A into columns $A = [\mathrm{Col}_1(A) | \cdots | \mathrm{Col}_n(A)]$, and partition ξ into rows

$$\xi = \begin{bmatrix} y_1 \\ \vdots \\ y_n \end{bmatrix}$$

Now apply Theorem 3.10 in the special form listed in equation (3.6).   □

Corollary 3.12 says the product of A and ξ is a linear combination of the columns of A with scalars from ξ. This fact has important consequences for the theory of linear equations.

Corollary 3.13 Consider the following system of m equations in n unknowns $x_1, \ldots, x_n$

$$a_{11}x_1 + \cdots + a_{1n}x_n = b_1$$
$$\vdots \qquad\qquad \vdots \quad \vdots \qquad\qquad (*)$$
$$a_{m1}x_1 + \cdots + a_{mn}x_n = b_m$$

Let $A = (a_{ij})$ be the coefficient matrix of $(*)$. Set $B = [b_1 \ \cdots \ b_m]^t$. Then $(*)$ has a solution if and only if B is a linear combination of the columns of A.

Proof. The system $(*)$ has a solution if and only if there exists a column vector $\xi = [y_1 \ \cdots \ y_n]^t$ such that $A\xi = B$. But $A\xi = y_1 \mathrm{Col}_1(A) + \cdots + y_n \mathrm{Col}_n(A)$ by Corollary 3.12. Thus, $(*)$ has a solution ξ if and only if B is a linear combination of the columns of A.   □

We can generalize Corollary 3.12 as follows.

Corollary 3.14 Let $A \in M_{m \times n}$ and $B \in M_{n \times p}$. Then the jth column of AB is a linear combination of the columns of A with scalars from the jth column of B.

Proof. Let $B = (b_{ij})$. Partition $B = [\mathrm{Col}_1(B) | \cdots | \mathrm{Col}_p(B)]$ into columns, and leave A alone. Then Theorem 3.10 implies $AB = [A\,\mathrm{Col}_1(B) | \cdots | A\,\mathrm{Col}_p(B)]$. Using Corollary 3.12, we have the following equation for $j = 1, \ldots, p$.

3.15 $$\text{Col}_j(AB) = A\,\text{Col}_j(B) = \sum_{i=1}^{n} b_{ij}\text{Col}_i(A). \qquad \square$$

There is an analog of Corollary 3.12 for rows.

Corollary 3.16 Let $A \in M_{m \times n}$. Then for any row vector $\delta = [x_1 \cdots x_m]$, we have $\delta A = x_1\,\text{Row}_1(A) + \cdots + x_m\,\text{Row}_m(A)$.

Proof. The ith column of A^t is the transpose of the ith row of A. Hence using Theorem 2.35 and Corollary 3.12, we have

$$\delta A = (\delta A)^{tt} = (A^t \delta^t)^t = \left(\sum_{i=1}^{m} x_i\,\text{Col}_i(A^t) \right)^t$$

$$= \left(\sum_{i=1}^{m} x_i(\text{Row}_i(A))^t \right)^t = \sum_{i=1}^{m} x_i(\text{Row}_i(A))^{tt} = \sum_{i=1}^{m} x_i\,\text{Row}_i(A).$$

$$\square$$

Thus, δA is a linear combination of the rows of A. This result can be generalized as follows.

Corollary 3.17 Let $A \in M_{m \times n}$ and $B \in M_{n \times p}$. Then the ith row of AB is a linear combination of the rows of B with scalars from the ith row of A.

Proof. Let $A = (a_{ij})$. Partition A into rows and leave B alone. It follows from Theorem 3.10 that $\text{Row}_i(AB) = \text{Row}_i(A)B$. Thus, by Corollary 3.16, we have the following equations for each $i = 1, \ldots, m$.

3.18
$$\text{Row}_i(AB) = \text{Row}_i(A)B = \sum_{j=1}^{n} a_{ij}\,\text{Row}_j(B). \qquad \square$$

We close this section with two definitions which will be useful in Chapter II.

Definition 3.19 Let $A \in M_{m \times n}$. The set of all linear combinations of the columns of A is called the column space of A. We will denote the column space of A by $CS(A)$.

The columns of A are column vectors of size m. If the entries of A are real numbers, then

$$CS(A) = \left\{ \sum_{j=1}^{n} x_j\,\text{Col}_j(A) \,|\, x_1, \ldots, x_n \in \mathbb{R} \right\} \subseteq \mathbb{R}^m.$$

If A has complex entries then $CS(A) \subseteq \mathbb{C}^m$.

Definition 3.20   Let $A \in M_{m \times n}$. The set of all linear combinations of the rows of A is called the row space of A. We will denote the row space of A by $RS(A)$.

Each row of A is a row vector of size n. Thus, $RS(A) \subseteq M_{1 \times n}$. If, for instance, A has real entries then

$$RS(A) = \left\{ \sum_{i=1}^{m} x_i \, \text{Row}_i(A) \,|\, x_1, \ldots, x_m \in \mathbb{R} \right\} \subseteq M_{1 \times n}(\mathbb{R}).$$

It will be helpful to interpret some of the corollaries to Theorem 3.10 in terms of column and row spaces. Corollary 3.13 càn be rephrased as follows.

3.21   If $A \in M_{m \times n}$ and $B \in M_{m \times 1}$, then the equation $AX = B$ has a solution if and only if $B \in CS(A)$.

Corollaries 3.14 and 3.17 can be rewritten as follows.

3.22 $CS(AB) \subseteq CS(A)$

3.23 $RS(AB) \subseteq RS(B)$

Exercises for Section 3

3.1   Let

$$A = \left[\begin{array}{cc|c|cc} 1 & 2 & 4 & 1 & 0 \\ 3 & 1 & 0 & 0 & 1 \\ \hline 0 & 1 & 1 & -1 & 2 \\ 1 & 0 & 2 & 3 & 1 \\ -1 & 1 & -1 & 2 & 4 \end{array} \right] \quad \text{and} \quad B = \left[\begin{array}{c|cc} 1 & 1 & 2 \\ 2 & -1 & 0 \\ 4 & 2 & 2 \\ 1 & 0 & 1 \\ 3 & 0 & 1 \end{array} \right]$$

(a) Compute AB.
(b) List the sizes of the blocks in the partitions given above and verify that the hypotheses of Theorem 3.10 are satisfied in this case.
(c) Compute the product of A and B by block multiplication.
3.2   Consider the same two matrices A and B as in Exercise 3.1. Give a partition of B for which block multiplication of A and B is not well defined.

3.3   Let

$$A = \left[\begin{array}{cc|cc} 1 & 0 & -1 & 1 \\ 1 & 2 & 0 & 1 \\ \hline 2 & 0 & 4 & -1 \\ 3 & 1 & -2 & 0 \end{array}\right] \quad \text{and} \quad B = \left[\begin{array}{cc|cc} 5 & 0 & -1 & 1 \\ 2 & 1 & 1 & -1 \\ \hline 3 & 0 & 2 & 0 \\ 1 & 5 & 1 & 3 \end{array}\right]$$

(a) Compute AB.

(b) Compute AB by block multiplication.

3.4   Let

$$A = \begin{bmatrix} 1 & -1 & 0 & 0 & 0 \\ 1 & 2 & 0 & 0 & 0 \\ 0 & 0 & 1 & 2 & 3 \\ 0 & 0 & 1 & 0 & 1 \\ 0 & 0 & -1 & 1 & 2 \end{bmatrix}$$

and

$$B = \begin{bmatrix} 0 & 0 & 0 & 0 & 0 \\ 0 & 0 & 0 & 0 & 0 \\ 0 & 0 & 1 & 0 & 0 \\ 0 & 0 & 0 & 1 & 0 \\ 0 & 0 & 0 & 0 & 1 \end{bmatrix}$$

Find a good partition of A and B for which block multiplication gives the product AB in a simple way.

3.5   Consider the block diagonal matrix

$$A = \left[\begin{array}{cc|cc} A_{11} & 0 & \cdots & 0 \\ 0 & A_{22} & \cdots & 0 \\ \vdots & \vdots & \vdots & \vdots \\ 0 & 0 & \cdots & A_{rr} \end{array}\right]$$

Here $A_{11}, A_{22}, \ldots, A_{rr}$ are square matrices not necessarily of the same size. Use Theorem 3.10 to compute A^k for any positive integer k.

3.6   Let

$$A = \begin{bmatrix} 1 & 2 & 0 & 0 \\ 3 & 0 & 0 & 0 \\ 0 & 0 & 1 & 2 \\ 0 & 0 & 0 & 1 \end{bmatrix} \quad \text{and} \quad B = \begin{bmatrix} 1 & 2 & 0 & 0 & 0 \\ 3 & 4 & 0 & 0 & 0 \\ 0 & 0 & 1 & 0 & 1 \\ 0 & 0 & 3 & 1 & 2 \\ 0 & 0 & 2 & 0 & 1 \end{bmatrix}$$

Use your answer from Exercise 3.5 to compute the following expressions:
(a) A^3,
(b) B^3,
(c) $A^3 - 2A^2 + A$,
(d) $B^3 + B^2$.

3.7   Suppose each A_{ii}, $i = 1, \ldots, r$, is invertible in Exercise 3.5. Is A then invertible? If so, what is the inverse of A?

3.8   Suppose A and B are invertible. Is

$$\left[\begin{array}{c|c} A & C \\ \hline 0 & B \end{array} \right]$$

invertible for any matrix C? We assume that C has the correct size so that the notation makes sense. If this matrix is invertible, what is its inverse?

3.9   Give an example of two matrices A and B for which $CS(AB) \neq CS(A)$.

3.10   Give an example of two matrices A and B for which $RS(AB) \neq RS(B)$.

3.11   Use (3.21) to construct a 3×3 matrix A and a column vector B (of size 3) for which the linear system $AX = B$ has no solution.

3.12   Let

$$A = \begin{bmatrix} 1 & 1 & 1 \\ 0 & 1 & 2 \\ 1 & 0 & 0 \end{bmatrix}$$

Show that $AX = B$ has a solution for any column vector B (of size 3).

4. GAUSSIAN ELIMINATION

In this section, we return to one of the fundamental problems of linear algebra. How do we find all solutions to a system of linear equations? Let us repeat a portion of our discussion at the end of Section 1. Consider the following system of m equations in n unknowns $x_1, \ldots, x_n$.

4.1

$$a_{11}x_1 + a_{12}x_2 + \cdots + a_{1n}x_n = b_1$$
$$\vdots \qquad \vdots \qquad \vdots \qquad \vdots$$
$$a_{m1}x_1 + a_{m2}x_2 + \cdots + a_{mn}x_n = b_m$$

The procedure for finding all solutions (if there are any) to the equations in (4.1) is to reduce this system to an equivalent system whose augmented matrix is upper triangular. If $m \leqslant n$, such a system would have the following form.

4.2

$$c_{11}x_1 + c_{12}x_2 + \cdots \qquad \cdots \qquad \cdots \qquad \cdots + c_{1n}x_n = d_1$$
$$c_{22}x_2 + c_{23}x_3 + \cdots \qquad \cdots \qquad \cdots + c_{2n}x_n = d_2$$
$$\ddots \qquad \qquad \vdots \qquad \vdots$$
$$c_{rr}x_r + \cdots + c_{rn}x_n = d_r$$

In equation (4.2), $r \leqslant m$. Having obtained such a system, we then use the method of back substitution to compute the solutions. Roughly speaking, this means solve the last equation, $c_{rr}x_r + \cdots + c_{rn}x_n = d_r$, first. (We may have to assign values to some of the variables $x_r, \ldots, x_n$ arbitrarily here.) Substitute the answers from the last equation into the second from the last equation in (4.2), $c_{r-1r-1}x_{r-1} + \cdots + c_{r-1n}x_n = d_{r-1}$. Solve this equation for x_{r-1}, etc. The reader should review Example 1.16 for a concrete problem.

How do we reduce (4.1) to an equivalent system like (4.2)? More precisely, how do we manipulate the equations in (4.1) to produce a new system of equations whose augmented matrix is upper triangular and whose solutions are the same as the solutions to the original system of equations? We will see in this section that the manipulations needed are precisely the following elementary operations.

4.3 (a) Interchange two equations.

(b) Replace an equation with itself plus a scalar multiple of another equation.

(c) Multiply an equation by a nonzero scalar.

The word scalar here means an element of $F = \mathbb{R}$ (or $\mathbb{C}$). A scalar multiple of an equation is the equation obtained by multiplying the original equation by some scalar in F. A word to the experts is in order here. We will see that in fact operation (c) is not needed to reduce the equations in (4.1) to upper triangular form. However, (c) is useful, e.g., in computing row reduced echelon forms. Hence, we include (c) in our list.

In order to solve the system of equations in (4.1), a judicious sequence of operations from (4.3) is applied to the equations. The goal is to obtain a new system of equations in upper triangular form. Before describing this procedure, we need to simplify our notation. We saw in Section 1 that the variables $x_1, \ldots, x_n$ play no real role in the reduction procedure. The variables merely serve as place markers for the columns of coefficients to be manipulated. Let us make this point more precise. Let $[A\,|\,B]$ denote the augmented matrix of the system of equations in (4.1). Thus,

4.4
$$[A\,|\,B] = \begin{bmatrix} a_{11} & \cdots & a_{1n} & b_1 \\ \vdots & & \vdots & \vdots \\ a_{m1} & \cdots & a_{mn} & b_m \end{bmatrix}$$

When applying operations (a), (b), and (c) to the equations in (4.1), we might as well just apply the corresponding row operations to the augmented matrix $[A\,|\,B]$.

Suppose equation i is interchanged with equation j in (4.1) (operation (a) in (4.3)). Clearly, the new system of equations has the same solutions as the original system of equations. The augmented matrix of the new system of equations is obtained from $[A\,|\,B]$ by interchanging rows i and j.

Example 4.5   The system

$$\begin{aligned} 2x + 3y + 4z &= 10 \\ x - y - z &= 2 \\ x + z &= 3 \end{aligned}$$

has augmented matrix

$$\begin{bmatrix} 2 & 3 & 4 & 10 \\ 1 & -1 & -1 & 2 \\ 1 & 0 & 1 & 3 \end{bmatrix}$$

Suppose equation 1 and equation 3 are interchanged. The new system of equations is

$$\begin{aligned} x + z &= 3 \\ x - y - z &= 2 \\ 2x + 3y + 4z &= 10 \end{aligned}$$

The augmented matrix of this new system of equations is

$$\begin{bmatrix} 1 & 0 & 1 & 3 \\ 1 & -1 & -1 & 2 \\ 2 & 3 & 4 & 10 \end{bmatrix}$$

Suppose c $(\in F)$ times the ith equation of (4.1) is added to the jth equation (operation (b) in (4.3)). The new system of equations has precisely the same solutions as those of the original system. (This is a little less obvious than the corresponding assertion for operation (a), but is still easy to prove. See Exercise 4.6 at the end of this section.) The augmented matrix of the new system is obtained from $[A \mid B]$ by adding c times the ith row of $[A \mid B]$ to the jth row of $[A \mid B]$.

Example 4.6 Again consider the system

$$\begin{aligned} 2x + 3y + 4z &= 10 \\ x - y - z &= 2 \\ x + z &= 3 \end{aligned}$$

The augmented matrix of this system is

$$\begin{bmatrix} 2 & 3 & 4 & 10 \\ 1 & -1 & -1 & 2 \\ 1 & 0 & 1 & 3 \end{bmatrix}$$

Suppose 3 times equation 2 is added to equation 1. The new system becomes

$$5x \quad\quad + z = 16$$
$$x - y - z = 2$$
$$x \quad\quad + z = 3$$

with augmented matrix

$$\begin{bmatrix} 5 & 0 & 1 & 16 \\ 1 & -1 & -1 & 2 \\ 1 & 0 & 1 & 3 \end{bmatrix}$$

Suppose the ith equation of (4.1) is multiplied by a nonzero constant c. The solutions of the new system of equations are clearly the same as those of the original system. The augmented matrix of the new system of equations is obtained from $[A \mid B]$ by multiplying the ith row by c.

Example 4.7   Again consider

$$2x + 3y + 4z = 10$$
$$x - y - z = 2$$
$$x \quad\quad + z = 3$$

Suppose equation 1 is multiplied by 1/2. The new system of equations is

$$x + (3/2)y + 2z = 5$$
$$x - \quad\quad y - z = 2$$
$$x \quad\quad\quad + z = 3$$

with augmented matrix

$$\begin{bmatrix} 1 & 3/2 & 2 & 5 \\ 1 & -1 & -1 & 2 \\ 1 & 0 & 1 & 3 \end{bmatrix}$$

Thus, applying operations (a), (b), and (c) to the equations in (4.1) is equivalent to applying the corresponding row operations to the augmented matrix $[A \mid B]$. The matrices produced in this way tell us all we need to know about the current state of the equations. Dropping the variables and the equal signs from the reduction procedure is a considerable saving in notation. The three operations (a), (b), and (c) in (4.3) correspond to the following three row operations on $[A\,|\,B]$.

4.8 (a) Interchange two rows of the matrix.
    (b) Add a scalar times one row of the matrix to another row of the matrix.
    (c) Multiply a row of the matrix by a nonzero scalar.

The row operations in (4.8) are called elementary row operations. We have seen in Examples 4.5, 4.6, and 4.7 that each operation (a), (b), or (c) (from (4.3)) applied to the equations in (4.1) produces a new system of equations whose augmented matrix is obtained from $[A\,|\,B]$ by applying the corresponding elementary row operation (a), (b), or (c) to $[A\,|\,B]$.

If an $m \times n$ matrix Y can be obtained from an $m \times n$ matrix Y' by applying a finite number of elementary row operations to Y', then Y is said to be row equivalent to Y'. For example, the matrix

$$\begin{bmatrix} 1 & 0 & 1 & 3 \\ 1 & -1 & -1 & 2 \\ 2 & 3 & 4 & 10 \end{bmatrix}$$

in Example 4.5 is row equivalent to

$$\begin{bmatrix} 2 & 3 & 4 & 10 \\ 1 & -1 & -1 & 2 \\ 1 & 0 & 1 & 3 \end{bmatrix}$$

If a matrix Y is row equivalent to a matrix Y', we will write $Y \underset{r}{\sim} Y'$. Notice that any elementary row operation can be undone by applying another row operation of the same type. In particular, $Y \underset{r}{\sim} Y'$ if and only if $Y' \underset{r}{\sim} Y$. Thus, it does not make any difference which matrix comes first in the relationship. The words "row equivalent" are used only for matrices of the same size. Matrices of different sizes cannot be row equivalent.

It should be clear that if the augmented matrix $[A\,|\,B]$ of the linear system $AX = B$ is row equivalent to the augmented matrix $[C\,|\,D]$ of the system $CX = D$, then the two systems of equations $AX = B$ and $CX = D$

have precisely the same solutions. Thus, we have the following important principle.

4.9 Let $AX = B$ and $CX = D$ be two systems of linear equations. If $[A \,|\, B] \underset{r}{\sim} [C \,|\, D]$, then $A\xi = B$ if and only if $C\xi = D$.

In (4.9), ξ is any column vector of the appropriate size. Although the statement in (4.9) is intuitively obvious, we ask the reader to provide a detailed proof of this result in Exercise 4.6

The process which begins with the augmented matrix $[A \,|\, B]$ of a linear system of equations $AX = B$ and row reduces $[A \,|\, B]$ to an upper triangular matrix $[C \,|\, D]$ via elementary row operations is called Gaussian elimination. The statement in (4.9) implies that Gaussian elimination produces a new system of equations $CX = D$ whose solutions are precisely those of the original system $AX = B$. We have now outlined a method, called Gaussian elimination, for finding all solutions to the linear system of equations in (4.1). The three basic steps in the method are summarized as follows.

4.10 To find all solutions (if any) to the linear system of equations

$$
\begin{array}{ccc}
a_{11}x_1 + \cdots + a_{1n}x_n = b_1 \\
\vdots \qquad\qquad \vdots \quad \vdots \\
a_{m1}x_1 + \cdots + a_{mn}x_n = b_m
\end{array} \tag{$*$}
$$

by Gaussian elimination carry out the following three steps:

(1) Set up the augmented matrix

$$
[A \,|\, B] = \begin{bmatrix} a_{11} & \cdots & a_{1n} & b_1 \\ \vdots & & \vdots & \vdots \\ a_{m1} & \cdots & a_{mn} & b_m \end{bmatrix}
$$

of $(*)$.

(2) Apply elementary row operations to $[A \,|\, B]$ to reduce $[A \,|\, B]$ to a matrix $[C \,|\, D]$ in upper triangular form.

(3) Solve the system $CX = D$ by back substitution.

The matrix $[C \,|\, D]$ in step (3) above will be the augmented matrix of an upper triangular system of equations like those in (4.2). Of course, it is possible that A is sufficiently complicated so that $CX = D$ is still very hard.

Nonetheless, the equation $CX = D$ is usually easier to solve than $AX = B$, because $CX = D$ is upper triangular (we can use back substitution). Also, $CX = D$ may have more zero coefficients than $AX = B$.

We owe the reader a proof of the fact that step (2) in (4.10) can always be done. We will prove an even stronger result in Theorem 4.18. Before proceeding with this theorem, let us look at some examples of how the algorithm in (4.10) works.

Example 4.11 Solve

$$2x + 3y + 4z = 10$$
$$x - y - z = 2 \qquad (*)$$
$$x \qquad + z = 3$$

The augmented matrix of this system of equations is the 3×4 matrix

$$\begin{bmatrix} 2 & 3 & 4 & 10 \\ 1 & -1 & -1 & 2 \\ 1 & 0 & 1 & 3 \end{bmatrix}$$

This matrix can be reduced to an upper triangular matrix by the following elementary row operations from (4.8) as indicated

$$\begin{bmatrix} 2 & 3 & 4 & 10 \\ 1 & -1 & -1 & 2 \\ 1 & 0 & 1 & 3 \end{bmatrix} \underset{(a)}{\longmapsto} \begin{bmatrix} 1 & 0 & 1 & 3 \\ 1 & -1 & -1 & 2 \\ 2 & 3 & 4 & 10 \end{bmatrix}$$

$$\underset{(b)}{\longmapsto} \begin{bmatrix} 1 & 0 & 1 & 3 \\ 0 & -1 & -2 & -1 \\ 2 & 3 & 4 & 10 \end{bmatrix}$$

$$\underset{(b)}{\longmapsto} \begin{bmatrix} 1 & 0 & 1 & 3 \\ 0 & -1 & -2 & -1 \\ 0 & 3 & 2 & 4 \end{bmatrix}$$

$$\underset{(c)}{\longmapsto} \begin{bmatrix} 1 & 0 & 1 & 3 \\ 0 & 1 & 2 & 1 \\ 0 & 3 & 2 & 4 \end{bmatrix} \underset{(b)}{\longmapsto} \begin{bmatrix} 1 & 0 & 1 & 3 \\ 0 & 1 & 2 & 1 \\ 0 & 0 & -4 & 1 \end{bmatrix}$$

(upper triangular)

Notice that operation (c) was not necessary but was convenient.

The new system of equations to be solved is

$$x + 0y + z = 3$$
$$y + 2z = 1$$
$$-4z = 1$$

Solve these equations by back substitution. The last equation implies $z = -1/4$. Substituting $z = -1/4$ in the middle equation yields $y = 3/2$. Substituting $z = -1/4$ and $y = 3/2$ in equation 1 gives $x = 13/4$. Hence, $x = 13/4$, $y = 3/2$, and $z = -1/4$ is the (unique) solution to (∗).

Example 4.12   Solve

$$x + y + z = 2$$
$$x - y + 2z = 3 \qquad\qquad (*)$$
$$2x \quad\quad + 3z = 5$$

The augmented matrix of (∗) is

$$\begin{bmatrix} 1 & 1 & 1 & 2 \\ 1 & -1 & 2 & 3 \\ 2 & 0 & 3 & 5 \end{bmatrix}$$

This matrix can be reduced to an upper triangular matrix as follows:

$$\begin{bmatrix} 1 & 1 & 1 & 2 \\ 1 & -1 & 2 & 3 \\ 2 & 0 & 3 & 5 \end{bmatrix} \underset{(b)}{\longmapsto} \begin{bmatrix} 1 & 1 & 1 & 2 \\ 0 & -2 & 1 & 1 \\ 2 & 0 & 3 & 5 \end{bmatrix}$$

$$\underset{(b)}{\longmapsto} \begin{bmatrix} 1 & 1 & 1 & 2 \\ 0 & -2 & 1 & 1 \\ 0 & -2 & 1 & 1 \end{bmatrix}$$

$$\underset{(b)}{\longmapsto} \begin{bmatrix} 1 & 1 & 1 & 2 \\ 0 & -2 & 1 & 1 \\ 0 & 0 & 0 & 0 \end{bmatrix} \quad \text{(upper triangular)}$$

The new system of equations corresponding to this matrix is

$$x + y + z = 2$$
$$-2y + z = 1$$

There are two equations in three unknowns here. Solve this system by back substitution after making an arbitrary choice in the last equation. Suppose $z = c$. Here $c \in F$. Then the last equation implies $y = (c - 1)/2$. Substituting $z = c$ and $y = (c - 1)/2$ in the first equation gives $x = (5 - 3c)/2$. Thus, a complete set of solutions to (∗) is the set of all column vectors of the following form

$$\left\{ \begin{bmatrix} (5 - 3c)/2 \\ (c - 1)/2 \\ c \end{bmatrix} \middle| \; c \in \mathbb{R}(\text{or } \mathbb{C}) \right\}$$

We can also write this answer in the following way

$$\left\{ c \begin{bmatrix} -3/2 \\ 1/2 \\ 1 \end{bmatrix} + \begin{bmatrix} 5/2 \\ -1/2 \\ 0 \end{bmatrix} \middle| \; c \in \mathbb{R} \;(\text{or } \mathbb{C}) \right\}$$

Example 4.13   Solve

$$x + 2y = 1$$
$$x + 3y = 2$$
$$2x + 5y = 3 \qquad \qquad (\ast)$$
$$2x + 4y = 3$$

The augmented matrix of this system of equations is the 4×3 matrix

$$\begin{bmatrix} 1 & 2 & 1 \\ 1 & 3 & 2 \\ 2 & 5 & 3 \\ 2 & 4 & 3 \end{bmatrix}$$

This matrix can be reduced to an upper triangular matrix as follows

$$
\begin{bmatrix} 1 & 2 & 1 \\ 1 & 3 & 2 \\ 2 & 5 & 3 \\ 2 & 4 & 3 \end{bmatrix} \xrightarrow{\text{(b)}} \begin{bmatrix} 1 & 2 & 1 \\ 0 & 1 & 1 \\ 2 & 5 & 3 \\ 2 & 4 & 3 \end{bmatrix} \xrightarrow{\text{(b)}} \begin{bmatrix} 1 & 2 & 1 \\ 0 & 1 & 1 \\ 0 & 1 & 1 \\ 2 & 4 & 3 \end{bmatrix}
$$

$$
\xrightarrow{\text{(b)}} \begin{bmatrix} 1 & 2 & 1 \\ 0 & 1 & 1 \\ 0 & 1 & 1 \\ 0 & 0 & 1 \end{bmatrix} \xrightarrow{\text{(b)}} \begin{bmatrix} 1 & 2 & 1 \\ 0 & 1 & 1 \\ 0 & 0 & 0 \\ 0 & 0 & 1 \end{bmatrix}
$$

$$
\xrightarrow{\text{(a)}} \begin{bmatrix} 1 & 2 & 1 \\ 0 & 1 & 1 \\ 0 & 0 & 1 \\ 0 & 0 & 0 \end{bmatrix} \quad \text{(upper triangular)}
$$

The system of equations corresponding to the last matrix above is

$$
x + 2y = 1
$$
$$
y = 1
$$
$$
0 = 1
$$

This system of equations clearly has no solution. Hence, (∗) has no solution.

We now take up the question of row reducing a matrix to an upper triangular matrix using the three elementary row operations in (4.8). This is the central point of Gaussian elimination. We will show that a matrix can always be row reduced to a matrix in row reduced echelon form. A row reduced echelon form is a special upper triangular matrix. We need two definitions.

Definition 4.14 An $m \times n$ matrix E is said to be in echelon form if there exists an integer k with $0 \leqslant k \leqslant m$ and such that

(a) $\text{Row}_i(E) \neq O$ if $i \leqslant k$, and $\text{Row}_i(E) = O$ if $i > k$.

(b)   If the first (counting from left to right) nonzero entry in $\text{Row}_i(E)$ lies in column j_i, and the first nonzero entry of $\text{Row}_{i+1}(E)$ lies in column j_{i+1}, then $j_i < j_{i+1}$. Here $i = 1, \ldots, k - 1$.

Condition (a) in Definition 4.14 merely says that all zero rows (if any) of E are at the bottom of the matrix. The second condition, (b) of Definition 4.14 is really the important part of the definition. When the leading nonzero entries in the nonzero rows of E are connected by a broken line, then this line slopes down and to the right. Consider the following examples.

Example 4.15

$$A = \begin{bmatrix} * & \times & \times \\ 0 & * & \times \\ 0 & 0 & * \end{bmatrix}, \quad B = \begin{bmatrix} * & \times & \times & \times & \times \\ 0 & 0 & 0 & * & \times \\ 0 & 0 & 0 & 0 & * \\ 0 & 0 & 0 & 0 & 0 \end{bmatrix}$$

$$C = \begin{bmatrix} 0 & 0 & * & \times & \times \\ 0 & 0 & 0 & 0 & * \\ 0 & 0 & 0 & 0 & 0 \\ 0 & 0 & 0 & 0 & 0 \end{bmatrix}, \quad D = \begin{bmatrix} * & \times \\ 0 & * \\ 0 & 0 \\ 0 & 0 \end{bmatrix}$$

In these examples, $*$ represents a nonzero entry of the matrix, and $\times$ represents an entry whose value is not relevant to the example. A, B, C, and D are all in echelon form.

Notice that the index i in (a) of Definition 4.14 ranges over the row numbers of E. Therefore, $1 \leqslant i \leqslant m$. In, particular, certain parts of (a) may be vacuous for certain values of k. For instance, suppose E is the zero matrix. Then O satisfies Definition 4.14 with $k = 0$. There are no i for which $i \leqslant 0$, and consequently, the first part of (a) is ignored. It is true $\text{Row}_i(O) = 0$ whenever $i > 0$. Thus, the zero matrix satisfies (a). Statement (b) for $E = O$ is vacuous. Thus, the zero matrix is in echelon form. Suppose E is in echelon form and has no zero rows. Then $k = m$. The identity matrix I_m is an example of a matrix in echelon form with no zero rows.

Notice that the conditions in Definition 4.14 guarantee that a matrix in echelon form is an upper triangular matrix. It is also helpful to notice that the integer k in Definition 4.14 satisfies the inequality $k \leqslant \min\{m, n\}$.

We can now define a row reduced echelon form.

Definition 4.16 Let E be a matrix in echelon form. E is in row reduced echelon form if

(a) The first nonzero entry in each nonzero row of E is 1.
(b) In any column of E which contains the first nonzero entry of some nonzero row, that entry (necessarily a 1) is the only nonzero entry in that column.

Example 4.17 Consider the four matrices in Example 4.15. If these matrices are in row reduced echelon form, then they have the following forms

$$A = \begin{bmatrix} 1 & 0 & 0 \\ 0 & 1 & 0 \\ 0 & 0 & 1 \end{bmatrix}, \quad B = \begin{bmatrix} 1 & \times & \times & 0 & 0 \\ 0 & 0 & 0 & 1 & 0 \\ 0 & 0 & 0 & 0 & 1 \\ 0 & 0 & 0 & 0 & 0 \end{bmatrix}$$

$$C = \begin{bmatrix} 0 & 0 & 1 & \times & 0 \\ 0 & 0 & 0 & 0 & 1 \\ 0 & 0 & 0 & 0 & 0 \\ 0 & 0 & 0 & 0 & 0 \end{bmatrix}, \quad D = \begin{bmatrix} 1 & 0 \\ 0 & 1 \\ 0 & 0 \\ 0 & 0 \end{bmatrix}$$

If a matrix E is in row reduced echelon form, then E is in echelon form. In particular, E is an upper triangular matrix. In our next theorem, we argue that any matrix can be row reduced, using the three elementary row operations in (4.8), to a matrix in row reduced echelon form. This result could be applied to the augmented matrix $[A \,|\, B]$ of a system of linear equations $AX = B$. Thus, step (2) in (4.10) is always possible.

Theorem 4.18 Let A be any $m \times n$ matrix. Then A is row equivalent to an $m \times n$ matrix in row reduced echelon form.

Proof. If $A = O$, then A is already in row reduced echelon form. Hence, we can assume A has a nonzero entry. Suppose j is the number of the first column (counting from left to right) which contains a nonzero entry of A. Applying an elementary row operation of type (a) to A, we have

$$A \underset{r}{\sim} \begin{bmatrix} 0 & \cdots & 0 & a_{1j} & \cdots & a_{1n} \\ \vdots & & \vdots & \vdots & & \vdots \\ 0 & \cdots & 0 & a_{mj} & \cdots & a_{mn} \end{bmatrix} = A_1 \quad \text{and} \quad a_{1j} \neq 0$$

Since $a_{1j} \neq 0$, we can multiply row 1 of A_1 by a_{1j}^{-1} and get

$$A \underset{r}{\sim} \begin{bmatrix} 0 & \cdots & 0 & 1 & * & \cdots & * \\ 0 & \cdots & 0 & a_{2j} & \cdots & \cdots & a_{2n} \\ \vdots & & \vdots & \vdots & & & \vdots \\ 0 & \cdots & 0 & a_{mj} & \cdots & \cdots & a_{mn} \end{bmatrix} = A_2.$$

The $*$ here in A_2 denotes an entry whose precise value is not relevant to this proof. We can now add $(-a_{kj})$ times row 1 of A_2 to row k of A_2 $(k = 2, \ldots, m)$. We get

$$A \underset{r}{\sim} \begin{bmatrix} 0 & \cdots & 0 & 1 & * & \cdots & * \\ 0 & \cdots & 0 & 0 & c_{2j+1} & \cdots & c_{2n} \\ \vdots & & \vdots & \vdots & \vdots & & \vdots \\ 0 & \cdots & 0 & 0 & c_{mj+1} & \cdots & c_{mn} \end{bmatrix} = A_3$$

Set

$$C = \begin{bmatrix} c_{2j+1} & \cdots & c_{2n} \\ \vdots & & \vdots \\ c_{mj+1} & \cdots & c_{mn} \end{bmatrix}$$

If $C = O$, then A_3 is in row reduced echelon form and the proof of the theorem is complete. Suppose C has a nonzero entry. We can repeat the same process just gone through on rows 2 through m of A_3. Notice that any row operations applied to rows 2 through m of A_3 cannot change any entry in the first j columns. Suppose l is the number of the first column of A_3 in which the submatrix C has a nonzero entry. Then $j < l$. Following the same procedure we applied to A (now applied to rows 2 through m of A_3); we get

$$A \underset{r}{\sim} \begin{bmatrix} 0 & \cdots & 0 & 1 & * & \cdots & * & * & \cdots & \cdots & * \\ 0 & \cdots & 0 & 0 & 0 & \cdots & 0 & 1 & * & \cdots & * \\ 0 & \cdots & 0 & 0 & 0 & \cdots & 0 & 0 & & & \\ \vdots & & \vdots & \vdots & \vdots & & \vdots & \vdots & & D & \\ 0 & \cdots & 0 & 0 & 0 & \cdots & 0 & 0 & & & \end{bmatrix} = A_4$$

In A_4, D is a submatrix of the following form

$$D = \begin{bmatrix} d_{3l+1} & \cdots & d_{3n} \\ \vdots & & \vdots \\ d_{ml+1} & \cdots & d_{mn} \end{bmatrix}$$

Now a suitable multiple of row 2 of A_4 added to row 1 of A_4 gives

$$A \underset{r}{\sim} \begin{bmatrix} 0 & \cdots & 0 & 1 & * & \cdots & * & 0 & * & \cdots & * \\ 0 & \cdots & 0 & 0 & 0 & \cdots & 0 & 1 & * & \cdots & * \\ 0 & \cdots & 0 & 0 & 0 & \cdots & 0 & 0 & & & \\ \vdots & & \vdots & \vdots & \vdots & & \vdots & \vdots & & D & \\ 0 & \cdots & 0 & 0 & 0 & \cdots & 0 & 0 & & & \end{bmatrix}$$

We can clearly continue this procedure and obtain a row reduced echelon form row equivalent to A in at most m steps. □

Thus, every matrix is row equivalent to a matrix in row reduced echelon form. We will show in Chapter II that the row reduced echelon form obtained in the reduction procedure in the proof of Theorem 4.18 is unique. Thus, if $A \underset{r}{\sim} E$, and $A \underset{r}{\sim} E'$ with both E and E' in row reduced echelon form, then $E = E'$. Hence, it makes sense to refer to *the* row reduced echelon form of A. By this, we mean the row reduced echelon form E for which $E \underset{r}{\sim} A$.

Let us say a few words about the strategy one should employ when using Gaussian elimination. Suppose we want to solve a linear system of equations $AX = B$. We know from Theorem 4.18 that $[A \,|\, B]$ can be row reduced to a matrix E in row reduced echelon form. The matrix E has important theoretical applications, but as a problem solving device it takes too long to compute. It is faster to reduce $[A \,|\, B]$ to an upper triangular matrix $[C \,|\, D]$ and back substitute in the equation $CX = D$. Thus, in Examples 4.11, 4.12, and 4.13, we do not reduce the matrix $[A \,|\, B]$ to its row reduced echelon form. Instead, we modify the procedure in Theorem 4.18 in the obvious way and reduce $[A \,|\, B]$ to an upper triangular matrix. Notice the proof of Theorem 4.18 shows that row operations of type (c) are not needed to reduce a matrix to an upper triangular matrix. Also notice that a matrix can be row equivalent to many different upper triangular matrices. In particular (unlike the row reduced echelon form of A), there may be many matrices in echelon form which are row equivalent to A.

We finish this section with some consequences of Theorem 4.18.

Definition 4.19 A homogeneous (linear) system of m equations in n unknowns $x_1, \ldots, x_n$ is any linear system of the following form

$$a_{11}x_1 + \cdots + a_{1n}x_n = 0$$
$$\vdots \qquad \vdots \quad \vdots \qquad (*)$$
$$a_{m1}x_1 + \cdots + a_{mn}x_n = 0$$

Let $A = (a_{ij})$ denote the matrix of coefficients in $(*)$ above. The homogeneous system of equations $AX = O$ always has at least one solution, namely $X = O$. The column vector O, i.e., $x_1 = 0, \ldots, x_n = 0$, is called the trivial solution of $AX = O$. For example, if A is invertible, then Corollary 2.31 implies $AX = O$ has only the trivial solution $X = O$. A column vector ξ (of size n) is called a nontrivial solution of $AX = O$ if $\xi \neq O$, and $A\xi = O$.

If $m < n$, then $(*)$ always has a nontrivial solution. This remark is a simple consequence of Theorem 4.18 and the method of back substitution. We begin with the following special case.

Lemma 4.20 Let

$$a_{11}x_1 + a_{12}x_2 + \cdots \qquad \cdots \qquad \cdots + a_{1n}x_n = 0$$
$$a_{22}x_2 + \cdots \qquad \cdots \qquad \cdots + a_{2n}x_n = 0$$
$$\ddots \qquad \qquad \vdots \quad \vdots \qquad (*)$$
$$a_{mm}x_m + \cdots + a_{mn}x_n = 0$$

be an upper triangular system of equations with $m < n$. Then $(*)$ has a nontrivial solution.

Proof. If $a_{11} = 0$, then $x_1 = 1, x_2 = \cdots = x_n = 0$ is a nontrivial solution to $(*)$. Hence, we can assume $a_{11} \neq 0$.

Suppose there exists an integer r such that $1 \leqslant r \leqslant m - 1$, $a_{11}, \ldots, a_{rr}$ are nonzero, and $a_{r+1\,r+1} = 0$. Then $x_{r+1} = 1, x_{r+2} = \cdots = x_n = 0$ satisfy equations $r + 1$ through m in $(*)$. Back substitute the values $x_{r+1} = 1$, $x_{r+2} = \cdots = x_n = 0$ in the rth equation in $(*)$. The rth equation becomes:

4.21 $$a_{rr}x_r + a_{rr+1} = 0$$

Since $a_{rr} \neq 0$, (4.21) can be solved for x_r. Hence, $x_r = -a_{rr+1}/a_{rr}$. Now substitute $x_r = -a_{rr+1}/a_{rr}$, $x_{r+1} = 1$, $x_{r+2} = \cdots = x_n = 0$ in equation $r - 1$ in $(*)$. Since $a_{r-1\,r-1} \neq 0$, this equation can be solved for x_{r-1}. Continuing to back substitute, a solution to $(*)$ is constructed in which $x_{r+1} = 1$. In particular, $(*)$ has a nontrivial solution.

Finally, suppose all the coefficients $a_{11}, \ldots, a_{mm}$ are nonzero. Set $x_{m+1} = \cdots = x_n = 1$. Notice that there is at least one variable here since $m < n$. Substitute these values in the mth equation of (*). Since $a_{mm} \neq 0$, this equation can be solved for x_m. Back substitute to the beginning. Thus, (*) has a solution in which $x_{m+1} = \cdots = x_n = 1$. □

Theorem 4.18 implies that any $m \times n$ matrix A is row equivalent to an upper triangular matrix E (in fact a matrix in row reduced echelon form). By (4.9), the solutions to $AX = O$ are precisely the same as the solutions to $EX = O$. If $m < n$, then Lemma 4.20 implies $EX = O$ has a nontrivial solution. Thus, $AX = O$ has a nontrivial solution. We have now proven the following theorem.

Theorem 4.22 A homogeneous system of m equations in n unknowns has a nontrivial solution if $m < n$.

Notice that Theorem 4.22 is not necessarily true if $m \geqslant n$. We have already pointed out that $AX = O$ has only the trivial solution if A is invertible.

Corollary 4.23 Let $A \in M_{m \times n}$. Suppose $A \underset{r}{\sim} E$ and E is a matrix in row reduced echelon form. Then the homogeneous system of equations $AX = O$ has a nontrivial solution if and only if the number of nonzero rows in E is less than n.

Proof. Let k denote the number of nonzero rows of E. If $A = O$, then $E = O$, and the corollary is trivial. Hence, we can assume $A \neq O$. In particular, $k \geqslant 1$. We have seen that the definition of a row reduced echelon form implies $k \leqslant \min\{m, n\}$.

Suppose $k = n$. Then

$$E = \left[\frac{I_n}{O}\right]$$

In this case, the only solution to $EX = O$ is $X = O$. Thus, by (4.9), the only solution to $AX = O$ is $X = O$.

Suppose $k < n$. Then

$$E = \left[\frac{U}{O}\right]$$

Here U is a $k \times n$ upper triangular matrix. Since $k < n$, Theorem 4.22

implies there exists a nonzero column vector ξ of size n such that $U\xi = O$. But then $E\xi = O$. Hence by (4.9), $A\xi = O$. ▫

If $m = n$ in Corollary 4.23, we get the following interesting result about square matrices.

Corollary 4.24 Let $A \in M_{n \times n}$. The homogeneous system of equations $AX = O$ has only $X = O$ as a solution if and only if $A \underset{r}{\sim} I_n$.

The proof of this corollary is left an an exercise.

Exercises for Section 4

4.1   Solve the following systems of equations using Gaussian elimination:

(a) $2x + 4y = -2$
$\quad\;\; x - \; y = \;\;\; 5$

(b) $2x_1 + 3x_2 - 4x_3 = -25$
$\quad\; x_1 + \; x_2 - \;\; x_3 = \;\; -6$
$\quad -x_1 + 2x_2 + 4x_3 = \;\;\;\; 12$

(c) $\;\; x_1 + \;\; x_2 - \;\; x_3 = 1$
$\quad 2x_1 + \;\; x_2 + 2x_3 = 2$
$\quad 8x_1 + 5x_2 + 4x_3 = 7$

(d) $2x_1 - 3x_2 + \;\; x_3 - x_4 = -5$
$\quad\; x_1 - \;\; x_2 - \;\; x_3 - x_4 = \;\;\;\; 0$
$\quad\; x_1 \qquad\quad + 2x_3 + x_4 = -1$
$\quad\; x_1 + \;\; x_2 + \;\; x_3 + x_4 = \;\;\;\; 2$

(e) $x + y + \;\; z + w = 1$
$\quad x - y + 2z + w = 2$
$\qquad\qquad\; z - w = 3$

(f) $\;\; x + \;\; y + z = 0$
$\quad\; x + 2y - z = 1$
$\quad 4x + 6y \qquad = 2$
$\quad 2x + 3y \qquad = 1$

4.2   Show the effect of an elementary row operation can be undone by applying another elementary row operation of the same type.

4.3   Show that $\underset{r}{\sim}$ is an equivalence relation. This means the following identities are true:

(a) $A \underset{r}{\sim} A$.

(b) $A \underset{r}{\sim} B$ if and only if $B \underset{r}{\sim} A$.

(c) If $A \underset{r}{\sim} B$ and $B \underset{r}{\sim} C$, then $A \underset{r}{\sim} C$.

In these identities, A, B, and C are all $m \times n$ matrices.

4.4 Determine which of the following matrices are in echelon form:

$$A = \begin{bmatrix} 1 & 1 & -1 & 4 \\ 0 & 1 & 2 & 1 \\ 0 & 0 & 0 & 2 \end{bmatrix}, \quad B = \begin{bmatrix} 1 & 1 & 0 \\ 0 & 2 & 0 \\ 0 & 1 & 0 \\ 0 & 4 & 1 \end{bmatrix},$$

$$C = \begin{bmatrix} -1 & 0 & 1 & 4 & 5 \\ 0 & 2 & 1 & 3 & 1 \\ 0 & 0 & 0 & 2 & 1 \end{bmatrix}, \quad D = \begin{bmatrix} 1 & 3 & 4 & 1 & 2 \\ 0 & 2 & 1 & 3 & 1 \\ 0 & 1 & 4 & 1 & 3 \\ 0 & 0 & 1 & 2 & 1 \end{bmatrix}$$

4.5 Reduce each matrix in Exercise 4.4 to a matrix in row reduced echelon form.

4.6 Prove the assertion in (4.9).

4.7 List all possible 2×2 and 3×3 matrices in row reduced echelon form. Use the same notation as in Example 4.17.

4.8 Prove Corollary 4.24.

4.9 Suppose A and B are two matrices for which every solution to $AX = O$ is a solution to $BX = O$ and vice versa. Does it follow that A and B are row equivalent?

4.10 Each matrix listed below is the coefficient matrix of a homogeneous system of equations. Find all solutions to the system.

(a) $\begin{bmatrix} 2 & 4 & 2 & 2 \\ 0 & 1 & 2 & 2 \\ 0 & 0 & 1 & 0 \end{bmatrix}$

(b) $\begin{bmatrix} 1 & 1 & 1 \\ 0 & 0 & 1 \\ 0 & 0 & 1 \end{bmatrix}$

(c) $\begin{bmatrix} 1 & 0 & 0 & 0 & 0 \\ 0 & 0 & 1 & 2 & 0 \\ 0 & 0 & 1 & 2 & 3 \end{bmatrix}$

(d) $\begin{bmatrix} 1 & -1 \\ 0 & 1 \\ 1 & 1 \end{bmatrix}$

4.11 Exhibit two nonzero 2×2 matrices which are not row equivalent.

4.12 Suppose

$$A = \begin{bmatrix} a & b \\ c & d \end{bmatrix}$$

with $ad - bc \neq 0$. Show $A \underset{r}{\sim} I_2$.

4.13 If the word "row" is replaced with "column" in (4.8), we get what are called elementary column operations. Re-prove Exercise 4.2 for elementary column operations.

4.14 Let $A, B \in M_{m \times n}$. Then $A \underset{c}{\sim} B$ (A is column equivalent to B) if A can be obtained from B by applying a finite number of elementary column operations to B. Show that $\underset{c}{\sim}$ is an equivalence relation.

4.15 If $\underset{r}{\sim}$ is replaced with $\underset{c}{\sim}$ in statement (4.9), does the same conclusion hold?

4.16 Develop the definitions of column echelon form and column reduced echelon form, and then prove the analog of Theorem 4.18 for columns.

4.17 Let $A \in M_{m \times n}$. What is the simplest matrix we can obtain from A by performing both column and row operations on A?

4.18 Is the analog of Theorem 4.22 true for nonhomogeneous equations, i.e., suppose $A \in M_{m \times n}$ with $m < n$ and B is a nonzero column vector of size m, is there necessarily a solution to $AX = B$? Justify your answer.

5. ELEMENTARY MATRICES AND INVERSES

In this section, we will show that the elementary row operations listed in (4.8) can be performed by multiplying on the left by what are called elementary matrices. Thus, Gaussian elimination can be performed within the framework of matrix arithmetic as a series of left multiplications.

Definition 5.1 An elementary matrix (of size $m \times m$) is a matrix obtained from the identity matrix I_m by performing a single elementary row operation on I_m.

Notice that an elementary matrix is square. Since there are three different types of elementary row operations, there are three different types of elementary matrices. We now introduce the notation used in the rest of this book for these three types of elementary matrices.

Definition 5.2 E_{ij} will denote the elementary matrix obtained from I_m by interchanging rows i and j of I_m.

In Definition 5.2, m is any positive integer, and $1 \leqslant i, j \leqslant m$. If $i = j$, then $E_{ij} = I_m$. Notice we have no room in our notation for the size of E_{ij}. This will cause no confusion in the sequel, since the reader will always be able to determine the size of an elementary matrix from the context in which it is used. E_{ij} obviously corresponds to (a) of (4.8). An example of E_{ij} for a fixed i and j such that $1 \leqslant i < j \leqslant m$ is as follows.

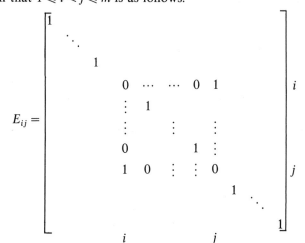

The i and j at the side and bottom of (5.3) indicate what rows and columns have been changed. The blank spaces represent 0 entries.

Definition 5.4 $E_{ij}(c)$ will denote the elementary matrix obtained from I_m by adding c times row j of I_m to row i of I_m.

Clearly, $E_{ij}(c)$ corresponds to (b) of (4.8). An example of $E_{ij}(c)$ when $i < j$ is as follows.

5.5

$$
E_{ij}(c) = \begin{bmatrix}
1 & 0 & \cdots & \cdots & \cdots & \cdots & \cdots & \cdots & \cdots & \cdots & 0 \\
\vdots & & & & & & & & & & \vdots \\
0 & \cdots & 0 & 1 & 0 & \cdots & 0 & c & 0 & \cdots & 0 \\
\vdots & & & & & & & & & & \vdots \\
0 & \cdots & \cdots & \cdots & \cdots & \cdots & \cdots & 1 & 0 & \cdots & 0 \\
\vdots & & & & & & & & & & \vdots \\
0 & \cdots & \cdots & \cdots & \cdots & \cdots & \cdots & \cdots & 0 & 1
\end{bmatrix}
\begin{matrix} \\ \\ i \\ \\ j \\ \\ \end{matrix}
$$

$$j$$

Definition 5.6 $E_i(c)$ will denote the elementary matrix obtained from I_m by multiplying row i of I_m by the nonzero constant c.

The matrix $E_i(c)$ corresponds to (c) of (4.8). An example of $E_i(c)$ is the following diagonal matrix.

5.7
$$
E_i(c) = \begin{bmatrix} 1 & 0 & \cdots & \cdots & \cdots & 0 \\ \vdots & & & & & \vdots \\ 0 & \cdots & c & \cdots & & 0 \\ \vdots & & & & & \vdots \\ 0 & \cdots & \cdots & \cdots & 0 & 1 \end{bmatrix} \begin{matrix} \\ \\ i \\ \\ \end{matrix}
$$

Notice that c is allowed to be zero in Definition 5.4. In this case, $E_{ij}(0) = I_m$. The constant c is not allowed to be zero in Definition 5.6. Also notice that some of these elementary matrices turn out to be the same matrix for various choices of i, j, and c. There is no uniqueness of representation here. Our first theorem says multiplying a matrix A on the left by E_{ij}, $E_{ij}(c)$ or $E_i(c)$ performs the corresponding row operation on A.

Theorem 5.8   Let $A \in M_{m \times n}$.

(a) $E_{ij}A$ is the $m \times n$ matrix obtained from A by interchanging rows i and j of A.

(b) $E_{ij}(c)A$ is the matrix obtained from A by adding c times row j of A to row i of A.

(c) $E_i(c)A$ is the matrix obtained from A by multiplying row i of A by c.

Proof. Partition A into columns, $A = [A_1 | \cdots | A_n]$. If E is any elementary matrix, then Theorem 3.10 implies $EA = [EA_1 | \cdots | EA_n]$. In particular, it suffices to prove the theorem for a single column vector $\xi = [x_1 \cdots x_m]^t$. The equations in (5.3), (5.5), and (5.7), immediately imply the following identities.

5.9
$$
E_{ij}\xi = [x_1 \quad \cdots \quad x_j \cdots x_i \cdots x_m]^t
$$
$$
\phantom{E_{ij}\xi = [x_1 \quad \cdots \quad} \underset{i}{} \quad \underset{j}{}
$$
$$
E_{ij}(c)\xi = [x_1 \cdots \underbrace{x_i + cx_j}_{i} \quad \cdots \quad x_m]^t
$$
$$
E_i(c)\xi = [x_1 \quad \cdots \quad cx_i \quad \cdots \quad x_m]^t
$$

In the first equation in (5.9), we have assumed $i < j$. We obviously get a similar result if $j < i$. The indices i and j below the equation indicate what rows are changing. The identities in (5.9) complete the proof of Theorem 5.8.   □

It follows from Theorem 5.8 that each elementary matrix is invertible. In fact, we can prove a little more.

Corollary 5.10   Each elementary matrix is invertible. The inverse of each type is given by the following formula:

(a) $E_{ij}^{-1} = E_{ij}$.
(b) $E_{ij}(c)^{-1} = E_{ij}(-c)$.
(c) $E_i(c)^{-1} = E_i(1/c)$.

Proof.   Set $A = E_{ij}$ in (a) of (5.8). Then $E_{ij}E_{ij}$ is the matrix obtained from E_{ij} by interchanging rows i and j of E_{ij}. Thus, $E_{ij}E_{ij} = I_m$. In particular, E_{ij} is invertible with inverse E_{ij}. Similar proofs can be given for the other two types of elementary matrices.   □

Notice that the inverse of an elementary matrix is another elementary matrix of the same type. It follows from Theorem 5.8 that Gaussian elimination is performed by multiplying on the left by a finite number of elementary matrices. Let us record this fact in the following corollary.

Corollary 5.11   Two $m \times n$ matrices A and B are row equivalent if and only if there exists a finite number of elementary matrices $E_1, \ldots, E_p$ such that $E_p E_{p-1} \cdots E_1 A = B$.

Consider the following example.

Example 5.12   In Example 4.11, we saw

$$A = \begin{bmatrix} 2 & 3 & 4 & 10 \\ 1 & -1 & -1 & 2 \\ 1 & 0 & 1 & 3 \end{bmatrix} \underset{r}{\sim} \begin{bmatrix} 1 & 0 & 1 & 3 \\ 0 & 1 & 2 & 1 \\ 0 & 0 & -4 & 1 \end{bmatrix} = B$$

The elementary matrices used to perform the indicated row operations in (4.11) are as follows: $E_1 = E_{13}$, $E_2 = E_{21}(-1)$, $E_3 = E_{31}(-2)$, $E_4 = E_2(-1)$, and $E_5 = E_{32}(-3)$. The reader can easily check that $E_{32}(-3)E_2(-1)E_{31}(-2)E_{21}(-1)E_{13}A = B$.

Elementary matrices are used to prove one of the fundamental results about invertible matrices.

Theorem 5.13 Let $A \in M_{n \times n}$. Then the following four conditions are equivalent:

(a) There exists an $n \times n$ matrix B such that $BA = I_n$.
(b) The linear system of equations $AX = O$ has only the trivial solution $X = O$.
(c) A is a product of elementary matrices.
(d) A is invertible.

Proof. We will prove the theorem by showing (a) $\Rightarrow$ (b) $\Rightarrow$ (c) $\Rightarrow$ (d) $\Rightarrow$ (a).

(a) $\Rightarrow$ (b): Suppose ξ is a solution to $AX = O$, i.e., $A\xi = O$. Then $\xi = I_n\xi = (BA)\xi = B(A\xi) = BO = O$. Thus, the only solution to $AX = O$ is the trivial solution $X = O$.

(b) $\Rightarrow$ (c): Suppose $AX = O$ has only the trivial solution $X = O$. By Corollary 4.24, $A \underset{r}{\sim} I_n$. By Corollary 5.11, there exists a finite number of elementary matrices $E_1, \ldots, E_p$ such that $A = E_p \cdots E_1 I_n = E_p \cdots E_1$. Hence, A is a product of elementary matrices.

(c) $\Rightarrow$ (d): We have seen that any elementary matrix is invertible. Products of invertible matrices are invertible by Theorem 2.30. Thus, if A is a product of elementary matrices, then A is invertible.

(d) $\Rightarrow$ (a): If A is invertible, then $A^{-1}A = I_n$. $\square$

Suppose $A \in M_{n \times n}$. A has a left inverse if there exists an $n \times n$ matrix B such that $BA = I_n$. In this case, B is called a left inverse of A. Similarly, A has a right inverse if $AB = I_n$. Theorem 5.13 implies that if A has a left inverse B, then A is invertible. But then, $B = BI_n = B(AA^{-1}) = (BA)A^{-1} = I_nA^{-1} = A^{-1}$. Thus, any left inverse of A is nothing but the (unique) inverse of A. The inverse of A is certainly a right inverse of A, and so we we conclude that any left inverse of A is also a right inverse of A. Conversely, one can also show any right inverse of A is a left inverse of A. So, there is no difference between a right inverse, a left inverse or the inverse of A.

Some of the ideas in the proof of Theorem 5.13 lead to an algorithm for computing the inverse of an invertible matrix A. Let us make this point clear. Suppose A is an invertible matrix. Then the equation $AX = O$ has only the trivial solution $X = O$. In particular, Corollary 4.24 implies A is row equivalent to I_n. Hence, there exist elementary matrices $E_1, \ldots, E_p$ such that $E_p \cdots E_1 A = I_n$. Thus, the matrix $E_p E_{p-1} \cdots E_1$ is a left inverse of A. Our comments in the last paragraph imply

that $E_p E_{p-1} \quad \cdots \quad E_1$ is the inverse of A. In particular, $E_p E_{p-1} \cdots E_1$
$[A \,|\, I_n] = [E_p \cdots E_1 A \,|\, E_p \cdots E_1 I_n] = [I_n \,|\, E_p \cdots E_1] = [I_n \,|\, A^{-1}]$. This last
equation suggests the following algorithm for computing A^{-1}.

5.14   To compute the inverse of an invertible matrix A:

(1)   Form the $n \times (2n)$ partitioned matrix $[A \,|\, I_n]$.
(2)   Apply row operations to $[A \,|\, I_n]$ so that A on the left in $[A \,|\, I_n]$ is
       reduced to I_n. Then I_n on the right in $[A \,|\, I_n]$ will be reduced to A^{-1}.

In symbols, step (2) above is $[A \,|\, I_n] \underset{r}{\sim} [I_n \,|\, A^{-1}]$. It follows from
Corollary 4.24 and Theorem 5.13 that A is invertible if and only if $A \underset{r}{\sim} I_n$.
In particular, step (2) in the algorithm (5.14) can only be performed for an
invertible matrix A. Consider the following example.

Example 5.15   Let

$$A = \begin{bmatrix} 2 & -1 & 0 \\ 1 & 2 & 1 \\ -1 & 0 & 3 \end{bmatrix}$$

Using the same notation as in (4.8), we have the following row reductions:

$$\begin{bmatrix} 2 & -1 & 0 & | & 1 & 0 & 0 \\ 1 & 2 & 1 & | & 0 & 1 & 0 \\ -1 & 0 & 3 & | & 0 & 0 & 1 \end{bmatrix}$$

$$\underset{(a)}{\longmapsto} \begin{bmatrix} -1 & 0 & 3 & | & 0 & 0 & 1 \\ 1 & 2 & 1 & | & 0 & 1 & 0 \\ 2 & -1 & 0 & | & 1 & 0 & 0 \end{bmatrix}$$

$$\underset{(c)}{\longmapsto} \begin{bmatrix} 1 & 0 & -3 & | & 0 & 0 & -1 \\ 1 & 2 & 1 & | & 0 & 1 & 0 \\ 2 & -1 & 0 & | & 1 & 0 & 0 \end{bmatrix}$$

$$\underset{(b)}{\longmapsto} \begin{bmatrix} 1 & 0 & -3 & | & 0 & 0 & -1 \\ 0 & 2 & 4 & | & 0 & 1 & 1 \\ 0 & -1 & 6 & | & 1 & 0 & 2 \end{bmatrix}$$

$$\xrightarrow[(c)]{}\begin{bmatrix} 1 & 0 & -3 & 0 & 0 & -1 \\ 0 & 1 & 2 & 0 & 1/2 & 1/2 \\ 0 & 1 & -6 & -1 & 0 & -2 \end{bmatrix}$$

$$\xrightarrow[(b)]{}\begin{bmatrix} 1 & 0 & -3 & 0 & 0 & -1 \\ 0 & 1 & 2 & 0 & 1/2 & 1/2 \\ 0 & 0 & -8 & -1 & -1/2 & -5/2 \end{bmatrix}$$

$$\xrightarrow[(c)]{}\begin{bmatrix} 1 & 0 & -3 & 0 & 0 & -1 \\ 0 & 1 & 2 & 0 & 1/2 & 1/2 \\ 0 & 0 & 1 & 1/8 & 1/16 & 5/16 \end{bmatrix}$$

$$\xrightarrow[(b)]{}\begin{bmatrix} 1 & 0 & 0 & 3/8 & 3/16 & -1/16 \\ 0 & 1 & 0 & -1/4 & 3/8 & -1/8 \\ 0 & 0 & 1 & 1/8 & 1/16 & 5/16 \end{bmatrix}$$

Therefore

$$A^{-1} = \begin{bmatrix} 3/8 & 3/16 & -1/16 \\ -1/4 & 3/8 & -1/8 \\ 1/8 & 1/16 & 5/16 \end{bmatrix}$$

We can also use Theorem 5.13 in problems involving polynomial interpolation. For this discussion, we will assume $F = \mathbb{R}$. Recall that a (real) polynomial $p(t)$ of degree n is an expression of the form $p(t) = a_0 + a_1 t + \cdots + a_n t^n$. Here $a_0 \ldots, a_n$ are fixed real numbers, with $a_n \neq 0$, and t is a real variable. In this book, we will let t denote a real (or when appropriate a complex) variable. This will cause no confusion with our previous notation A^t (the transpose of a matrix A) since the contexts are so different. Notice that p defines a function from $\mathbb{R}$ to $\mathbb{R}$ given by $p(c) = a_0 + a_1 c + \cdots + a_n c^n$ for all $c \in \mathbb{R}$.

The only fact we will need about polynomials is the following theorem from algebra. If $p(t)$ is a nonzero polynomial, then the number of roots of $p(t)$ [in $\mathbb{C}$] cannot exceed the degree of p. We will use this result in proving the basic interpolation theorem.

Theorem 5.16 Suppose $a_0, \ldots, a_n$ are $n + 1$ distinct real numbers. Then for any real numbers $b_0, \ldots, b_n$, there exists a polynomial $p(t)$ whose degree is at most n and such that $p(a_i) = b_i$ for all $i = 0, \ldots, n$.

Proof.   Form the following $(n + 1) \times (n + 1)$ matrix V.

5.17

$$V = \begin{bmatrix} 1 & a_0 & a_0^2 & \cdots & a_0^n \\ 1 & a_1 & a_1^2 & \cdots & a_1^n \\ \vdots & \vdots & \vdots & & \vdots \\ 1 & a_n & a_n^2 & \cdots & a_n^n \end{bmatrix}$$

V is called the Vandermonde matrix (for $a_0, \ldots, a_n$). We claim V is an invertible matrix. By Theorem 5.13, it is enough to argue the only solution to $VX = O$ is $X = O$. Suppose $VX = O$ has a nonzero solution $\xi = [c_0 \cdots c_n]^t$. Then $V\xi = O$, and at least one c_i is not zero. Let $q(t) = c_0 + c_1 t + \cdots + c_n t^n$. Since some c_i is not zero, $q(t)$ is a nonzero polynomial of degree at most n. $V\xi = O$ is precisely the statement $q(a_i) = 0$ for $i = 0, \ldots, n$. Thus, $q(t)$ has at least $n + 1$ distinct roots $a_0, \ldots, a_n$. This is impossible. We conclude that $VX = O$ has only the trivial solution $X = O$. In particular, V is invertible.

Now set $\beta = [b_0 \cdots b_n]^t$. Since V is invertible, $VX = \beta$ has the unique solution $V^{-1}\beta = [z_0 \cdots z_n]^t$. Set $p(t) = z_0 + \cdots + z_n t^n$. Then $V(V^{-1}\beta) = \beta$ implies $p(a_i) = b_i$ for all $i = 0, \ldots, n$. Since the degree of $p(t)$ is at most n, our proof is complete. $\square$

Since the Vandermonde matrix is invertible, the equation $VX = \beta$ has only one solution. In particular, there is only one polynomial $p(t)$ of degree at most n and such that $p(a_i) = b_i$ for all $i = 0, \ldots, n$. This polynomial is usually called the interpolating polynomial for the data $(a_0, b_0), \ldots, (a_n, b_n)$. The interpolating polynomial is the unique polynomial of degree at most n whose graph passes through the points $(a_0, b_0), \ldots, (a_n, b_n)$.

Example 5.18   Find the interpolating polynomial for the data $(-1, 1)$, $(0, 1)$, and $(1, 2)$.

In this problem, $n = 2$, $a_0 = -1$, $a_1 = 0$, and $a_2 = 1$. The Vandermonde matrix for these values is

$$\begin{bmatrix} 1 & -1 & 1 \\ 1 & 0 & 0 \\ 1 & 1 & 1 \end{bmatrix} = V$$

The inverse of V is computed using the algorithm in (5.14).

$$V^{-1} = \begin{bmatrix} 0 & 1 & 0 \\ -1/2 & 0 & 1/2 \\ 1/2 & -1 & 1/2 \end{bmatrix}$$

The solution to

$$VX = \begin{bmatrix} 1 \\ 1 \\ 2 \end{bmatrix}$$

is

$$V^{-1}\begin{bmatrix} 1 \\ 1 \\ 2 \end{bmatrix} = \begin{bmatrix} 1 \\ 1/2 \\ 1/2 \end{bmatrix}$$

Hence, the interpolating polynomial is

$$p(t) = 1 + (1/2)t + (1/2)t^2$$

One final application of Theorem 5.13 is worth mentioning here. Invertible matrices are nothing but products of elementary matrices. Consequently, Corollary 5.11 can be re-worded.

Theorem 5.19 Let $A, B \in M_{m \times n}$. Then $A \underset{r}{\sim} B$ if and only if there exists an $m \times m$ invertible matrix P such that $PA = B$.

Recall that the row space, $RS(A)$, of an $m \times n$ matrix A is the set of all linear combinations of the rows of A. Thus, $RS(A) \subseteq M_{1 \times n}$. Matrices which are row equivalent have the same row space.

Corollary 5.20 Let A and B be two $m \times n$ matrices. If A is row equivalent to B, then $RS(A) = RS(B)$.

Proof. Suppose $A \underset{r}{\sim} B$. By Theorem 5.19, there exists an invertible matrix P such that $PA = B$. Then (3.23) implies $RS(B) = RS(PA) \subseteq RS(A)$. If $A \underset{r}{\sim} B$, then $B \underset{r}{\sim} A$. Thus, reversing the roles of A and B in this proof gives $RS(A) \subseteq RS(B)$. Therefore, $RS(A) = RS(B)$. □

The converse of Corollary 5.20 is also true. We will give a proof of this fact in Chapter II.

We close this section with a few remarks about column operations. If $A \in M_{m \times n}$ and E is an elementary matrix of size $n \times n$, then multiplying A on the right by E performs a column operation on A. Suppose A is partitioned into rows

$$A = \begin{bmatrix} \text{Row}_1(A) \\ \vdots \\ \text{Row}_m(A) \end{bmatrix}$$

Then by Theorem 3.10

$$AE = \begin{bmatrix} \text{Row}_1(A)E \\ \vdots \\ \text{Row}_m(A)E \end{bmatrix}$$

Thus, the action of E on A is determined by what E does to row vectors. If we multiply a row vector by an elementary matrix of each type, we get the equations analogous to those in (5.9). Hence, we have

Theorem 5.21   Let $A \in M_{m \times n}$.

(a) AE_{ij} is the $m \times n$ matrix obtained from A in interchanging columns i and j of A.

(b) $AE_{ij}(c)$ is the $m \times n$ matrix obtained from A by adding c times column i of A to column j of A.

(c) $AE_i(c)$ is the $m \times n$ matrix obtained from A by multiplying column i of A by c.

Notice that statement (b) in Theorem 5.21 is slightly different from the corresponding row statement in Theorem 5.8(b). At any rate, any elementary column operation on A can be performed by multiplying A on the right with a suitable elementary matrix. In particular, we have the analogs of Corollary 5.11, Theorem 5.19, and Corollary 5.20.

Theorem 5.22   Let $A, B \in M_{m \times n}$.

(a) $A \underset{c}{\sim} B$ if and only if there exists a finite sequence of elementary matrices $E_1, \ldots, E_p$ such that $AE_pE_{p-1} \cdots E_1 = B$.

(b) $A \underset{c}{\sim} B$ if and only if there exists an invertible matrix Q such that $AQ = B$.

(c)   If $A \underset{c}{\sim} B$, then $CS(A) = CS(B)$.

Sometimes it is convenient to do both row and column operations on a matrix. This leads to the following definition.

Definition 5.23 Let A, $B \in M_{m \times n}$. A and B are equivalent if there exists invertible matrices P and Q such that $PAQ = B$.

Thus, A and B are equivalent if B can be obtained from A by applying a finite number of row and column operations to A. If A and B are equivalent, we will write $A \approx B$. A simple argument shows $\approx$ is an equivalence relation. This means the following.

5.24 (a) $A \approx A$.
     (b) $A \approx B$ if and only if $B \approx A$.
     (c) If $A \approx B$ and $B \approx C$, then $A \approx C$.

What is the simplest matrix equivalent to a given matrix A? This question is easy to answer. By Theorem 5.19, there exists an invertible matrix P such that $PA = E$ is a matrix in row reduced echelon form. If E has k nonzero rows, then using elementary column operations, E can be reduced to the block matrix

$$\left[\begin{array}{c|c} I_k & O \\ \hline O & O \end{array}\right].$$

Thus, we have the following theorem.

Theorem 5.25 Let A be a nonzero $m \times n$ matrix. Then there exists a positive integer k with $k \leqslant \min\{m, n\}$ and invertible matrices P and Q such that

$$PAQ = \left[\begin{array}{c|c} I_k & O \\ \hline O & O \end{array}\right]$$

Exercises for Section 5

5.1   Complete the proofs of (b) and (c) in Corollary 5.10.
5.2   For each matrix A listed below, find a sequence of elementary matrices $E_1, \ldots, E_p$ such that $E_p \cdots E_1 A$ is in row reduced echelon form.

(a)
$$A = \begin{bmatrix} 2 & 1 \\ 3 & 4 \end{bmatrix}$$

(b)
$$A = \begin{bmatrix} 2 & 3 & 4 \\ -1 & 1 & 1 \\ 1 & 0 & 1 \end{bmatrix}$$

(c)
$$A = \begin{bmatrix} 5 & 0 & 1 & 16 \\ 1 & -1 & -1 & 2 \\ 1 & 0 & 1 & 3 \end{bmatrix}$$

5.3   Use Theorem 5.13 to show that

$$A = \begin{bmatrix} a & b \\ c & d \end{bmatrix}$$

is invertible if and only if $ad - bc \neq 0$.

5.4   Let $A \in M_{m \times n}$. Suppose $i_1, \ldots, i_r$ are positive integers for which $1 \leqslant i_1 < \cdots < i_r \leqslant m$. Describe $E_{i_1 i_r} E_{i_1 i_{r-1}} \cdots E_{i_1 i_2} A$.

5.5   Use the algorithm in (5.14) to compute the inverses of the following matrices:

(a)
$$\begin{bmatrix} 1 & 0 & 0 \\ 2 & 1 & 0 \\ 1 & -1 & 3 \end{bmatrix}$$

(b)
$$\begin{bmatrix} -2 & 1 & 3 \\ 0 & -1 & 1 \\ 1 & 2 & 0 \end{bmatrix}$$

(c)
$$\begin{bmatrix} 1 & 2 & 0 \\ 0 & 1 & -1 \\ 0 & 0 & 2 \end{bmatrix}$$

(d)
$$\begin{bmatrix} 1 & 0 & -1 & 1 \\ 0 & 1 & 2 & 1 \\ -1 & 1 & 1 & 0 \\ 1 & 2 & 1 & -2 \end{bmatrix}$$

5.6   Suppose A is invertible. Use the algorithm in (5.14) to prove the following theorems:
(a) If A is lower triangular, so is A^{-1}

      (b) If A is upper triangular, so is A^{-1}

      (c) If A is diagonal, so is A^{-1}

5.7   Is Theorem 5.16 true for complex polynomials $p(t) = z_0 + \cdots + z_n t^n$ $(z_0, \ldots, z_n \in \mathbb{C})$ and complex points $(a_0, b_0), \ldots, (a_n, b_n)$?

5.8   Find the interpolating polynomial for the data indicated.

      (a) $(-1, 2)$, $(0, 1)$, and $(1, 3)$.

      (b) $(-1, 7)$, $(0, 5)$, $(1, 7)$, and $(2, 19)$.

      (c) $(1, 5)$, $(2, 7)$, and (3.9).

5.9   If $a_0, \ldots, a_n$ are all distinct, then there is only one polynomial $p(t)$ of degree at most n passing through the points $(a_0, b_0), \ldots, (a_n, b_n)$. Is this statement still true if p is allowed to have degree bigger than n?

5.10   Complete the details of the proof of Theorem 5.21.

5.11   Give a detailed proof of Theorem 5.25.

5.12   Show that

$$\begin{bmatrix} 5 & 1 & 6 & 13 \\ 1 & 0 & 1 & 2 \\ 1 & 1 & 2 & 5 \end{bmatrix} \approx \begin{bmatrix} 1 & 0 & 0 & 0 \\ 0 & 1 & 0 & 0 \\ 0 & 0 & 0 & 0 \end{bmatrix}$$

5.13   Show that

$$\begin{bmatrix} 5 & 1 & 6 & 13 \\ 1 & 0 & 1 & 2 \\ 1 & 1 & 2 & 5 \end{bmatrix} \not\approx \begin{bmatrix} 6 & 0 & -1 & 4 \\ 1 & 1 & 2 & 3 \\ 2 & 3 & 1 & 2 \end{bmatrix}$$

5.14   Suppose A is invertible. Show that if A is symmetric, then A^{-1} is also symmetric.

5.15   Show that $A \underset{c}{\sim} B$ if and only if $A^t \underset{r}{\sim} B^t$.

5.16   Let $A, B \in M_{n \times n}$. A and B are similar if there exists an invertible matrix P such that $PAP^{-1} = B$. If A is similar to B, we will write $A \sim B$. Show that $\sim$ is an equivalence relation.

5.17   Show that

$$\begin{bmatrix} -4 & 15 \\ -2 & 7 \end{bmatrix} \sim \begin{bmatrix} 1 & 0 \\ 0 & 2 \end{bmatrix}$$

5.18   Find constants c_1, c_2, c_3, and c_4 such that $f(t) = c_1 + c_2 \sin(t) + c_3 \cos(t) + c_4 \cos(2t)$ has the following values:

$$f(-\pi/2) = 1, \quad f(0) = 2, \quad f(\pi/4) = -1, \quad f(\pi/2) = 3$$

5.19   Prove the converse of Theorem 2.30(a). If AB is invertible, then A and B are both invertible.

6. *LU* FACTORIZATIONS

In this section, we examine the process of Gaussian elimination more carefully. We have seen that row operations are performed on a matrix A by multiplying A on the left by elementary matrices. Suppose A is to be reduced to an upper triangular matrix U. This is the most important step in solving linear equations (see (2) of (4.10)). We had pointed out that elementary matrices of type $E_i(c)$ are not needed in this procedure. We therefore concentrate on elementary matrices of the type E_{ij} and $E_{ij}(c)$. Let us introduce the following definition.

Definition 6.1 An elementary matrix of the type E_{ij} is called a transposition. Any finite product of transpositions is called a permutation matrix.

Example 6.2

$$E_{12} = \begin{bmatrix} 0 & 1 & 0 \\ 1 & 0 & 0 \\ 0 & 0 & 1 \end{bmatrix} \quad \text{and} \quad E_{13} = \begin{bmatrix} 0 & 0 & 1 \\ 0 & 1 & 0 \\ 1 & 0 & 0 \end{bmatrix}$$

are transpositions.

$$P = E_{12}E_{13} = \begin{bmatrix} 0 & 1 & 0 \\ 0 & 0 & 1 \\ 1 & 0 & 0 \end{bmatrix}$$

is a permutation matrix.

Notice that P in Example 6.2 is obtained from I_3 by permuting, i.e., interchanging, the rows of I_3. Any permutation matrix is obtained from I_m by permuting the rows of I_m. This follows from Exercise 5.4. We will allow i and j to be equal in E_{ij}. Thus, the identity matrix I_m is a transposition, and hence a permutation matrix. When row reducing a matrix A to an upper triangular matrix U, it may be necessary to interchange rows of A. Thus, it may be necessary to multiply A on the left by a permutation matrix.

When row reducing A to an upper triangular matrix, we need to produce zeros below a given nonzero entry in a given column. We had seen in Section 4 that this is accomplished by applying to A a sequence of elementary row operations of type (b) from (4.8). In other words, we multiply A on the left by a sequence of elementary matrices of the following form

$$E_{mi}(c_m)E_{m-1i}(c_{m-1}) \cdots E_{i+2i}(c_{i+2})E_{i+1i}(c_{i+1})$$

Let us analyze this product carefully. Consider the following $m \times m$ matrix L.

6.3

$$L = \begin{bmatrix} 1 & 0 & \cdots & 0 & 0 & \cdots & \cdots & 0 \\ \vdots & & & \vdots & & & & \vdots \\ 0 & \cdots & \cdots & 1 & 0 & \cdots & \cdots & 0 \\ 0 & \cdots & \cdots & c_{i+1} & 1 & 0 & \cdots & 0 \\ \vdots & & & \vdots & & & & \vdots \\ 0 & \cdots & \cdots & c_m & 0 & \cdots & \cdots & 1 \end{bmatrix} \begin{matrix} \\ \\ i \\ i+1 \\ \\ \end{matrix}$$

$$i$$

In equation (6.3), the constants $c_{i+1}, \ldots, c_m$ lie in the ith column of L below the (i, i)th entry. Thus, L has the same entries as I_m except that $[L]_{i+1\,i} = c_{i+1}, \ldots, , [L]_{mi} = c_m$. The index i here can be any positive integer between 1 and $m-1$, and the constants $c_{i+1}, \ldots, c_m$ any scalars from $F (= \mathbb{R}$ or $\mathbb{C})$. If $c_{i+1} = \cdots = c_m = 0$, then L is just I_m. In any case, L is always a lower triangular, square matrix for any choice of i and $c_{i+1}, \ldots, c_m$. Our first observation about L is the following identity.

6.4 $$L = E_{mi}(c_m)E_{m-1i}(c_{m-1}) \cdots E_{i+2i}(c_{i+2})E_{i+1i}(c_{i+1})$$

Equation (6.4) follows directly from Theorem 5.8(b). $E_{i+1i}(c_{i+1})$ is obtained from I_m by adding c_{i+1} times the ith row of I_m to row $i+1$. Thus, $E_{i+1i}(c_{i+1})$ looks like I_m except for column i which has the following form.

6.5 $$\mathrm{Col}_i(E_{i+1i}(c_{i+1})) = [0 \quad \cdots \quad 0 \quad 1 \quad c_{i+1} \quad 0 \quad \cdots \quad 0]^t$$
$$\phantom{\mathrm{Col}_i(E_{i+1i}(c_{i+1})) = [0 \quad \cdots \quad 0 \quad } i \quad\; i+1$$

$E_{i+2i}(c_{i+2})E_{i+1i}(c_{i+1})$ is the $m \times m$ matrix obtained from $E_{i+1i}(c_{i+1})$ by adding c_{i+2} times row i of $E_{i+1i}(c_{i+1})$ to row $i+2$ of $E_{i+1i}(c_{i+1})$. Thus, $E_{i+2i}(c_{i+2})E_{i+1i}(c_{i+1})$ looks like I_m except for column i which has the following form

6.6

$$\mathrm{Col}_i(E_{i+2i}(c_{i+2})E_{i+1i}(c_{i+1})) = [0 \quad \cdots \quad 0 \quad 1 \quad c_{i+1} \quad c_{i+2} \quad 0 \quad \cdots \quad 0]^t$$
$$\phantom{\mathrm{Col}_i(E_{i+2i}(c_{i+2})E_{i+1i}(c_{i+1})) = [0 \quad \cdots \quad 0 \quad } i \quad\; i+1 \quad i+2$$

Continuing with this argument, we derive the identity in (6.4).

The equation in (6.4) implies any $m \times m$ matrix L of the type given in equation (6.3) is a product of elementary matrices. In particular, L is invertible. If A is any $m \times n$ matrix, then LA is easily computed. We begin with a column vector $\xi = [x_1 \cdots x_m]^t$. Suppose L is the matrix in (6.3). Then we have

6.7 $\qquad L\xi = [x_1 \quad \cdots \quad x_i \quad x_{i+1} + c_{i+1}x_i \quad \cdots \quad x_m + c_m x_i]^t$

We can easily generalize the result in (6.7) by using Theorem 3.10. If A is partitioned into columns $A = [A_1 | \cdots | A_n]$, then $LA = [LA_1 | \cdots | LA_n]$. Switching to rows, we have

6.8 If

$$A = \begin{bmatrix} R_1 \\ \vdots \\ R_m \end{bmatrix}$$

then

$$LA = \begin{bmatrix} R_1 \\ \vdots \\ R_i \\ R_{i+1} + c_{i+1}R_i \\ \vdots \\ R_m + c_m R_i \end{bmatrix}$$

In equation (6.8), A is partitioned into rows $R_1, \ldots, R_m$. Thus, multiplying A by L has the effect of adding c_{i+1} times row i of A to row $i + 1$ of A, c_{i+2} times row i of A to row $i + 2$ of A and so on out to c_m times row i of A to row m of A. This is precisely the sort of thing done in Gaussian elimination to produce zeros below a given nonzero entry. Consider the following example.

Example 6.9 Let

$$A = \begin{bmatrix} 1 & 0 & 1 & 2 \\ 2 & 0 & 1 & 3 \\ -1 & 1 & 2 & 1 \end{bmatrix}$$

If

$$L = \begin{bmatrix} 1 & 0 & 0 \\ -2 & 1 & 0 \\ 1 & 0 & 1 \end{bmatrix}$$

then

$$LA = \begin{bmatrix} 1 & 0 & 1 & 2 \\ 0 & 0 & -1 & -1 \\ 0 & 1 & 3 & 3 \end{bmatrix} \quad \text{and} \quad E_{23}(LA) = \begin{bmatrix} 1 & 0 & 1 & 2 \\ 0 & 1 & 3 & 3 \\ 0 & 0 & -1 & -1 \end{bmatrix}$$

Thus, multiplication by $E_{23}L$ converts A to an upper triangular matrix.

We can now describe Gaussian elimination in terms of transpositions and matrices of the type in (6.3).

Theorem 6.10 Let $A \in M_{m \times n}$. There exist a finite number of transpositions $E_1, \ldots, E_k$ and a finite number of matrices $L_1, \ldots, L_k$ of the type given in equation (6.3) such that $L_k E_k \cdots L_1 E_1 A$ is upper triangular.

Proof. We follow the proof of Theorem 4.18 using transpositions to perform row operations of the type (a) of (4.8), and matrices like L in (6.3) to produce columns of zeros. We can assume $m > 1$.

If $A = O$, then A is already upper triangular, and there is nothing to do. We can set $E_1 = L_1 = I_m$. Hence, we can assume $A \neq O$. Suppose j is the number of the first column of A which contains a nonzero entry. Then after performing a suitable interchange of rows, i.e., after multiplying by a suitable transposition E_1, we have

6.11

$$E_1 A = \begin{bmatrix} 0 & \cdots & 0 & a_{1j} & * & \cdots & * \\ \vdots & & \vdots & \vdots & & & \vdots \\ 0 & \cdots & 0 & a_{mj} & * & \cdots & * \end{bmatrix}$$

Here $a_{1j} \neq 0$. The $*$ denote entries of $E_1 A$ whose precise values are not relevant to the proof. Now let L_1 be the following $m \times m$ matrix.

6.12

$$L_1 = \begin{bmatrix} 1 & 0 & 0 & \cdots & & 0 \\ -a_{2j}/a_{1j} & 1 & 0 & \cdots & & 0 \\ \vdots & \vdots & \vdots & & & \vdots \\ -a_{mj}/a_{1j} & 0 & 0 & \cdots & 0 & 1 \end{bmatrix}$$

Using equation (6.8), we have

6.13
$$L_1 E_1 A = \begin{bmatrix} 0 & \cdots & 0 & a_{1j} & * & \cdots & * \\ 0 & \cdots & 0 & 0 & & & \\ 0 & \cdots & 0 & 0 & & A' & \end{bmatrix}$$

If the submatrix A' in (6.13) is zero, then $L_1 E_1 A$ is upper triangular, and the proof is finished. Suppose A' has a nonzero entry. Let l denote the number of the first column of $L_1 E_1 A$ in which A' has a nonzero entry. Then $j < l$, and there exists a transposition E_2 such that $E_2(L_1 E_1 A)$ has the following form.

6.14
$$E_2 L_1 E_1 A = \begin{bmatrix} 0 & \cdots & 0 & a_{1j} & * & \cdots & * & * & * & \cdots & * \\ 0 & \cdots & 0 & 0 & 0 & \cdots & 0 & b_{2l} & * & \cdots & * \\ \vdots & & \vdots & \vdots & \vdots & & \vdots & \vdots & \vdots & & \vdots \\ 0 & \cdots & 0 & 0 & 0 & \cdots & 0 & b_{ml} & * & \cdots & * \end{bmatrix}$$

In equation (6.14), $b_{2l} \neq 0$. Now let L_2 be the following $m \times m$ matrix.

6.15
$$L_2 = \begin{bmatrix} 1 & 0 & 0 & \cdots & 0 \\ 0 & 1 & 0 & \cdots & 0 \\ 0 & -b_{3l}/b_{2l} & 1 & \cdots & 0 \\ \vdots & \vdots & \vdots & & \vdots \\ 0 & -b_{ml}/b_{2l} & 0 & \cdots & 0 & 1 \end{bmatrix}$$

Again using (6.8), we have

6.16
$$L_2 E_2 L_1 E_1 A = \begin{bmatrix} 0 & \cdots & 0 & a_{1j} & * & \cdots & * & * & * & \cdots & * \\ 0 & \cdots & 0 & 0 & 0 & \cdots & 0 & b_{2l} & * & \cdots & * \\ 0 & \cdots & 0 & 0 & 0 & \cdots & 0 & 0 & & & \\ \vdots & & \vdots & \vdots & \vdots & & \vdots & \vdots & & A' & \\ 0 & \cdots & 0 & 0 & 0 & \cdots & 0 & 0 & & & \end{bmatrix}$$

Obviously at most $m - 1$ applications of this argument reduces A to an upper triangular matrix.   □

The most interesting case of Theorem 6.10 occurs when no transpositions are needed to reduce A to an upper triangular matrix U. In this case, there exist lower triangular matrices $L_{m-1}, \ldots, L_1$ as in (6.3) such that

$L_{m-1} \cdots L_1 A = U$. It follows from the proof of Theorem 6.10, that each L_i has the following form.

6.17

$$L_1 = \begin{bmatrix} 1 & 0 & \cdots & \cdots & 0 \\ l_{21} & 1 & 0 & \cdots & 0 \\ \vdots & \vdots & & & \vdots \\ l_{m1} & 0 & \cdots & 0 & 1 \end{bmatrix},$$

$$L_2 = \begin{bmatrix} 1 & 0 & 0 & 0 & \cdots & 0 \\ 0 & 1 & 0 & 0 & \cdots & 0 \\ 0 & l_{32} & 1 & 0 & \cdots & 0 \\ \vdots & \vdots & \vdots & & & \vdots \\ 0 & l_{m2} & 0 & \cdots & 0 & 1 \end{bmatrix} \cdots L_{m-1} = \begin{bmatrix} 1 & 0 & \cdots & & 0 & 0 \\ \vdots & & & & \vdots & \vdots \\ 0 & \cdots & & & 1 & 0 \\ 0 & \cdots & & & l_{mm-1} & 1 \end{bmatrix}$$

(If $m = 1$, A is already upper triangular. We assume $m > 1$ in the rest of this section.)

The reader should understand that some of the L_i listed in (6.17) could be I_m to allow for zero columns or the case $m > n$ in the reduction. We had noted any lower triangular matrix of the type in (6.3) is invertible. Equation (6.8) implies the following formula for an inverse.

6.18

$$\begin{bmatrix} 1 & 0 & \cdots & 0 & 0 & \cdots & 0 \\ \vdots & & & \vdots & \vdots & & \vdots \\ 0 & \cdots & 0 & 1 & 0 & \cdots & 0 \\ 0 & \cdots & 0 & c_{i+1} & 1 & \cdots & 0 \\ \vdots & & & \vdots & \vdots & & \vdots \\ 0 & \cdots & 0 & c_m & 0 & \cdots & 1 \end{bmatrix}^{-1}$$

$$= \begin{bmatrix} 1 & 0 & \cdots & 0 & 0 & \cdots & 0 \\ \vdots & & & \vdots & \vdots & & \vdots \\ 0 & \cdots & 0 & 1 & 0 & \cdots & 0 \\ 0 & \cdots & 0 & -c_{i+1} & 1 & \cdots & 0 \\ \vdots & & & \vdots & \vdots & & \vdots \\ 0 & \cdots & 0 & -c_m & 0 & \cdots & 1 \end{bmatrix}$$

Now if $L_{m-1} \cdots L_1 A = U$, then $A = L_1^{-1} L_2^{-1} \cdots L_{m-1}^{-1} U$. Equation (6.18) implies $L_1^{-1} \cdots L_{m-1}^{-1}$ is the following product of lower triangular matrices.

6.19

$$L_1^{-1} L_2^{-1} \cdots L_{m-1}^{-1} = \begin{bmatrix} 1 & 0 & 0 & \cdots & 0 \\ -l_{21} & 1 & 0 & \cdots & 0 \\ \vdots & \vdots & \vdots & & \vdots \\ -l_{m1} & 0 & 0 & \cdots & 1 \end{bmatrix} \begin{bmatrix} 1 & 0 & 0 & \cdots & 0 \\ 0 & 1 & 0 & \cdots & 0 \\ 0 & -l_{32} & 1 & \cdots & 0 \\ \vdots & \vdots & \vdots & & \vdots \\ 0 & -l_{m2} & 0 & \cdots & 1 \end{bmatrix}$$

$$\cdots \begin{bmatrix} 1 & \cdots & 0 & 0 \\ \vdots & & \vdots & \vdots \\ 0 & \cdots & 1 & 0 \\ 0 & \cdots & -l_{mm-1} & 1 \end{bmatrix}$$

We analyze the product in equation (6.19) by first computing $L_{m-2}^{-1} L_{m-1}^{-1}$, then $L_{m-3}^{-1} (L_{m-2}^{-1} L_{m-1}^{-1})$, etc. Using equation (6.8) for these products, we have the following equation.

6.20

$$L_1^{-1} L_2^{-1} \cdots L_{m-1}^{-1} = \begin{bmatrix} 1 & 0 & \cdots & \cdots & 0 & 0 \\ -l_{21} & 1 & \cdots & \cdots & 0 & 0 \\ -l_{31} & -l_{32} & 1 & \cdots & 0 & 0 \\ \vdots & \vdots & & & \vdots & \vdots \\ -l_{m1} & -l_{m2} & \cdots & \cdots & -l_{mm-1} & 1 \end{bmatrix}$$

Thus, A is a product of a lower triangular matrix $L_1^{-1} \cdots L_{m-1}^{-1}$ and an upper triangular matrix U. Let us record these facts in the following theorem.

Theorem 6.21 Let $A \in M_{m \times n}$. Suppose A can be row reduced to an upper triangular matrix U without using transpositions. Then $A = LU$ with L a lower $m \times m$ matrix of the form

$$L = \begin{bmatrix} 1 & 0 & \cdots & \cdots & 0 & 0 \\ -l_{21} & 1 & \cdots & \cdots & 0 & 0 \\ -l_{31} & -l_{32} & 1 & \cdots & 0 & 0 \\ \vdots & \vdots & & & \vdots & \vdots \\ -l_{m1} & -l_{m2} & \cdots & \cdots & -l_{mm-1} & 1 \end{bmatrix}$$

Furthermore the columns of L determine the lower triangular matrices L_1, ..., L_{m-1} given in equation (6.17) for which $L_{m-1} \cdots L_1 A = U$.

The factorization of A obtained in Theorem 6.21 is called an *LU*-factorization of A. If A can be written as a product of two matrices $A = LU$ with L an $m \times m$, lower triangular matrix having 1s on its diagonal, and U an $m \times n$ upper triangular matrix, then the product LU is called an *LU*-factorization of A. If A can be row reduced to an upper triangular matrix without using transpositions, then Theorem 6.21 guarantees A has an *LU*-factorization. Although *LU*-factorizations are not in general unique (see Exercise 6.6), we will see that such factorizations can be very useful.

Example 6.22 Let

$$A = \begin{bmatrix} 1 & 4 & 0 & 0 \\ 2 & 1 & 1 & 1 \\ 3 & 5 & 2 & 3 \end{bmatrix}$$

Set

$$L_1 = \begin{bmatrix} 1 & 0 & 0 \\ -2 & 1 & 0 \\ -3 & 0 & 1 \end{bmatrix}$$

Then

$$L_1 A = \begin{bmatrix} 1 & 4 & 0 & 0 \\ 0 & -7 & 1 & 1 \\ 0 & -7 & 2 & 3 \end{bmatrix}$$

Set

$$L_2 = \begin{bmatrix} 1 & 0 & 0 \\ 0 & 1 & 0 \\ 0 & -1 & 1 \end{bmatrix}$$

Then

$$L_2 L_1 A = \begin{bmatrix} 1 & 4 & 0 & 0 \\ 0 & -7 & 1 & 1 \\ 0 & 0 & 1 & 2 \end{bmatrix} = U$$

an upper triangular matrix. Then

$$L_1^{-1}L_2^{-1}U = \begin{bmatrix} 1 & 0 & 0 \\ 2 & 1 & 0 \\ 3 & 1 & 1 \end{bmatrix} \begin{bmatrix} 1 & 4 & 0 & 0 \\ 0 & -7 & 1 & 1 \\ 0 & 0 & 1 & 2 \end{bmatrix}$$

is an LU-factorization of A.

We have seen in the proof of Theorem 6.10 that at most $m - 1$ transpositions and matrices of the type in equation (6.3) are needed to reduce A to an upper triangular matrix U. In Example 6.22 $m - 1 = 2$, and it requires two L_i to reduce A to U. In our next example, $m > n$, and it requires fewer than $(m - 1)$ L_i to reduce A to an upper triangular matrix.

Example 6.23 Let

$$A = \begin{bmatrix} 1 & 2 \\ 1 & 3 \\ 2 & 1 \\ 1 & 4 \end{bmatrix}$$

Set

$$L_1 = \begin{bmatrix} 1 & 0 & 0 & 0 \\ -1 & 1 & 0 & 0 \\ -2 & 0 & 1 & 0 \\ -1 & 0 & 0 & 1 \end{bmatrix}$$

Then

$$L_1 A = \begin{bmatrix} 1 & 2 \\ 0 & 1 \\ 0 & -3 \\ 0 & 2 \end{bmatrix}$$

Next set

$$L_2 = \begin{bmatrix} 1 & 0 & 0 & 0 \\ 0 & 1 & 0 & 0 \\ 0 & 3 & 1 & 0 \\ 0 & -2 & 0 & 1 \end{bmatrix}$$

Then

$$L_2 L_1 A = \begin{bmatrix} 1 & 2 \\ 0 & 1 \\ 0 & 0 \\ 0 & 0 \end{bmatrix} = U$$

an upper triangular matrix. Thus, an LU-factorization of A is as follows:

$$\begin{bmatrix} 1 & 2 \\ 1 & 3 \\ 2 & 1 \\ 1 & 4 \end{bmatrix} = \begin{bmatrix} 1 & 0 & 0 & 0 \\ 1 & 1 & 1 & 0 \\ 2 & -3 & 1 & 0 \\ 1 & 2 & 0 & 1 \end{bmatrix} \begin{bmatrix} 1 & 2 \\ 0 & 1 \\ 0 & 0 \\ 0 & 0 \end{bmatrix}$$

In this example, $m - 1 = 3$, and it takes two L_i to reduce A to an upper triangular matrix U.

The next question to think about is: What happens if transpositions are needed in order to row reduce A to an upper triangular matrix? It turns out that an LU-factorization is still available provided we are willing to permute the rows of A first. We need the following lemma.

Lemma 6.24 Let L be the $m \times m$ matrix given in equation (6.3). Let E be a transposition of the form

$$E = \left[\begin{array}{c|c} I_i & O \\ \hline O & Q \end{array} \right]$$

where Q is an $(m - i) \times (m - i)$ transposition. Set

$$Q \begin{bmatrix} c_{i+1} \\ \vdots \\ c_m \end{bmatrix} = \begin{bmatrix} y_{i+1} \\ \vdots \\ y_m \end{bmatrix}.$$

Then $EL = L'E$ where L' is the $m \times m$ matrix

$$L' = \begin{bmatrix} 1 & 0 & \cdots & 0 & 0 & \cdots & 0 \\ \vdots & & & \vdots & \vdots & & \vdots \\ 0 & \cdots & \cdots & 1 & 0 & \cdots & 0 \\ 0 & \cdots & \cdots & y_{i+1} & 1 & \cdots & 0 \\ \vdots & & & \vdots & \vdots & & \vdots \\ 0 & \cdots & \cdots & y_m & 0 & \cdots & 1 \end{bmatrix}$$

Proof. This is just a simple exercise using Theorem 3.10. Partition

$$L = \left[\begin{array}{c|c} I_i & O \\ \hline Z & I_{m-i} \end{array}\right]$$

Here Z is the $(m-i) \times i$ matrix

$$\begin{bmatrix} 0 & \cdots & 0 & c_{i+1} \\ \vdots & & \vdots & \vdots \\ 0 & \cdots & 0 & c_m \end{bmatrix}$$

Then

$$EL = \left[\begin{array}{c|c} I_i & O \\ \hline O & Q \end{array}\right]\left[\begin{array}{c|c} I_i & O \\ \hline Z & I_{m-i} \end{array}\right] = \left[\begin{array}{c|c} I_i & O \\ \hline QZ & QI_{m-i} \end{array}\right] = \left[\begin{array}{c|c} I_i & O \\ \hline Z' & Q \end{array}\right]$$

Here

$$Z' = QZ = \begin{bmatrix} 0 & \cdots & 0 & y_{i+1} \\ \vdots & & \vdots & \vdots \\ 0 & \cdots & 0 & y_m \end{bmatrix}$$

On the other hand

$$L'E = \left[\begin{array}{c|c} I_i & O \\ \hline Z' & I_{m-i} \end{array}\right]\left[\begin{array}{c|c} I_i & O \\ \hline O & Q \end{array}\right] = \left[\begin{array}{c|c} I_i & O \\ \hline Z' & Q \end{array}\right]$$

Thus, $EL = L'E$.   □

Thus, if L is a lower triangular matrix of the type in equation (6.3), and E is a transposition which interchanges two rows below the ith row, then $EL = L'E$ where L' is another lower triangular matrix of the same type as L. We can use this idea to generalize our discussion about LU-factorizations.

Theorem 6.25   Let $A \in M_{m \times n}$. Then there exist a permutation matrix P and lower triangular matrices $L_{m-1}, \ldots, L_1$ as in equation (6.17) such that $L_{m-1} \cdots L_1(PA)$ is an upper triangular matrix.

Proof. We had seen in the proof of Theorem 6.10 that there exist transpositions $E_1, \ldots, E_{m-1}$ and lower triangular matrices $L'_1, \ldots, L'_{m-1}$ such that

6.26

$$L'_{m-1}E_{m-1} \cdots L'_1 E_1 A = U \quad \text{(an upper triangular matrix)}$$

Furthermore, each L'_i can be chosen to have the form given in equation (6.17). Thus, for each $i = 1, \ldots, m-1$

6.27
$$L'_i = \left[\begin{array}{c|c} I_i & O \\ \hline Z_i & I_{m-i} \end{array} \right]$$

where

$$Z_i = \begin{bmatrix} 0 & \cdots & 0 & l_{i+1\,i} \\ \vdots & & \vdots & \vdots \\ 0 & \cdots & 0 & l_{mi} \end{bmatrix}$$

Each E_i in equation (6.26) can be taken to be a transposition of the following form.

6.28
$$E_i = \left[\begin{array}{c|c} I_{i-1} & O \\ \hline O & Q_i \end{array} \right]$$

where Q_i is a transposition of size $(m - i + 1) \times (m - i + 1)$.

The idea of this proof is to use Lemma 6.24 to push the E_i in equation (6.26) past the L'_j. Lemma 6.24 implies $E_2 L'_1 = L''_1 E_2$ where L''_1 is a lower triangular matrix having the same form as L'_1. Making this substitution in equation (6.26), we have

6.29
$$L'_{m-1} E_{m-1} \cdots L'_3 E_3 L'_2 L''_1 E_2 E_1 A = U$$

Now again by Lemma 6.24, $E_3 L'_2 = L''_2 E_3$, and $E_3 L''_1 = L'''_1 E_3$. These new matrices L''_2 and L'''_1 have exactly the same form as L'_2 and L''_1, respectively. Making these substitutions in equation (6.29), we have

6.30
$$L'_{m-1} E_{m-1} \cdots L'_3 L''_2 L'''_1 E_3 E_2 E_1 A = U$$

It is now clear that repeated applications of this argument eventually result in the following factorization.

6.31
$$L_{m-1} L_{m-2} \cdots L_1 E_{m-1} E_{m-2} \cdots E_1 A = U$$

In equation (6.31), the L_i have the form given in equation (6.17), and the E_i are the same transpositions which appear in (6.26). Set $P = E_{m-1} \cdots E_1$. Then P is a permutation matrix and $L_{m-1} \cdots L_1(PA) = U$. $\square$

Thus, if we are willing to permute the rows of A first, the new matrix PA will have an LU-factorization.

Corollary 6.32 Let $A \in M_{m \times n}$. There exists a permutation matrix P such that PA has an LU-factorization.

Proof. By Theorem 6.25, there exist a permutation matrix P and lower triangular matrices $L_{m-1}, \ldots, L_1$ as in equation (6.17) such that $L_{m-1} \cdots L_1 PA = U$ is an upper triangular matrix. Then using equations (6.18) through (6.20), we have $PA = (L_1^{-1} \cdots L_{m-1}^{-1})U$ is an LU-factorization of PA. □

Let us return to Example 6.9 for an illustration of these ideas.

Example 6.33 Let

$$A = \begin{bmatrix} 1 & 0 & 1 & 2 \\ 2 & 0 & 1 & 3 \\ -1 & 1 & 2 & 1 \end{bmatrix}$$

We had seen in Example 6.9 that a transposition E_{23} was needed to reduce A to an upper triangular matrix. Set $P = E_{23}$. Then

$$PA = \begin{bmatrix} 1 & 0 & 1 & 2 \\ -1 & 1 & 2 & 1 \\ 2 & 0 & 1 & 3 \end{bmatrix}$$

Reduce PA with

$$L = \begin{bmatrix} 1 & 0 & 0 \\ 1 & 1 & 0 \\ -2 & 0 & 1 \end{bmatrix}$$

$$LPA = \begin{bmatrix} 1 & 0 & 0 \\ 1 & 1 & 0 \\ -2 & 0 & 1 \end{bmatrix} \begin{bmatrix} 1 & 0 & 1 & 2 \\ -1 & 1 & 2 & 1 \\ 2 & 0 & 1 & 3 \end{bmatrix} = \begin{bmatrix} 1 & 0 & 1 & 2 \\ 0 & 1 & 3 & 3 \\ 0 & 0 & -1 & -1 \end{bmatrix} = U$$

Thus, the LU-factorization of PA is as follows:

$$PA = \begin{bmatrix} 1 & 0 & 0 \\ -1 & 1 & 0 \\ 2 & 0 & 1 \end{bmatrix} \begin{bmatrix} 1 & 0 & 1 & 2 \\ 0 & 1 & 3 & 3 \\ 0 & 0 & -1 & -1 \end{bmatrix}.$$

There are many computer algorithms now available for finding an LU-factorization of A. Most of these algorithms will automatically pass to PA if needed. The reader should always ask the computer to multiply L and U together. In this way, one can easily tell if A or some permutation of the rows of A has been factored.

We finish this section with a discussion of why LU-factorizations are useful in applications. Suppose we want to solve a large number of problems of the form $AX = B$. Here A is a fixed $m \times n$ matrix, the same in every problem. B is a column vector of size m which varies from problem to problem. In this situation, it is senseless to keep row reducing $[A \mid B]$ over and over as B varies. It is more efficient to find an LU-factorization of some permutation of A, say PA, and proceed as follows.

6.34   To solve $AX = B$:

(a)   Find an LU-factorization $PA = LU$.
(b)   Replace $AX = B$ with $PAX = PB$, i.e., with $LUX = PB$.
(c)   Solve $LY = PB$ by forward substitution.
(d)   Solve $UX = Y$ by back substitution.

The important points here are (c) and (d). Since both L and U are triangular matrices, the equations $LY = PB$ and $UX = Y$ are easily solved. Consider the following example.

Example 6.35   Suppose

$$\begin{bmatrix} 1 & 4 & 0 \\ 2 & 1 & 1 \\ 3 & 5 & 2 \end{bmatrix} \begin{bmatrix} x_1 \\ x_2 \\ x_3 \end{bmatrix} = \begin{bmatrix} 1 \\ 4 \\ 8 \end{bmatrix}.$$

The steps in the algorithm (6.34) are as follows:

(a)
$$\begin{bmatrix} 1 & 4 & 0 \\ 2 & 1 & 1 \\ 3 & 5 & 2 \end{bmatrix} = \begin{bmatrix} 1 & 0 & 0 \\ 2 & 1 & 0 \\ 3 & 1 & 1 \end{bmatrix} \begin{bmatrix} 1 & 4 & 0 \\ 0 & -7 & 1 \\ 0 & 0 & 1 \end{bmatrix}$$

is an LU-factorization of A. Notice that no transpositions were needed to reduce A to an upper triangular matrix.

(c)   Solve the following lower triangular system of equations by forward substitution

$$\begin{bmatrix} 1 & 0 & 0 \\ 2 & 1 & 0 \\ 3 & 1 & 1 \end{bmatrix}\begin{bmatrix} y_1 \\ y_2 \\ y_3 \end{bmatrix} = \begin{bmatrix} 1 \\ 4 \\ 8 \end{bmatrix}$$

Thus, $y_1 = 1$, $y_2 = 2$, and $y_3 = 3$.

(d)   Solve the following upper triangular system of equations by back substitution

$$\begin{bmatrix} 1 & 4 & 0 \\ 0 & -7 & 1 \\ 0 & 0 & 1 \end{bmatrix}\begin{bmatrix} x_1 \\ x_2 \\ x_3 \end{bmatrix} = \begin{bmatrix} 1 \\ 2 \\ 3 \end{bmatrix}$$

Thus, $x_3 = 3$, $x_2 = 1/7$, and $x_1 = 3/7$.

Exercises for Section 6

6.1   Find an LU-factorization of each of the following matrices:

(a) $\begin{bmatrix} 0 & 2 & 1 \\ 1 & 1 & 2 \\ 3 & 4 & 1 \end{bmatrix}$

(b) $\begin{bmatrix} 1 & 0 & 1 & 1 & 1 \\ 2 & 1 & 1 & 4 & 6 \\ 3 & 1 & 2 & 0 & 1 \end{bmatrix}$

(c) $\begin{bmatrix} 2 & 1 & 1 \\ 4 & 1 & 1 \\ -2 & 2 & 1 \end{bmatrix}$

(d) $\begin{bmatrix} 1 & 3 & 2 \\ 2 & 6 & 8 \\ 2 & 8 & 1 \end{bmatrix}$

6.2   Use the algorithm in (6.34) to solve $AX = B$ when A is each of the matrices in Exercise 6.1 and $B = [1 \quad 0 \quad 1]^t$.

6.3   Prove the formula listed in equation (6.18).

6.4   Show the product $L_1^{-1} \cdots L_{m-1}^{-1}$ listed in (6.19) is indeed the matrix listed in (6.20).

6.5   In Exercise 6.4, does the product $L_{m-1}^{-1} \cdots L_1^{-1}$ have any particularly nice form?

6.6   Show that an $m \times n$ matrix A can have two different LU-factorizations.

6.7   Let $A \in M_{m \times m}$. Suppose A has an LU-factorization, $A = LU$, with the diagonal entries of U nonzero. Thus, $[U]_{ii} \neq 0$ for all $i = 1, \ldots,$ m. Show there exists an $m \times m$ diagonal matrix D, and an upper triangular matrix U_1 such that
       (a) $[U_1]_{ii} = 1$ for all $i = 1, \ldots, m$
       (b) $A = LDU_1$.
       The decomposition in Exercise 6.7 is called an LDU-factorization of A.

6.8   In Exercise 6.7 above, suppose a diagonal entry of U is zero. Does A have an LDU-factorization?

6.9   Find an LDU-factorization (if they exist) of the three square matrices in Exercise 6.1. You may have to apply a permutation to A first.

6.10   Show that LDU-factorizations (when they exist) are unique. To be more specific, prove the following theorem. Let $A \in M_{m \times m}$. Suppose $A = L_1 D_1 U_1 = L_2 D_2 U_2$ where L_1 and L_2 are lower $m \times m$ triangular matrices with unit diagonals, U_1 and U_2 are upper $m \times m$ triangular matrices with unit diagonals, and D_1 and D_2 are diagonal matrices with no zeros on their diagonals. Show $L_1 = L_2$, $D_1 = D_2$, and $U_1 = U_2$. (It may be wise to review the results of Exercise 2.15.)

6.11   Suppose $A \in M_{m \times m}$ has an LDU-factorization. If A is symmetric, what does this factorization look like?

6.12   Suppose

$$A = \begin{bmatrix} 1 & 3 & 5 \\ 3 & 12 & 18 \\ 5 & 18 & 30 \end{bmatrix}$$

       Find the LDU-factorization of A.

6.13   Suppose

$$A = \begin{bmatrix} a & b \\ c & d \end{bmatrix}$$

       has an LDU-factorization. What does it look like?

6.14   Suppose P is an $m \times m$ permutation matrix. Show P contains exactly one 1 in each row and column.

6.15   List all possible 2×2 and 3×3 permutation matrices. How many $m \times m$ permutation matrices are there?

6.16   Suppose P is an $m \times m$ permutation matrix. Show $P^t = P^{-1}$.

II
Vector Spaces

1. DEFINITIONS AND EXAMPLES

In this chapter, we study abstract vector spaces and the linear transformations between them. As in Chapter I, much of our discussion has nothing to do with whether the scalars are real or complex. Consequently, we will let F denote either $\mathbb{R}$ or $\mathbb{C}$ and study vector spaces over F.

Definition 1.1 A vector space V over F is a nonempty set V together with two functions, $(\alpha, \beta) \mapsto \alpha + \beta$ from $V \times V$ to V and $(x, \alpha) \mapsto x\alpha$ from $F \times V$ to V, which satisfy the following axioms:

V1. $\alpha + \beta = \beta + \alpha$ for all α, $\beta \in V$.
V2. $\alpha + (\beta + \gamma) = (\alpha + \beta) + \gamma$ for all α, β, $\gamma \in V$.
V3. There exists an element $o \in V$ such that $o + \alpha = \alpha$ for all $\alpha \in V$.
V4. For any $\alpha \in V$, there exists a $\beta \in V$ such that $\alpha + \beta = o$.
V5. $(xy)\alpha = x(y\alpha)$ for all x, $y \in F$ and $\alpha \in V$.
V6. $x(\alpha + \beta) = x\alpha + x\beta$ for all $x \in F$ and α, $\beta \in V$.
V7. $(x + y)\alpha = x\alpha + y\alpha$ for all x, $y \in F$ and $\alpha \in V$.
V8. $1\alpha = \alpha$ for all $\alpha \in V$.

The first thing that should be noticed about this definition is that a vector space is really three objects, a nonempty set V, a function from $V \times V$ to V, and a function from $F \times V$ to V. These functions must satisfy the eight

86

axioms listed in Definition 1.1. It is conceivable that the same set V might have several different functions from $V \times V$ to V and $F \times V$ to V satisfying the eight axioms in Definition 1.1. Thus, to completely describe a vector space whose underlying set is V, we must carefully define the two functions $V \times V \mapsto V$ and $F \times V \mapsto V$ and verify that these two functions satisfy V1 through V8.

Suppose $(V, V \times V \mapsto V, F \times V \mapsto V)$ is a vector space over F. When the two functions $V \times V \mapsto V$ and $F \times V \mapsto V$ are understood from the context or when it is not important to know the exact form of these two functions, we will drop them from our notation and simply refer to V as the vector space over F.

Suppose V is a vector space over F. The elements in the set V are called vectors. We will usually (but not always) denote vectors in V with small Greek letters α, β, γ, etc. The numbers in F ($= \mathbb{R}$ or $\mathbb{C}$) are called scalars, and will be denoted by small Roman letters x, y, z, etc. If $F = \mathbb{R}$, then V is called a real vector space. If $F = \mathbb{C}$, then V is called a complex vector space.

Since V is a vector space over F, there exist two functions $V \times V \mapsto V$ and $F \times V \mapsto V$ which satisfy the eight axioms in Definition 1.1. The function $V \times V \mapsto V$ is always called (vector) addition. The value of this function on an ordered pair $(\alpha, \beta) \in V \times V$ is denoted by $\alpha + \beta$. Axiom V1 says vector addition is commutative. V2 says addition is associative. Notice that vector addition is a binary operation on V. This means the operation of addition is applied to only two vectors at a time. If three vectors α, β, and $\gamma \in V$, are to be added, first form $\alpha + \beta$ and then $(\alpha + \beta) + \gamma$, or form $\beta + \gamma$ and then $\alpha + (\beta + \gamma)$. Axiom V2 guarantees that either procedure results in the same answer. Hence, we can write $\alpha + \beta + \gamma$ for the sum of these three vectors with no fear of ambiguity. A simple induction argument as in Exercise I.2.19 shows that any finite sum of vectors $\alpha_1, \ldots, \alpha_n \in V$ can be written unambiguously as $\alpha_1 + \cdots + \alpha_n$. Any placement of parentheses in $\alpha_1 + \cdots + \alpha_n$ will result in the same answer. Consequently, we will drop all parentheses when dealing with finite sums in V.

The vector o which appears in V3 is called the zero vector of V. It is clearly unique. For suppose V contains another vector o' which satisfies V3. Then using V1 and V3, we have $o = o' + o = o + o' = o'$. Hence, o is the unique vector satisfying V3.

The vector β in V4 for which $\alpha + \beta = o$ is called the negative of α. The negative of α is also unique. For suppose $\alpha + \beta = o$ and $\alpha + \gamma = o$. Then $\gamma = o + \gamma = (\alpha + \beta) + \gamma = (\beta + \alpha) + \gamma = \beta + (\alpha + \gamma) = \beta + o = \beta$. It is a good exercise using the axioms to explain why each equality here is true.

We will denote the negative of a vector α by $-\alpha$. Thus, $-\alpha$ is the unique vector in V for which $\alpha + (-\alpha) = o$.

The function $F \times V \mapsto V$ is called scalar multiplication. The value of scalar multiplication on the ordered pair $(x, \alpha) \in F \times V$ is always denoted by $x\alpha$. Notice that $x\alpha$ is a vector in V. Axioms V5 through V8 describe how scalar multiplication behaves with respect to addition in V and multiplication in F. Note that the plus sign is being used with a double meaning in axiom V7. The plus sign on the left side of the equation in V7 is addition of scalars in F. The plus sign on the right in V7 is vector addition in V. The statements in V6 and V7 are both called distributive laws. Some of the more important facts about scalar multiplication are summarized in the following lemma.

Lemma 1.2   Let V be a vector space over F

(a) $xo = o$ for any $x \in F$.
(b) $0\alpha = o$ for any $\alpha \in V$.
(c) $(-1)\alpha = -\alpha$ for all $\alpha \in V$.
(d)   If $x\alpha = o$, then $x = 0$ or $\alpha = o$.

Proof.   (a) $xo = x(o + o) = xo + xo$. Adding $-(xo)$ to both sides of this equation, gives $o = xo + -(xo) = xo + xo + -(xo) = xo$. Thus, any scalar multiple of the zero vector is again the zero vector.

(b) In statement (b), 0 is the zero scalar in F. We have $0\alpha = (0 + 0)\alpha = 0\alpha + 0\alpha$. Adding -0α to both sides of this equation gives $o = 0\alpha + -(0\alpha) = 0\alpha + 0\alpha + -(0\alpha) = 0\alpha$.

(c) $o = 0\alpha = (-1 + 1)\alpha = (-1)\alpha + 1\alpha = (-1)\alpha + \alpha$. Since the negative of α is unique, we conclude that $(-1)\alpha = -\alpha$.

(d) Suppose $x\alpha = o$ and $x \neq 0$. Then $\alpha = 1\alpha = (1/x)x\alpha = (1/x)o = o$.

$\square$

We will make use of these identities whenever it is convenient to do so. Let us now consider some examples of vector spaces which were present in Chapter I.

Example 1.3   The set $M_{m \times n}(\mathbb{R})$ is a vector space over $\mathbb{R}$ when vector addition and scalar multiplication are defined as in Definitions I.2.1 and I.2.4. In this example, the vectors are $m \times n$ matrices. Theorem I.2.6 is precisely the statement that matrix addition and scalar multiplication satisfy axioms V1 through V8. Notice that the zero vector of $M_{m \times n}$ is the zero matrix O of size $m \times n$.

We can also consider the complex vector space $M_{m \times n}(\mathbb{C})$. Again vector addition and scalar multiplication are given by Definitions I.2.1 and I.2.4.

There is a special case of Example 1.3 which will be very important in the sequel.

Example 1.4 $F^n = \{[x_1 \cdots x_n]^t \mid x_i \in F\}$ is a vector space over F with addition and scalar multiplication defined as follows

$$\begin{bmatrix} x_1 \\ \vdots \\ x_n \end{bmatrix} + \begin{bmatrix} y_1 \\ \vdots \\ y_n \end{bmatrix} = \begin{bmatrix} x_1 + y_1 \\ \vdots \\ x_n + y_n \end{bmatrix} \quad \text{and} \quad x \begin{bmatrix} x_1 \\ \vdots \\ x_n \end{bmatrix} = \begin{bmatrix} xx_1 \\ \vdots \\ xx_n \end{bmatrix}$$

Since $F^n = M_{n \times 1}(F)$, Example 1.3 implies F^n is a vector space over F with these operations. A typical vector in F^n is a column vector of size n. Notice that the zero vector in F^n is the column vector $O = [0 \cdots 0]^t$.

If $F = \mathbb{R}$ in Example 1.4, then $\mathbb{R}^n$ is just euclidean n-space, a space familiar from the calculus. If $F = \mathbb{C}$ in Example 1.4, then the vector space $\mathbb{C}^n$ is called complex n-space.

Every $m \times n$ matrix A has several vector spaces associated with it. Here are the examples we considered in Chapter I.

Example 1.5 Let $A = [A_1 \mid \cdots \mid A_n] \in M_{m \times n}(F)$. Here A is partitioned into columns $A_1, \ldots, A_n \in F^m$. Recall the column space, $CS(A)$, of A is the following subset of F^m

$$CS(A) = \{x_1 A_1 + \cdots + x_n A_n \mid x_1, \ldots, x_n \in F\}$$

Thus, $CS(A)$ is just all linear combinations of the columns of A.

F^m is a vector space over F when addition and scalar multiplication are defined as in Example 1.4. We can restrict these operations to the subset $CS(A)$. If $\alpha = x_1 A_1 + \cdots + x_n A_n$ and $\beta = y_1 A_1 + \cdots + y_n A_n$ are two vectors in $CS(A)$, then, using Theorem I.2.6, we have

$$\alpha + \beta = (x_1 + y_1)A_1 + \cdots + (x_n + y_n)A_n \in CS(A).$$

Thus, the sum of two vectors from $CS(A)$ is another vector in $CS(A)$. Similarly,

$$x\alpha = (xx_1)A_1 + \cdots + (xx_n)A_n \in CS(A).$$

Thus, any scalar multiple of a vector in $CS(A)$ is again a vector in $CS(A)$.

Clearly, $0 = 0A_1 + \cdots + 0A_n \in CS(A)$. Thus, $CS(A)$ satisfies axiom V3. If $\alpha = x_1 A_1 + \cdots + x_n A_n \in CS(A)$, then

$$-\alpha = (-x_1)A_1 + \cdots + (-x_n)A_n \in CS(A).$$

In particular, $CS(A)$ satisfies V4. The remaining axioms are all satisfied in $CS(A)$ since they are true in the larger vector space F^m.

We can also put a natural vector space structure on the row space $RS(A)$ of an $m \times n$ matrix A.

Example 1.6 Let $A \in M_{m \times n}(F)$. Partition A into rows

$$A = \begin{bmatrix} R_1 \\ \vdots \\ R_m \end{bmatrix}$$

Then $RS(A) = \{x_1 R_1 + \cdots + x_m R_m \mid x_1, \ldots, x_m \in F\}$. Thus, $RS(A)$ is the subset of $M_{1 \times n}(F)$ consisting of all linear combinations of $R_1, \ldots, R_m$.
$M_{1 \times n}(F)$ is a vector space over F when addition and scalar multiplication are defined as follows:

$$[x_1, \ldots, x_n] + [y_1, \ldots, y_n] = [x_1 + y_1, \ldots, x_n + y_n]$$

and $x[x_1, \ldots, x_n] = [xx_1, \ldots, xx_n]$. This is just a special case of Example 1.3. If we restrict these operations to the set $RS(A)$, the same reasoning as in Example 1.5 shows $RS(A)$ is a vector space over F.

Example 1.7 Let $A \in M_{m \times n}(F)$. Set $NS(A) = \{\xi \in F^n \mid A\xi = 0\}$. $NS(A)$ is called the null space of A. Clearly, $NS(A)$ is all solutions to the homogeneous system of equations $AX = 0$.

We saw in Example 1.4 that F^n is a vector space over F. If we restrict the operations of column vector addition and column vector scalar multiplication to the set $NS(A)$, the null space of A becomes a vector space in its own right. To see this, suppose $\xi_1, \xi_2 \in NS(A)$. Then $\xi_1 + \xi_2 \in F^n$ and $A(\xi_1 + \xi_2) = A\xi_1 + A\xi_2 = 0 + 0 = 0$. Thus, $\xi_1 + \xi_2 \in NS(A)$. In particular, the sum of any two vectors in $NS(A)$ is again a vector in $NS(A)$. Similarly, $A(x\xi_1) = xA\xi_1 = x0 = 0$. Thus, $x\xi_1 \in NS(A)$, i.e., any scalar multiple of a vector in $NS(A)$ is again in $NS(A)$. The zero vector $0 \in F^n$ lies in $NS(A)$ since $A0 = 0$. Therefore, $NS(A)$ satisfies V3. If $\xi \in NS(A)$, then $-\xi = (-1)\xi \in NS(A)$. In particular, $NS(A)$ satisfies V4. The remaining axioms of Definition 1.1 are true in $NS(A)$ since they hold in F^n.

There are many interesting examples of real vector spaces that are familiar from elementary calculus.

Example 1.8 Suppose I is a nonempty subset of $\mathbb{R}$. Let $\mathbb{R}^I$ denote the set of real-valued functions on I. Thus, $\mathbb{R}^I = \{f : I \mapsto \mathbb{R} \mid f \text{ is a function on } I\}$. Two functions in $\mathbb{R}^I$ are added or a function in $\mathbb{R}^I$ is multiplied by a real number pointwise. Thus, if f and g are functions in $\mathbb{R}^I$, then $f + g$ is the real-valued function on I whose value at every $x \in I$ is given by $(f + g)(x) = f(x) + g(x)$. Similarly, if $c \in \mathbb{R}$, then cf is the real-valued function on I defined by $(cf)(x) = cf(x)$ for all $x \in I$.

These operations are the usual definitions of addition and scalar multiplication of functions on I. The fact that these operations satisfy axioms V1 through V8 is a familiar result from the calculus. Thus, $\mathbb{R}^I$ with these operations is a real vector space.

Notice that a typical vector in $\mathbb{R}^I$ is a real valued function on I. The zero vector of $\mathbb{R}^I$ is the function which is identically zero on I.

There are many interesting subsets of $\mathbb{R}^I$ which are also vector spaces with addition and scalar multiplication defined as in Example 1.8.

Example 1.9 Let I be an open interval in $\mathbb{R}$. Set $C(I) = \{f \in \mathbb{R}^I \mid f \text{ is continuous on } I\}$. Set $C^k(I) = \{f \in \mathbb{R}^I \mid f \text{ is } k \text{ times differentiable on } I\}$. Then $C(I) \supset C^1(I) \supset C^2(I) \supset \cdots$ are all vector spaces over $\mathbb{R}$.

Those functions, which are Riemann integrable, also form a real vector space.

Example 1.10 Let $I = [a, b]$ be a closed interval in $\mathbb{R}$. Let $\mathbb{R}(I) = \{f \in \mathbb{R}^I \mid f \text{ is Riemann integrable on } I\}$. When addition and scalar multiplication are defined as in Example 1.8, $\mathbb{R}(I)$ is a vector space over $\mathbb{R}$.

There is one last example which we will refer to often in the sequel.

Example 1.11 Let $\mathbb{R}[t](\mathbb{C}[t])$ denote the set of all polynomials $a_0 + a_1 t + a_2 t^2 + \cdots + a_n t^n$ with coefficients $a_0, \ldots, a_n \in \mathbb{R}(\mathbb{C})$. Two polynomials are added by adding coefficients of like powers of t. Thus, $(1 + t) + (2 - 4t + 3t^2) = 3 - 3t + 3t^2$. A polynomial is multiplied by a scalar c by multiplying the coefficients of the polynomial by c. For example, $6(1 + 2t + 3t^2) = 6 + 12t + 18t^2$. The fact that these two operations satisfy axioms V1 through V8 is familiar from elementary algebra.

Thus, $\mathbb{R}[t]$ ($\mathbb{C}[t]$) is a real (complex) vector space. A typical vector in $\mathbb{R}[t]$ is a polynomial in t. The zero vector in $\mathbb{R}[t]$ is the constant 0.

There are subsets of $\mathbb{R}[t]$ (and $\mathbb{C}[t]$) which are vector spaces in their own right with the same definitions of addition and scalar multiplication as in Example 1.11. We will let $\mathscr{P}_n$ denote the set of polynomials in $\mathbb{R}[t]$ whose degree is at most n. It is easy to see that $\mathscr{P}_n$ is a vector space over $\mathbb{R}$. See Exercise 1.4.

We have now introduced eleven examples of vector spaces, $M_{m \times n}$, F^n, $CS(A)$, $RS(A)$, $NS(A)$, $\mathbb{R}^I$, $C(I)$, $C^k(I)$, $\mathbb{R}(I)$, $\mathbb{R}[t]$ (or $\mathbb{C}[t]$), and $\mathscr{P}_n$. In the rest of this book, we will use this notation to mean the vector space described in the appropriate example above. Thus, $M_{m \times n}$ is the vector space over F whose vector addition and scalar multiplication are described in Example 1.3, F^n is the vector space over F whose addition and scalar multiplication are described in Example 1.4, etc.

Some of the most interesting vector spaces are subspaces of larger vector spaces.

Definition 1.12   Let W be a nonempty subset of a vector space V. W is a subspace of V if W is itself a vector space over F with the same operations as those on V.

Definition 1.12 is the usual definition of a subspace given in most modern textbooks. It can be a bit confusing to the careful reader. Hence, we take this opportunity to carefully explain what the statement "W is itself a vector space over F with the same operations as those on V" means in Definition 1.12.

Suppose vector addition and scalar multiplication in the vector space V are denoted by $\alpha + \beta$ and $x\alpha$, respectively. Then, vector addition is the function from $V \times V$ to V which sends the ordered pair (α, β) to $\alpha + \beta$. Scalar multiplication is the function from $F \times V$ to V which sends the ordered pair (x, α) to $x\alpha$. These two functions satisfy the eight axioms given in Definition 1.1.

If W is a nonempty subset of V, we can consider the subset $W \times W$ of $V \times V$ given by $W \times W = \{(\alpha, \beta) \in V \times V \mid \alpha, \beta \in W\}$. We can also consider the subset $F \times W$ of $F \times V$ given by $F \times W = \{(x, \alpha) \in F \times V \mid x \in F, \alpha \in W\}$. It makes perfectly good sense to restrict vector addition to the subset $W \times W$ (i.e., only add vectors from W), and restrict scalar multiplication to the subset $F \times W$ (i.e., only multiply vectors in W by scalars). If W is an arbitrary subset of V, then of course there is no reason why the sum of two vectors from W is another vector in W. Likewise, a scalar multiple of some vector in W may not be another vector in W. The statement "W is itself a vector space over F with the same operations as those on V" means the restriction of vector addition to the subset $W \times W$ maps $W \times W$ into W, the restriction of scalar multiplication to the subset $F \times W$ maps $F \times W$ into W, and W is a vector space with these two operations. Thus, if W is a subspace of V, then $\alpha + \beta \in W$ whenever $\alpha, \beta \in W$, and $x\alpha \in W$ whenever $\alpha \in W$ and $x \in F$.

Again suppose V is a vector space over F with vector addition denoted by $\alpha + \beta$, and scalar multiplication denoted by $x\alpha$. A subset W of V is said to be closed under vector addition if $\alpha + \beta \in W$ whenever $\alpha, \beta \in W$. The set W is said to be closed under scalar multiplication if $x\alpha \in W$ whenever $x \in F$ and $\alpha \in W$. Our discussion in the last paragraph implies a subspace of V is a nonempty subset of V which is closed under vector addition and scalar multiplication.

The converse of this statement is also true. Suppose W is a nonempty subset of V which is closed under vector addition and scalar multiplication. Then the restriction of vector addition (from $V \times V$ to $W \times W$) maps $W \times W$ to W. The restriction of scalar multiplication (from $F \times V$ to $F \times W$) maps $F \times W$ to W. These operations on W satisfy axioms V1 through V8. To see this, first note that axioms V1, V2, and V5 through V8 are automatically satisfied. (If one of these axioms fails for vectors in W, then it fails in V since $W \subseteq V$.) Let $\alpha \in W$. Since W is closed under scalar multiplication, $0\alpha = o \in W$. Thus, W contains the zero vector of V. Since $o + \beta = \beta$ for all $\beta \in V$, this is surely true for $\beta \in W$. Thus, W satisfies V3. Again let $\alpha \in W$. Using Lemma 1.2, we have $-\alpha = (-1)\alpha \in W$ since W is closed under scalar multiplication. Hence, W contains the negative of any vector in W. In particular, W satisfies V4. Thus, W satisfies all eight axioms in Definition 1.1 and consequently, is a vector space over F. Therefore, W is a subspace of V. We have now proven the following theorem.

Theorem 1.13 Let W be a nonempty subset of V. Then W is a subspace of V if and only if $\alpha + \beta \in W$, and $x\alpha \in W$ whenever $\alpha, \beta \in W$ and $x \in F$.

Notice that Theorem 1.13 can be stated more succinctly as follows. A nonempty subset W of V is a subspace of V if and only if $x\alpha + y\beta \in W$ whenever $\alpha, \beta \in W$ and $x, y \in F$.

We have already seen several examples of subspaces. If $A \in M_{m \times n}$, then $CS(A)$ is a subspace of F^m, $NS(A)$ is a subspace of F^n and $RS(A)$ is a subspace of $M_{1 \times n}$. The vector spaces $\mathbb{R}[t]$, $C(I)$, $C^k(I)$, and $\mathbb{R}(I)$ are all subspaces of $\mathbb{R}^I$. $\mathscr{P}_n$ is a subspace of $\mathbb{R}[t]$.

Not every interesting subset of V is a subspace of V. Consider the following example.

Example 1.14   Is $W = \{A \in M_{2 \times 2} \mid A \text{ is nonsingular}\}$ a subspace of $M_{2 \times 2}$? To answer this question, we check if W is closed under addition and scalar multiplication in $M_{2 \times 2}$. We quickly see that W is not closed under either operation.

$$\begin{bmatrix} 1 & 0 \\ 0 & 1 \end{bmatrix} \quad \text{and} \quad \begin{bmatrix} 1 & 0 \\ 0 & -1 \end{bmatrix}$$

are in W, but

$$\begin{bmatrix} 1 & 0 \\ 0 & 1 \end{bmatrix} + \begin{bmatrix} 1 & 0 \\ 0 & -1 \end{bmatrix} = \begin{bmatrix} 2 & 0 \\ 0 & 0 \end{bmatrix}$$

is not in W. Also

$$0 \begin{bmatrix} 1 & 0 \\ 0 & 1 \end{bmatrix}$$

is not in W. Hence, W is not a subspace of $M_{2 \times 2}$.

Every vector space V has at least two subspaces, namely $\{o\}$ and V. A subspace W of V is said to be proper if $W \neq \{o\}$ and $W \neq V$. For example, if

$$A = \begin{bmatrix} 1 & 0 \\ 0 & 0 \end{bmatrix}$$

then $CS(A)$ is a proper subspace of F^2.

One useful method for producing subspaces of a given vector space V is to form the so-called linear span of a subset S of V. Suppose $\alpha_1, \ldots, \alpha_n$ are vectors in V and $x_1, \ldots, x_n$ are scalars in F. The vector $x_1 \alpha_1 + \cdots + x_n \alpha_n$ is called a linear combination of the vectors $\alpha_1, \ldots, \alpha_n$.

Definition 1.15   Let S be a subset of a vector space V. The set of all linear combinations of vectors from S will be called the linear span of S. We will let $L(S)$ denote the linear span of S.

In symbols, $L(S) = \{\sum_{i=1}^{n} x_i \alpha_i \mid \alpha_i \in S, x_i \in F, n > 0\}$. We allow S to be the empty subset of V in Definition 1.15. If $S = \varnothing$, then we define $L(\varnothing) = (o)$.

We have seen a few examples of linear spans of subsets. If $A \in M_{m \times n}$, then $CS(A)$ is the linear span of the set $S = \{\text{Col}_1(A), \ldots, \text{Col}_n(A)\}$ in F^m. $RS(A)$ is the linear span of the set $S = \{\text{Row}_1(A), \ldots, \text{Row}_m(A)\}$ in $M_{1 \times n}$.

If the set S in Definition 1.15 is finite, say $S = \{\alpha_1, \ldots, \alpha_n\}$, then we will write $L(\alpha_1, \ldots, \alpha_n)$ for $L(S)$. For example, $CS(A) = L(\text{Col}_1(A), \ldots, \text{Col}_n(A))$.

There are a few elementary rules about forming $L(S)$. We summarize these rules in our next theorem.

Theorem 1.16   Let V be a vector space over F.
(a)   For any subset $S \subseteq V$, $L(S)$ is a subspace of V containing S.
(b)   If $S_1 \subseteq S_2 \subseteq V$, then $L(S_1) \subseteq L(S_2) \subseteq V$.
(c)   If $\alpha \in L(S)$, then there exists a finite subset $S_1 \subseteq S$ such that $\alpha \in L(S_1)$.
(d) $L(L(S)) = L(S)$.
(e)   If $\beta \in L(S \cup \{\alpha\})$ and $\beta \notin L(S)$, then $\alpha \in L(S \cup \{\beta\})$.

Proof.   All of these statements are easy to prove. We give arguments for (a) and (e) and leave the rest to the reader.

(a) Let $\alpha, \beta \in L(S)$. Then $\alpha = x_1\alpha_1 + \cdots + x_n\alpha_n$ for some vectors $\alpha_1, \ldots, \alpha_n \in S$ and scalars $x_1, \ldots, x_n \in F$. Similarly, $\beta = y_1\beta_1 + \cdots + y_m\beta_m$ for $\beta_1, \ldots, \beta_m \in S$, and $y_1, \ldots, y_m \in F$. Now if $x, y \in F$, then

$$x\alpha + y\beta = (xx_1)\alpha_1 + \cdots + (xx_n)\alpha_n + (yy_1)\beta_1 + \cdots + (yy_m)\beta_m.$$

If some of the β_k are the same as some of the α_j, the coefficients of these vectors can be added together. At any rate, $x\alpha + y\beta$ is certainly some linear combination of vectors from S. Thus, by Theorem 1.13 $L(S)$ is a subspace of V.

If $\alpha \in S$, then $\alpha = 1\alpha$ is a linear combination of vectors from S. Thus, $\alpha \in L(S)$. In particular, $S \subseteq L(S)$.

(e) Suppose $\beta \in L(S \cup \{\alpha\}) - L(S)$. Then β is a linear combination of vectors from $S \cup \{\alpha\}$. Thus, $\beta = x_1\alpha_1 + \cdots + x_n\alpha_n$ with $\alpha_1, \ldots, \alpha_n \in S \cup \{\alpha\}$. We can assume here that no $x_i\alpha_i$ is zero, and that the vectors $\alpha_1, \ldots, \alpha_n$ are all distinct. If $\alpha \neq \alpha_i$ for $i = 1, \ldots, n$, then $\beta \in L(S)$ which is contrary to our assumption. Therefore, $\alpha = \alpha_i$ for some i. We can assume with no loss in generality that $\alpha = \alpha_1$. Since $x_1\alpha_1 \neq o$, $x_1 \neq 0$ by Lemma 1.2(d). Then

$$\alpha = (x_1^{-1})\beta + (-x_1^{-1}x_2)\alpha_2 + \cdots + (-x_1^{-1}x_n)\alpha_n \in L(S \cup \{\beta\}) \qquad \square$$

Theorem 1.16(a) implies that every subset S is contained in a subspace $L(S)$. If S is already a subspace of V, then Theorem 1.13 implies $L(S) = S$. $L(S)$ is clearly the smallest subspace of V which contains S, i.e., if W is a subspace of V and $S \subseteq W$, then $L(S) \subseteq W$.

The statement in Theorem 1.16(e) is often called the exchange principle. Two vectors α and β can be exchanged in the relation $\beta \in L(S \cup \{\alpha\})$

provided $\beta \notin L(S)$. This principle will be used in the next section to show that any two bases of V contain the same number of vectors.

Vector spaces V such that $V = L(\alpha_1, \ldots, \alpha_n)$ for some set $\{\alpha_1, \ldots, \alpha_n\} \subseteq V$ will get special treatment in this book.

Definition 1.17 Let V be a vector space over F. V is a finite dimensional vector space (over F) if there exist vectors $\alpha_1, \ldots, \alpha_n \in V$ such that $V = L(\alpha_1, \ldots, \alpha_n)$.

If V is not finite dimensional, then V is called an infinite dimensional vector space. Let us consider our previous examples.

Example 1.18 $M_{m \times n}$ is a finite dimensional vector space over F. In fact, a little more is true. For each $i = 1, \ldots, m$ and $j = 1, \ldots, n$. let A_{ij} be the $m \times n$ matrix whose entries are defined by the following equations.

1.19
$$[A_{ij}]_{pq} = \begin{cases} 1 & \text{if } p = i \text{ and } q = j \\ 0 & \text{if } (p, q) \neq (i, j) \end{cases}$$

In equation (1.19), p ranges from 1 to m, and q ranges from 1 to n. Thus, A_{ij} is the $m \times n$ matrix having 1 in its (i,j)th entry and zeros elsewhere. If $B = (b_{ij})$ is any $m \times n$ matrix in $M_{m \times n}$, then

1.20
$$B = \sum_{i=1}^{m} \sum_{j=1}^{n} b_{ij} A_{ij}$$

Set $S = \{A_{ij} | i = 1, \ldots, m; \ j = 1, \ldots, n\}$. Equation (1.20) implies $M_{m \times n} = L(S)$. In particular, $M_{m \times n}$ is a finite dimensional vector space over F.

The set $\{A_{ij} | i = 1, \ldots, m; j = 1, \ldots, n\}$ defined in Example 1.18 is called the set of *matrix units* of $M_{m \times n}$. We note that the representation given in equation (1.20) is unique. This means if

$$\sum_{i=1}^{m} \sum_{j=1}^{n} b_{ij} A_{ij} = \sum_{i=1}^{m} \sum_{j=1}^{n} a_{ij} A_{ij}$$

for scalars a_{ij} and b_{ij}, then $a_{ij} = b_{ij}$ for all $i = 1, \ldots, m$ and $j = 1, \ldots, n$. Thus, every $m \times n$ matrix is a unique linear combination of the matrix units of $M_{m \times n}$.

There is one special case of Example 1.18 which is worth emphasizing here.

Example 1.21 F^n is a finite dimensional vector space over F. Since $F^n = M_{n \times 1}$, this fact follows from Example 1.18. We take this opportunity to introduce some notation which we will use for the rest of the book. The matrix units for F^n are the column vectors

$$\varepsilon_1 = \begin{bmatrix} 1 \\ \vdots \\ 0 \end{bmatrix}, \quad \varepsilon_2 = \begin{bmatrix} 0 \\ 1 \\ \vdots \\ 0 \end{bmatrix}, \dots, \varepsilon_n = \begin{bmatrix} 0 \\ \vdots \\ 1 \end{bmatrix}$$

Thus, ε_i, is the column vector of size n having 1 in its ith entry and zeros elsewhere. If $\alpha = [x_1 \ \cdots \ x_n]^t \in F^n$, then clearly

1.22
$$\alpha = x_1 \varepsilon_1 + \cdots + x_n \varepsilon_n$$

Furthermore, the representation of α as a linear combination of $\varepsilon_1, \dots, \varepsilon_n$ is obviously unique. Thus, $F^n = L(\varepsilon_1, \dots, \varepsilon_n)$. The set $\{\varepsilon_1, \dots, \varepsilon_n\}$ is called the canonical basis of F^n. We will have more to say about bases in the next section.

If $A \in M_{m \times n}$, then $CS(A)$ and $RS(A)$ are finite dimensional vector spaces by definition. $NS(A)$ is also a finite dimensional vector space. This remark will easily follow from some of our theorems in the next section.

Not all vector spaces are finite dimensional. It is easy to see that $\mathbb{R}[t]$ is not finite dimensional over $\mathbb{R}$.

Example 1.23 The vector space $\mathbb{R}[t]$ is not finite dimensional over $\mathbb{R}$. To see this, let $p_1(t) \dots, p_n(t)$ be any finite number of polynomials in $\mathbb{R}[t]$. Suppose $d >$ degree of $p_i(t)$ for $i = 1, \dots, n$. Then clearly $t^d \notin L(p_1, \dots, p_n)$. In particular, $\mathbb{R}[t] \neq L(p_1, \dots, p_n)$ for any finite set of vectors $p_1, \dots, p_n$. Thus, $\mathbb{R}[t]$ is infinite dimensional.

Using various counting arguments, we can show that all of the vector spaces $\mathbb{R}^I$, $C(I)$, $C^k(I)$, and $\mathbb{R}(I)$ are infinite dimensional vector spaces over $\mathbb{R}$ when I is an interval of positive length.

Exercises for Section 1

1.1   Show that the field $\mathbb{C}$ (with the usual addition and multiplication of complex numbers) is a vector space over both $\mathbb{R}$ and $\mathbb{C}$. Is $\mathbb{C}$ a finite dimensional vector space over $\mathbb{R}$?

1.2   Let $I = (0, 1) \subseteq \mathbb{R}$. Decide which of the subsets listed below are subspaces of $\mathbb{R}^I$.
      (a) $S = \{f \in \mathbb{R}^I \mid f(1/2) = 0\}$.
      (b) $S = \{f \in \mathbb{R}^I \mid f(1/2) = 1\}$.
      (c) $S = \{f \in C^1(I) \mid f'(1/2) = 0\}$ (f' = the derivative of f).
      (d) $S = \{f \in C(I) \mid \lim_{x \to 1} f(x) = 0\}$.
      (e) $S = \{f \in C(I) \mid \lim_{x \to 1} f(x) = 1\}$.
      (f) $S = \{f \in C^2(I) \mid |f'(1/2)| + |f''(1/2)| = 1\}$.

1.3   Which of the following subsets of $M_{2 \times 3}$ are subspaces?
      (a) $S = \{A = (a_{ij}) \mid a_{11} = 0\}$.
      (b) $S = \{A = (a_{ij}) \mid |a_{11}| + |a_{12}| = 0\}$.
      (c) $S = \{A = (a_{ij}) \mid a_{11} a_{12} a_{13} = 0\}$.

1.4   Let $\mathscr{P}_n$ denote the subset of $\mathbb{R}[t]$ consisting of all polynomials of degree at most n. Thus,

$$\mathscr{P}_n = \{a_0 + a_1 t + \cdots + a_n t^n \in \mathbb{R}[t] \mid a_0, \ldots, a_n \in \mathbb{R}\}.$$

Show that $\mathscr{P}_n$ is a finite dimensional subspace of $\mathbb{R}[t]$.

1.5   Let $W = \{A \in M_{n \times n} \mid A = A^t\}$. Show that W is a subspace of $M_{n \times n}$. Find a finite set S such that $W = L(S)$.

1.6   Let $W = \{A \in M_{n \times n}(\mathbb{C}) \mid A^* = A\}$. Is W a subspace of $M_{n \times n}(\mathbb{C})$?

1.7   Let $W = \{A \in M_{n \times n}(\mathbb{R}) \mid A^t = -A\}$. Show that W is a subspace of $M_{n \times n}(\mathbb{R})$. Find a finite set S such that $L(S) = W$.

1.8   Consider the differential equation

$$y'' + 2xy' + (x + 1)y = 0 \qquad\qquad (*)$$

Show that the set of $f(x) \in C^2((0, 1))$ such that $f(x)$ is a solution to $(*)$ is a subspace of $C^2((0, 1))$.

1.9   Let $C([0, 1])$ denote the set of all continuous, real valued functions on the closed interval $[0, 1]$. Show that $C([0, 1])$ is a vector space over $\mathbb{R}$ when addition and scalar multiplication are defined as in Example 1.8.

1.10 Let $W = \{f \in C([0, 1]) \mid \int_0^1 f(t) \, dt = 0\}$. Show that W is a subspace of $C([0, 1])$.

1.11 Let $[a_1 \cdots a_n]^t \in \mathbb{R}^n - \{0\}$. Show that

$$W = \{[x_1 \ \cdots \ x_n]^t \in \mathbb{R}^n \mid x_1 a_1 + \cdots + x_n a_n = 0\}$$

is a proper subspace of $\mathbb{R}^n$.

1.12 Show that $\mathbb{R}$ is a vector space over $\mathbb{R}$. Does $\mathbb{R}$ have any proper subspaces? Do the same problem for $\mathbb{C}$.

1.13 Suppose W_1 and W_2 are subspaces of a vector space V. Show that $W_1 \cap W_2$ is a subspace of V.

1.14 Give an example in $\mathbb{R}^2$ which shows that the union, $W_1 \cup W_2$, of two subspaces W_1 and W_2 need not be a subspace.

1.15 Let W_1 and W_2 be subspaces of a vector space V. Show $W_1 \cup W_2$ is a subspace of V if and only if $W_1 \subseteq W_2$ or $W_2 \subseteq W_1$.

1.16 Let S be a subset of a vector space V. Show that $L(S)$ is the intersection of all subspaces of V which contain S.

1.17 Let $\delta = (1, 0, 0) \in M_{1 \times 3}(\mathbb{R})$. Show that δ is not in the linear span of α, β, γ where $\alpha = (1, 1, 1)$, $\beta = (0, 1, -1)$, and $\gamma = (1, 0, 2)$.

1.18 Let W_1 and W_2 be subspaces of a vector space V. Show that $W_1 + W_2 = \{\alpha + \beta \mid \alpha \in W_1, \beta \in W_2\}$ is a subspace of V.

1.19 In Exercise 1.18, show $W_1 + W_2$ is the smallest subspace of V containing W_1 and W_2.

1.20 Let S_1 and S_2 be subsets of a vector space V. Show $L(S_1 \cup S_2) = L(S_1) + L(S_2)$.

1.21 Let $A \in M_{m \times n}$ and $B \in F^m$. Is $W = \{\xi \in F^n \mid A\xi = B\}$ a subspace of F^n?

1.22 Describe all subspaces of $\mathbb{R}^2$ and $\mathbb{R}^3$. Draw pictures.

1.23 Find a nonempty subset of $\mathbb{R}^2$ which is closed under vector addition but is not a subspace of $\mathbb{R}^2$.

1.24 Find a nonempty subset of $\mathbb{R}^2$ which is closed under scalar multiplication but is not a subspace of $\mathbb{R}^2$.

2. FINITE DIMENSIONAL VECTOR SPACES, BASES, AND DIMENSION

Throughout this section, V will denote a vector space over F. As usual, F can be either $\mathbb{R}$ or $\mathbb{C}$. We are primarily interested in proving the basic theorems about bases and dimension for finite dimensional vector spaces. But, for the time being, V can be any vector space over F.

Definition 2.1 Let $\alpha_1, \ldots, \alpha_n$ be vectors in V. The vectors $\alpha_1, \ldots, \alpha_n$ are said to be linearly dependent (over F) if there exist scalars $x_1, \ldots, x_n \in F$ such that $x_1\alpha_1 + \cdots + x_n\alpha_n = o$, and at least one x_i is not zero.

A linear combination of the form $x_1\alpha_1 + \cdots + x_n\alpha_n$ in which at least one of the scalars x_i is not zero is called a nontrivial linear combination of α_1, $\ldots$, α_n. Thus, vectors $\alpha_1, \ldots, \alpha_n \in V$ are linearly dependent if some nontrivial linear combination of $\alpha_1, \ldots, \alpha_n$ is zero.

Example 2.2   Let

$$\alpha = \begin{bmatrix} 1 \\ 0 \\ 1 \end{bmatrix}, \quad \beta = \begin{bmatrix} -1 \\ 1 \\ 2 \end{bmatrix}, \quad \gamma = \begin{bmatrix} -1 \\ 3 \\ 8 \end{bmatrix}$$

in F^3. Then α, β, γ are linearly dependent since $2\alpha + 3\beta - \gamma = 0$.
On the other hand

$$\varepsilon_1 = \begin{bmatrix} 1 \\ 0 \\ 0 \end{bmatrix}, \quad \varepsilon_2 = \begin{bmatrix} 0 \\ 1 \\ 0 \end{bmatrix}, \quad \varepsilon_2 = \begin{bmatrix} 0 \\ 0 \\ 1 \end{bmatrix}$$

are not linearly dependent. For suppose, $x_1\varepsilon_1 + x_2\varepsilon_2 + x_3\varepsilon_3 = 0$. Then

$$0 = \begin{bmatrix} 0 \\ 0 \\ 0 \end{bmatrix} = x_1\varepsilon_1 + x_2\varepsilon_2 + x_3\varepsilon_3 = \begin{bmatrix} x_1 \\ x_2 \\ x_3 \end{bmatrix}$$

Thus, $x_1 = x_2 = x_3 = 0$. Consequently, the canonical basis $\underline{\varepsilon} = \{\varepsilon_1, \varepsilon_2, \varepsilon_3\}$ of F^3 consists of vectors which are not linearly dependent.

Notice that Definition 2.1 is really a statement about the set of vectors $\{\alpha_1, \ldots, \alpha_n\}$. Clearly the order in which the vectors appear has nothing to do with whether they are linearly dependent or not. In particular, if $\alpha_1, \ldots, \alpha_n$ are linearly dependent, then any permutation of the α_i are linearly dependent.

There are several remarks about linear dependence which follow directly from the definition. We gather these together in our first theorem.

Theorem 2.3   Let $\alpha_1, \ldots, \alpha_n \in V$.

(a) If some $\alpha_i = o$, then $\alpha_1, \ldots, \alpha_n$ are linearly dependent.
(b) If $\alpha_i = \alpha_j$ for some $i \neq j$, then $\alpha_1, \ldots, \alpha_n$ are linearly dependent.
(c) $\alpha_1, \ldots, \alpha_n$ are linearly dependent if and only if some α_i is in the linear span of the others, i.e., $\alpha_i \in L(\alpha_1, \ldots, \alpha_{i-1}, \alpha_{i+1}, \ldots, \alpha_n)$.
(d) If $\alpha_1, \ldots, \alpha_n \in F^m$, then $\alpha_1, \ldots, \alpha_n$ are linearly dependent if and only if the equation $[\alpha_1 | \cdots | \alpha_n] X = O$ has a nontrivial solution.

Before giving a proof of Theorem 2.3, let us introduce some convenient notation. We will let $\alpha_1, \ldots, \hat{\alpha}_i, \ldots, \alpha_n$ denote the sequence $\alpha_1, \ldots, \alpha_{i-1}, \alpha_{i+1}, \ldots, \alpha_n$. The caret $\hat{\ }$ indicates which vector is missing from the list.

In Theorem 2.3(d), $[\alpha_1 \mid \cdots \mid \alpha_n]$ is the $m \times n$ matrix whose columns are the column vectors $\alpha_1, \ldots, \alpha_n$.

Proof of Theorem 2.3. (a) Suppose $\alpha_i = o$, for some i with $1 \leqslant i \leqslant n$. Then, $0\alpha_1 + \cdots + 0\alpha_{i-1} + 1\alpha_i + 0\alpha_{i+1} + \cdots + 0\alpha_n = o$ is a nontrivial $(x_i = 1)$ linear combination of $\alpha_1, \ldots, \alpha_n$ which is zero. Thus, any collection of vectors containing zero is automatically linearly dependent.

(b) Suppose two of the vectors, say α_i and α_j, are the same. We can assume $1 \leqslant i < j \leqslant n$. Then

$$0\alpha_1 + \cdots + 0\alpha_{i-1} + 1\alpha_i + 0\alpha_{i+1} + \cdots$$
$$+ 0\alpha_{j-1} + (-1)\alpha_j + 0\alpha_{j+1} + \cdots + 0\alpha_n = o$$

is a nontrivial linear combination of $\alpha_1, \ldots, \alpha_n$ which is zero. Therefore, if the vectors are not all distinct, then they are automatically linearly dependent.

(c) Suppose $\alpha_i \in L(\alpha_1, \ldots, \hat{\alpha}_i, \ldots, \alpha_n)$ where $1 \leqslant i \leqslant n$. Then there exist scalars $x_1, \ldots, \hat{x}_i, \ldots, x_n$ such that

$$\alpha_i = x_1\alpha_1 + \cdots + x_{i-1}\alpha_{i-1} + x_{i+1}\alpha_{i+1} + \cdots + x_n\alpha_n.$$

Clearly

$$x_1\alpha_1 + \cdots + x_{i-1}\alpha_{i-1} + (-1)\alpha_i + x_{i+1}\alpha_{i+1} + \cdots + x_n\alpha_n = o$$

is a nontrivial linear combination of $\alpha_1, \ldots, \alpha_n$ which is zero.

Conversely, suppose $\alpha_1, \ldots, \alpha_n$ are linearly dependent. Then there exist scalars $x_1, \ldots, x_n \in F$, not all zero, such that $x_1\alpha_1 + \cdots + x_n\alpha_n = o$. Suppose $x_i \neq 0$. Then

$$\alpha_i = (-x_i^{-1}x_1)\alpha_1 + \cdots + (-x_i^{-1}x_{i-1})\alpha_{i-1} + (-x_i^{-1}x_{i+1})\alpha_{i+1} + \cdots$$
$$+ (-x_i^{-1}x_n)\alpha_n \in L(\alpha_1, \ldots, \hat{\alpha}_i, \ldots, \alpha_n).$$

(d) If $X = [x_1 \cdots x_n]^t$, then we have seen in Corollary I.3.12 that $[\alpha_1 \mid \cdots \mid \alpha_n]X = x_1\alpha_1 + \cdots + x_n\alpha_n$. Thus, the homogeneous system of equations $[\alpha_1 \mid \cdots \mid \alpha_n]X = O$ has a nontrivial solution precisely when the column vectors $\alpha_1, \ldots, \alpha_n$ are linear dependent.

$\square$

Theorem 2.3(d) is the important link between the notion of linear dependence and the ideas in Chapter I. If $A \in M_{m \times n}$, then the homogeneous system of equations $AX = O$ has a nontrivial solution precisely when the columns of A are linearly dependent. Theorem 2.3(d) is also the most important computational device for deciding when column vectors

$\alpha_1, \ldots, \alpha_n$ are linearly dependent. Arrange the vectors $\alpha_1, \ldots, \alpha_n$ in a matrix $[\alpha_1 \mid \cdots \mid \alpha_n]$ and compute the solutions to the homogeneous system of equations $[\alpha_1 \mid \cdots \mid \alpha_n]X = O$. The vectors $\alpha_1, \ldots, \alpha_n$ are linearly dependent if and only if this system has a nontrivial solution.

Example 2.4   Again consider

$$\alpha = \begin{bmatrix} 1 \\ 0 \\ 1 \end{bmatrix}, \quad \beta = \begin{bmatrix} -1 \\ 1 \\ 2 \end{bmatrix}, \quad \gamma = \begin{bmatrix} -1 \\ 3 \\ 8 \end{bmatrix}$$

in F^3. The matrix

$$A = [\alpha \mid \beta \mid \gamma] = \begin{bmatrix} 1 & -1 & -1 \\ 0 & 1 & 3 \\ 1 & 2 & 8 \end{bmatrix}$$

has

$$\xi = \begin{bmatrix} 2 \\ 3 \\ -1 \end{bmatrix}$$

as a nontrivial solution to $AX = O$. Consequently, α, β, γ are linearly dependent over F.

Theorem 2.3(d) has one important corollary which we mention here.

Corollary 2.5   Let $\alpha_1, \ldots, \alpha_n \in F^m$. If $n > m$, then $\alpha_1, \ldots, \alpha_n$ are linearly dependent.

Proof.   Let $A = [\alpha_1 \mid \cdots \mid \alpha_n]$. Since $n > m$, the homogeneous system of equations $AX = O$ has a nontrivial solution by Theorem I.4.22. Hence, α_1, $\ldots, \alpha_n$ are linearly dependent by Theorem 2.3(d).   □

Vectors which are not linearly dependent are said to be linearly independent. The formal definition is as follows.

Definition 2.6   Let $\alpha_1, \ldots, \alpha_n \in V$. The vectors $\alpha_1, \ldots, \alpha_n$ are said to be linearly independent (over F) if they are not linearly dependent.

It is clear from Definition 2.1 that the vectors $\alpha_1, \ldots, \alpha_n$ are linearly independent if and only if whenever some linear combination of $\alpha_1, \ldots, \alpha_n$ is zero, say $x_1\alpha_1 + \cdots + x_n\alpha_n = o$, then $x_1 = \cdots = x_n = 0$. Thus, if $\alpha_1, \ldots,$ α_n are linearly independent, the only linear combination of these vectors which can be zero is the trivial combination in which all of the scalars are zero.

We have already seen some examples of linearly independent vectors. The canonical basis $\underline{\varepsilon} = \{\varepsilon_1, \varepsilon_2, \varepsilon_3\}$ of F^3 in Example 2.2 is a set of linearly independent vectors. The same reasoning shows the canonical basis $\underline{\varepsilon} = \{\varepsilon_1, \ldots, \varepsilon_n\}$ of F^n (Example 1.21) is a set of linearly independent vectors. Similarly, the matrix units $\{A_{ij} \mid i = 1, \ldots, m; j = 1, \ldots, n\}$ in $M_{m \times n}$ are linearly independent over F.

We again point out that Definition 2.6 is really a statement about the set $\{\alpha_1, \ldots, \alpha_n\}$. If $\alpha_1, \ldots, \alpha_n$ are linearly independent, then clearly any permutation of these vectors is still a set of vectors which are linearly independent. Notice that Lemma 1.2(d) implies that any nonzero vector α is linearly independent over F. We also have the analog of Theorem 2.3(d).

2.7 Column vectors $\alpha_1, \ldots, \alpha_n \in F^m$ are linearly independent if and only if the homogeneous system of equations $[\alpha_1 \mid \cdots \mid \alpha_n]X = O$ has only the trivial solution $X = O$.

Vectors which are linearly independent and span V are very important in linear algebra.

Definition 2.8   Let V be a finite dimensional vector space over F. A basis of V is a finite set of vectors $\{\alpha_1, \ldots, \alpha_n\}$ such that

(a) $\alpha_1, \ldots, \alpha_n$ are linearly independent over F.
(b) $V = L(\alpha_1, \ldots, \alpha_n)$.

We have already seen several examples of bases. The canonical basis $\underline{\varepsilon} = \{\varepsilon_1, \ldots, \varepsilon_n\}$ of F^n is a basis of F^n. The matrix units $\{A_{ij} \mid i = 1, \ldots, m;$ $j = 1, \ldots, n\}$ is a basis of $M_{m \times n}$. Clearly, any polynomial of degree at most n can be written uniquely as a linear combination of the monomials $1, t, t^2,$ $\ldots, t^n$. Thus, the set $B = \{1, t, t^2, \ldots, t^n\}$ is a basis of $\mathscr{P}_n$.

Suppose $B = \{\alpha_1, \ldots, \alpha_n\}$ is a basis of V. If some of the vectors in the sequence $\alpha_1, \ldots, \alpha_n$ are interchanged, a new sequence of vectors $\gamma_1, \ldots, \gamma_n$ is formed. Each γ_i is equal to some α_j and vice versa. Since B is a basis of V, $\alpha_1, \ldots, \alpha_n$ satisfy (a) and (b) in Definition 2.8. These two statements do not depend on the order of the vectors. Thus, $\gamma_1, \ldots, \gamma_n$ also satisfy (a) and (b) in

Definition 2.8. In particular, the question "Is $B = \{\alpha_1, \ldots, \alpha_n\}$ a basis of V?" does not depend on the order of the vectors in B.

Since B is a basis of V, Definition 2.8(b) implies every vector $\beta \in V$ is a linear combination of $\alpha_1, \ldots, \alpha_n$. Thus, $\beta = x_1\alpha_1 + \cdots + x_n\alpha_n$ for some choice of scalars $x_1, \ldots, x_n \in F$. Definition 2.8(a) implies that this representation of β as a linear combination of $\alpha_1, \ldots, \alpha_n$ is unique. For suppose, $\beta = y_1\alpha_1 + \cdots + y_n\alpha_n$ for some scalars $y_1, \ldots, y_n \in F$. Then

$$o = \beta - \beta = (y_1\alpha_1 + \cdots + y_n\alpha_n) - (x_1\alpha_1 + \cdots + x_n\alpha_n)$$

$$= (y_1 - x_1)\alpha_1 + \cdots + (y_n - x_n)\alpha_n.$$

Since $\alpha_1, \ldots, \alpha_n$ are linearly independent, Definition 2.6 implies, $y_1 - x_1 = 0, \ldots, y_n - x_n = 0$. Therefore, $y_1 = x_1, \ldots, y_n = x_n$. Thus, if $B = \{\alpha_1, \ldots, \alpha_n\}$ is a basis of V, every vector in V can be written in one and only one way as a linear combination of $\alpha_1, \ldots, \alpha_n$. We will see that this property makes a basis very useful.

In the next four theorems, we will present the basic facts about bases. Before presenting these theorems, we need to say a few words about the zero vector space. Suppose $V = \{o\}$. Since the empty set $\varnothing$ is a linearly independent set of vectors, and $L(\varnothing) = \{o\}$, $\varnothing$ is a basis of $\{o\}$.

Our first theorem says that every set of linearly independent vectors can be expanded to a basis.

Theorem 2.9 Let V be a finite dimensional vector space over F. Let $\{\gamma_1, \ldots, \gamma_r\}$ be a set of linearly independent vectors in V. Then there exists a basis B of V such that $\{\gamma_1, \ldots, \gamma_r\} \subseteq B$.

Proof. Since V is finite dimensional over F, there exist vectors $\alpha_1, \ldots, \alpha_n$ in V such that $V = L(\alpha_1, \ldots, \alpha_n)$. Suppose $\{\alpha_1 \ldots, \alpha_n\} \subseteq L(\gamma_1, \ldots, \gamma_r)$. Then $V = L(\alpha_1, \ldots, \alpha_n) \subseteq L(L(\gamma_1, \ldots, \gamma_r)) = L(\gamma_1, \ldots, \gamma_r)$ by Theorem 1.16. But then $V = L(\gamma_1, \ldots, \gamma_r)$, and $\{\gamma_1, \ldots, \gamma_r\}$ is a already a basis of V. In this case, set $B = \{\gamma_1, \ldots, \gamma_r\}$, and the proof of the theorem is complete.

Suppose some $\alpha_i \notin L(\gamma_1 \ldots, \gamma_r)$. After relabeling the α_ks if need be, we can assume $\alpha_1 \notin L(\gamma_1, \ldots, \gamma_r)$. We claim the vectors $\alpha_1, \gamma_1, \ldots, \gamma_r$ are linearly independent. Suppose $x_1\alpha_1 + y_1\gamma_1 + \cdots + y_r\gamma_r = o$. If $x_1 \neq 0$, then this equation implies $\alpha_1 \in L(\gamma_1, \ldots, \gamma_r)$ which is impossible. Hence, $x_1 = 0$. Since $\gamma_1, \ldots, \gamma_r$ are linearly independent, we conclude that $y_1 = \cdots = y_r = 0$. Thus, $\alpha_1, \gamma_1, \ldots, \gamma_r$ are linearly independent.

Now suppose $\{\alpha_2, \ldots, \alpha_n\} \subseteq L(\alpha_1, \gamma_1, \ldots, \gamma_r)$. Then $\{\alpha_1, \ldots, \alpha_n\} \subseteq L(\alpha_1, \gamma_1, \ldots, \gamma_r)$. The same argument as in the first paragraph of this proof shows

$V = L(\alpha_1, \gamma_1, \ldots, \gamma_r)$. In particular, $B = \{\alpha_1, \gamma_1, \ldots, \gamma_r\}$ is a basis of V, and the proof of the theorem is complete.

If some $\alpha_j \notin L(\alpha_1, \gamma_1, \ldots, \gamma_r)$ with $2 \leq j \leq n$, we can repeat the same argument again. After relabeling $\alpha_2 \ldots, \alpha_n$ if need be, we get the vectors α_1, $\alpha_2, \gamma_1, \ldots, \gamma_r$ are linearly independent. If $\alpha_1, \alpha_2, \gamma_1, \ldots, \gamma_r$ span V, then $B = \{\alpha_1, \alpha_2, \gamma_1, \ldots, \gamma_r\}$ is the required basis of V. If these vectors do not span V, then the argument continues. Now the vectors $\alpha_1, \ldots, \alpha_n$ do span V and there are only finitely many of them. Hence, this argument terminates in a finite number of steps with a basis of V of the following form $B = \{\alpha_1, \ldots, \alpha_p, \gamma_1, \ldots, \gamma_r\}$. Here $p \leq n$. This completes the proof of the theorem. □

There is an obvious corollary to Theorem 2.9 which is our second fundamental theorem about bases.

Theorem 2.10   Every finite dimensional vector space has a basis.

Proof.   Suppose V is a finite dimensional vector space over F. If $V = \{o\}$, then $\varnothing$ is a basis of V. Suppose $V \neq \{o\}$. Then V contains a nonzero vector α. We have previously observed that any nonzero vector is linearly independent over F. By Theorem 2.9, V has a basis B which contains $\{\alpha\}$. In particular, V has a basis B. □

Theorem 2.9 says that any linearly independent set of vectors in V can be expanded to a basis of V. Our next result says that if a set of vectors span V, then that set must contain a basis of V.

Theorem 2.11   Suppose V is a finite dimensional vector space over F. If $V = L(\alpha_1, \ldots, \alpha_n)$, then $\{\alpha_1, \ldots, \alpha_n\}$ contains a basis of V.

Proof.   If the vectors $\alpha_1, \ldots, \alpha_n$ are linearly independent over F, then $\{\alpha_1, \ldots, \alpha_n\}$ is a basis of V, and the proof is complete. Suppose $\alpha_1, \ldots, \alpha_n$ are linearly dependent. Then by Theorem 2.3(c), $\alpha_i \in L(\alpha_1, \ldots, \hat{\alpha}_i, \ldots, \alpha_n)$ for some i with $1 \leq i \leq n$. Relabeling the α_ks if need be, we can assume $\alpha_1 \in L(\alpha_2, \ldots, \alpha_n)$. Then $V = L(\alpha_2, \ldots, \alpha_n)$. If $\alpha_2, \ldots, \alpha_n$ are linearly independent, then $\{\alpha_2, \ldots, \alpha_n\}$ is a basis of V.

If $\alpha_2, \ldots, \alpha_n$ are linearly dependent, then we can repeat the proof and throw out another α_k. After finitely many applications of this argument, we find a subset B of $\{\alpha_1, \ldots, \alpha_n\}$ (possibly empty) such that B is a basis of V. □

Our fourth theorem leads to the notion of dimension of a vector space V.

Theorem 2.12 Suppose V is a finite dimensional vector space over F. Then any two bases of V contain the same number of vectors.

Proof. Suppose $B = \{\beta_1, \ldots, \beta_m\}$ and $A = \{\alpha_1, \ldots, \alpha_n\}$ are two bases of V. We will use the exchange principle to argue $m = n$.

Suppose $m < n$. Since $\beta_1 \in L(\alpha_1, \ldots, \alpha_n)$, there exist scalars $x_1, \ldots, x_n \in F$ such that $\beta_1 = x_1\alpha_1 + \cdots + x_n\alpha_n$. Since β_1 is part of a basis of V, $\beta_1 \neq o$. Hence, some x_i is nonzero. Relabeling the α_ks if need be, we can assume $x_1 \neq 0$. Then $\beta_1 \in L(\alpha_1, \ldots, \alpha_n)$, but $\beta_1 \notin L(\alpha_2, \ldots, \alpha_n)$. For if $\beta_1 \in L(\alpha_2, \ldots, \alpha_n)$, then $\beta_1 = y_2\alpha_2 + \cdots + y_n\alpha_n$ for some y_i in F. We would then have $o = \beta_1 - \beta_1 = x_1\alpha_1 + (x_2 - y_2)\alpha_2 + \cdots + (x_n - y_n)\alpha_n$ is a nontrivial ($x_1 \neq 0$) linear combination of $\alpha_1, \ldots, \alpha_n$ which is zero. Since A is a basis of V, this is impossible. Hence, $\beta_1 \in L(\alpha_1, \ldots, \alpha_n) - L(\alpha_2, \ldots, \alpha_n)$.

We can now apply the exchange principle, Theorem 1.16(e), and conclude that $\alpha_1 \in L(\beta_1, \alpha_2, \ldots, \alpha_n)$. In particular, $V = L(\beta_1, \alpha_2, \ldots, \alpha_n)$. We had observed that β_1 is not a linear combination of $\alpha_2, \ldots, \alpha_n$. Hence it follows that $\beta_1, \alpha_2, \ldots, \alpha_n$ are linearly independent. Thus, $A_1 = \{\beta_1, \alpha_2, \ldots, \alpha_n\}$ is a basis of V.

We can now repeat the entire argument again with A replaced with A_1, and β_1 replaced with β_2. We have $\beta_2 \in L(\beta_1, \alpha_2, \ldots, \alpha_n)$. Hence, $\beta_2 = x_1\beta_1 + x_2\alpha_2 + \cdots + x_n\alpha_n$. Since $\beta_1, \beta_2 \in B$, β_1 and β_2 are linearly independent. Thus, some x_i is nonzero with $i \geqslant 2$. After relabeling the vectors $\alpha_2, \ldots, \alpha_n$, if need be, we can assume $x_2 \neq 0$. Then the same argument used above shows $A_2 = \{\beta_1, \beta_2, \alpha_3, \ldots, \alpha_n\}$ is a basis of V.

Repeating this argument m times, we construct a basis of V of the following form $A_m = \{\beta_1, \ldots, \beta_m, \alpha_{m+1}, \ldots, \alpha_n\}$. But B is a basis of V. Thus, $V = L(\beta_1, \ldots, \beta_m)$. In particular, α_{m+1} must be a nontrivial linear combination of $\beta_1, \ldots, \beta_m$. Thus, the vectors in A_m are linearly dependent. Since A_m is a basis of V, this is impossible. Hence, if $m < n$, we get a contradiction. We conclude that $m \geqslant n$.

Reversing the roles of A and B in this proof, we get $n \geqslant m$. Therefore, $m = n$, and the proof of the theorem is complete. $\qquad\square$

Notice that all four of our basic theorems about bases are for finite dimensional vector spaces. The theorems remain true for infinite dimensional vector spaces as well. A subset S of a (possibly infinite dimensional) vector space V is said to be linearly independent if every finite subset of distinct vectors from S are linearly independent. A linearly independent subset S of V is called a basis of V if $L(S) = V$. For example, $B = \{1, t, t^2, t^3, \ldots\}$ is clearly a basis of $\mathbb{R}[t]$. The four theorems for arbitrary vector spaces read as follows.

2.13 Suppose V is a vector space over F. Any linearly independent subset S of V is contained in a basis of V.

2.14 Every vector space V has a basis.

2.15 If $L(S) = V$, then S contains a basis of V.

2.16 Any two bases of V has the same cardinality.

The proofs of the assertions given in (2.13) through (2.16) are quite different from the finite dimensional arguments given in Theorems 2.9 through 2.12. We invite the reader to consult reference [2] for the more general arguments.

We can now introduce the notion of dimension.

Definition 2.17 Let V be a finite dimensional vector space over F. The dimension of V is the number of vectors in any basis of V.

Theorem 2.12 guarantees that this definition makes sense. Any two bases of V contain the same number of vectors. We will denote the dimension of V by the letters dim V. For any finite dimensional vector space V, dim $V < \infty$. Let us compute the dimensions of some of the vector spaces we have discussed.

Example 2.18 dim $M_{m \times n} = mn$. The matrix units $\{A_{ij} \mid i = 1, \ldots, m; j = 1, \ldots, n\}$ is a basis of $M_{m \times n}$ (Example 1.18).

Example 2.19 dim $F^n = n$. The canonical basis $\underline{\varepsilon} = \{\varepsilon_1, \ldots, \varepsilon_n\}$ is a basis of F^n (Example 1.21).

Example 2.20 dim $\mathscr{P}_n = n + 1$. Clearly, $\{1, t, \ldots, t^n\}$ is a basis of $\mathscr{P}_n$.

If V is not finite dimensional, then V has no finite basis. In this case, V is called an infinite dimensional vector space. If V is an infinite dimensional vector space, then we will write dim $V = \infty$. For example, dim $\mathbb{R}[t] = \infty$. If I is an interval of positive length in $\mathbb{R}$, then one can argue that dim $\mathbb{R}^I = \dim C(I) = \dim C^k(I) = \dim \mathbb{R}(I) = \infty$.

In our next theorem, we gather together some of the more elementary facts about dim V.

Theorem 2.21   Let V be a finite dimensional vector space over F.

(a)   If $V = \{o\}$, then dim $V = 0$.
(b)   If $V = L(\alpha_1, \ldots, \alpha_n)$, then dim $V \leqslant n$.
(c)   If W is a subspace of V, then dim $W \leqslant$ dim V. Furthermore, dim $W =$ dim V if and only if $W = V$.

Proof.   (a) If $V = \{o\}$, then $\varnothing$ is a basis of V. The number of vectors in $\varnothing$ is zero. Thus, dim $V = 0$.

(b) If $V = L(\alpha_1, \ldots, \alpha_n)$, then the set $\{\alpha_1, \ldots, \alpha_n\}$ contains a basis of V by Theorem 2.11. Hence, dim $V \leqslant n$.

(c) Suppose W is a subspace of V. By Theorem 2.10, W has a basis $B = \{\alpha_1, \ldots, \alpha_r\}$. The vectors in B are linearly independent and, consequently, may be expanded by Theorem 2.9 to a basis $B' = \{\alpha_1, \ldots, \alpha_r, \alpha_{r+1}, \ldots, \alpha_n\}$ of V. Therefore, dim $W = r \leqslant n =$ dim V.

Now suppose dim $W =$ dim V. With the same notation as in the last paragraph, we have $r = n$. Thus, B is already a basis of V. Then $W = L(\alpha_1, \ldots, \alpha_r) = V$. $\square$

One of the most important applications of Theorem 2.10 is the construction of coordinate maps on a finite dimensional vector space V. Suppose $B = \{\alpha_1, \ldots, \alpha_n\}$ is a basis of V. Then any vector $\gamma \in V$ can be written uniquely as a linear combination, $\gamma = x_1\alpha_1 + \cdots + x_n\alpha_n$, of the vectors in B. Thus, if the order of the vectors in B is fixed, each vector $\gamma \in V$ determines a unique column vector $[x_1, \ldots, x_n]^t \in F^n$. Let us introduce the following notation.

Definition 2.22   Let $B = \{\alpha_1, \ldots, \alpha_n\}$ be a basis of V. Let $\gamma \in V$. Then $[\gamma]_B = [x_1 \ \cdots \ x_n]^t$ if $x_1\alpha_1 + \cdots + x_n\alpha_n = \gamma$. Thus, $[\gamma]_B$ is the column vector in F^n whose entries are the unique scalars needed to represent γ as a linear combination of $\alpha_1, \ldots, \alpha_n$ (in that order).

Notice that unlike most of our other comments about bases, the order of the vectors in B is important here. If we interchange some of the vectors in B, then the column vector $[\gamma]_B$ changes. Consider the following example.

Example 2.23   Let $V = F^3$. Let $\underline{\varepsilon} = \{\varepsilon_1, \varepsilon_2, \varepsilon_3\}$ be the canonical basis of F^3. Set $B = \{\varepsilon_2, \varepsilon_3, \varepsilon_1\}$. Since B is the same set as $\underline{\varepsilon}$, B is a basis of F^3. Let

$\gamma = [a \; b \; c]^t \in F^3$. Then $[\gamma]_{\underline{\varepsilon}} = [a \; b \; c]^t$ since $\gamma = a\varepsilon_1 + b\varepsilon_2 + c\varepsilon_3$. On the other hand, $[\gamma]_B = [b \; c \; a]^t$ since $\gamma = b\varepsilon_2 + c\varepsilon_3 + a\varepsilon_1$.

The notation $[\gamma]_B$ is a bit ambiguous because the order of the vectors in B is not specified by the symbols. This will cause no real confusion in the sequel. One usually picks a fixed basis B and a fixed ordering of the vectors in B. If we change the ordering of the vectors in B, as in Example 2.23, we will change the name of the basis from B to something else.

The column vector $[\gamma]_B$ is called the B-skeleton of γ or the coordinates of γ with respect to $\alpha_1, \ldots, \alpha_n$. As γ varies over V, $[\gamma]_B$ varies over F^n. Thus, $[*]_B : V \mapsto F^n$ is a well-defined function from V to F^n. Our notation, $[*]_B : V \mapsto F^n$, means $[*]_B$ is the function from V to F^n whose value on any vector $\gamma \in V$ is the column vector $[\gamma]_B$. The function $[*]_B$ is called a coordinate map on V. Obviously different bases of V (or different orderings of the same basis of V) determine different coordinate maps on V. We will use the notation $[*]_B : V \mapsto F^n$ only when B is a basis of V and $n = \dim V$.

Example 2.24 Let $V = F^3$. Set $B = \{\alpha_1, \alpha_2, \alpha_3\}$ where

$$\alpha_1 = \begin{bmatrix} 2 \\ 1 \\ -1 \end{bmatrix}, \quad \alpha_2 = \begin{bmatrix} -1 \\ 2 \\ 0 \end{bmatrix}, \quad \alpha_3 = \begin{bmatrix} 0 \\ 1 \\ 3 \end{bmatrix}$$

We have seen in Example I.5.15 that the matrix

$$A = \begin{bmatrix} 2 & -1 & 0 \\ 1 & 2 & 1 \\ -1 & 0 & 3 \end{bmatrix}$$

is nonsingular. Thus, (2.7) implies the vectors α_1, α_2, α_3 are linearly independent. Since $\dim F^3 = 3$, B is a basis of F^3. If $\underline{\varepsilon} = \{\varepsilon_1, \varepsilon_2, \varepsilon_3\}$ is the canonical basis of F^3, then

$$[\varepsilon_1]_B = \begin{bmatrix} 3/8 \\ -1/4 \\ 1/8 \end{bmatrix}, \quad [\varepsilon_2]_B = \begin{bmatrix} 3/16 \\ 3/8 \\ 1/16 \end{bmatrix}, \quad [\varepsilon_3]_B = \begin{bmatrix} -1/16 \\ -1/8 \\ 5/16 \end{bmatrix}$$

These column vectors are computed by solving the equations $AX_i = \varepsilon_i$. Since A is invertible, $X_i = A^{-1}\varepsilon_i = \mathrm{Col}_i(A^{-1})$ for $i = 1, 2,$ and 3.

Any coordinate map $[*]_B : V \mapsto F^n$ satisfies three important properties.

2.25 (a) $[\gamma]_B = O$ if and only if $\gamma = o$.
(b) $[\gamma + \gamma']_B = [\gamma]_B + [\gamma']_B$ for all $\gamma, \gamma' \in V$.
(c) $[x\gamma]_B = x[\gamma]_B$ for all $x \in F$ and $\gamma \in V$.

The statement in (a) of (2.25) follows immediately from the fact that B is a basis for V. Suppose γ and γ' are two vectors in V. If $B = \{\alpha_1, \ldots, \alpha_n\}$, then $\gamma = x_1\alpha_1 + \cdots + x_n\alpha_n$ and $\gamma' = y_1\alpha_1 + \cdots + y_n\alpha_n$. In particular, $\gamma + \gamma' = (x_1 + y_1)\alpha_1 + \cdots + (x_n + y_n)\alpha_n$. Therefore,

$$[\gamma + \gamma']_B = [x_1 + y_1 \cdots x_n + y_n]^t = [x_1 \cdots x_n]^t + [y_1 \cdots y_n]^t$$
$$= [\gamma]_B + [\gamma']_B.$$

A similar proof can be given for (c) of (2.25).

Coordinate maps have all kinds of useful applications when studying abstract (finite dimensional) vector spaces. A coordinate map $[*]_B : V \mapsto F^n$ can be used to transform a problem in the abstract vector space V to a corresponding problem in the concrete space F^n. We can then use matrix arithmetic to solve the problem. For instance, suppose we want to decide if $\gamma_1, \ldots, \gamma_r$ are linearly independent in V. From the equations in (2.25), we have $x_1\gamma_1 + \cdots + x_r\gamma_r = o$ if and only if $x_1[\gamma_1]_B + \cdots + x_r[\gamma_r]_B = O$. Thus, the vectors $\gamma_1, \ldots, \gamma_r$ are linearly independent in V if and only if the column vectors $[\gamma_1]_B, \ldots, [\gamma_r]_B$ are linearly independent in F^n. Theorem 2.3(d) implies the column vectors $[\gamma_1]_B, \ldots, [\gamma_r]_B$ are linearly independent if and only if the homogeneous system of equations

$$[[\gamma_1]_B | \cdots | [\gamma_r]_B]X = O$$

has only the trivial solution $X = O$.

Consider the following example.

Example 2.26 Let $V = \mathscr{P}_2$. Show that the vectors $\gamma_1 = 2 + t - t^2$, $\gamma_2 = -1 + 2t$ and $\gamma_3 = t + 3t^2$ are linearly independent in $\mathscr{P}_2$.

A basis for $\mathscr{P}_2$ is $B = \{1, t, t^2\}$.

$$[\gamma_1]_B = \begin{bmatrix} 2 \\ 1 \\ -1 \end{bmatrix}, \quad [\gamma_2]_B = \begin{bmatrix} -1 \\ 2 \\ 0 \end{bmatrix}, \quad [\gamma_3]_B = \begin{bmatrix} 0 \\ 1 \\ 3 \end{bmatrix}$$

The matrix

$$A = [[\gamma_1]_B | [\gamma_2]_B | [\gamma_3]_B] = \begin{bmatrix} 2 & -1 & 0 \\ 1 & 2 & 1 \\ -1 & 0 & 3 \end{bmatrix}$$

is nonsingular by Example I.5.15. In particular, $AX = O$ has only the trivial solution $X = O$. Thus, $[\gamma_1]_B$, $[\gamma_2]_B$, $[\gamma_3]_B$ are linearly independent in F^3. Our remarks in the last paragraph now imply γ_1, γ_2, γ_3 are independent in $\mathscr{P}_2$.

It is often convenient to switch from one coordinate map to another. The equations in (2.25) can be used to work out the relationships between any two coordinate maps on V. Suppose $B = \{\alpha_1, \ldots, \alpha_n\}$ and $C = \{\beta_1, \ldots, \beta_n\}$ are two (ordered) bases for V. Let $M(C, B)$ denote the $n \times n$ matrix whose columns are defined by the following equation.

2.27
$$M(C, B) = [[\beta_1]_B \,|\, [\beta_2]_B \,|\, \cdots \,|\, [\beta_n]_B].$$

Thus, the ith column of $M(C, B)$ is the B-skeleton of the ith vector in C. We can now state the fundamental relationship between the two coordinate maps $[*]_B$ and $[*]_C$.

Theorem 2.28 $M(C, B)[\gamma]_C = [\gamma]_B$ for all $\gamma \in V$.

Proof. Let $\gamma \in V$. Since C is a basis of V, there exist unique scalars $x_1, \ldots, x_n$ in F such that $\gamma = x_1\beta_1 + \cdots + x_n\beta_n$. Using the equations in (2.25), we have

$$M(C, B)[\gamma]_C = M(C, B)[x_1\beta_1 + \cdots + x_n\beta_n]_C$$
$$= x_1 M(C, B)[\beta_1]_C + \cdots + x_n M(C, B)[\beta_n]_C.$$

Similarly, $[\gamma]_B = x_1[\beta_1]_B + \cdots + x_n[\beta_n]_B$. Hence, it suffices to prove that $M(C, B)[\beta_i]_C = [\beta_i]_B$ for all $i = 1, \ldots, n$.

Let i be a fixed integer with $1 \leqslant i \leqslant n$. Then

$$\beta_i = 0\beta_1 + \cdots + 0\beta_{i-1} + 1\beta_i + 0\beta_{i+1} + \cdots + 0\beta_n.$$

Consequently, $[\beta_i]_C = [0 \,\cdots\, 1 \,\cdots\, 0]^t = \varepsilon_i$, the ith vector of the canonical basis ε of F^n. Therefore,

$$M(C, B)[\beta_i]_C = M(C, B)\varepsilon_i = \mathrm{Col}_i(M(C, B)) = [\beta_i]_B.$$

This completes the proof of the theorem. $\qquad\qquad\qquad\qquad\square$

The matrix $M(C, B)$ given in equation (2.27) is called a change of basis matrix. Once we have computed $M(C, B)$, Theorem 2.28 allows us to switch freely from coordinates with respect to C to coordinates with respect to B. Let us return to Example 2.24 for an illustration of these ideas.

Example 2.29 In Example 2.24, let $C = \underline{\varepsilon} = \{\varepsilon_1, \varepsilon_2, \varepsilon_3\}$ be the canonical basis of F^3. The computations for $[\varepsilon_i]_B$ given in Example 2.24 imply

$$M(C, B) = \begin{bmatrix} 3/8 & 3/16 & -1/16 \\ -1/4 & 3/8 & -1/8 \\ 1/8 & 1/16 & 5/16 \end{bmatrix}$$

Thus, if

$$\gamma = \begin{bmatrix} x \\ y \\ z \end{bmatrix}$$

is any vector in F^3, Theorem 2.28 implies

$$[\gamma]_B = \begin{bmatrix} 3/8 & 3/16 & -1/16 \\ -1/4 & 3/8 & -1/8 \\ 1/8 & 1/16 & 5/16 \end{bmatrix} \begin{bmatrix} x \\ y \\ z \end{bmatrix}$$

For example, if

$$\gamma = \begin{bmatrix} 1 \\ 1 \\ 1 \end{bmatrix}$$

then

$$[\gamma]_B = \begin{bmatrix} 1/2 \\ 0 \\ 1/2 \end{bmatrix}$$

Exercises for Section 2

2.1   Let

$$\alpha = \begin{bmatrix} 1 \\ 2 \\ 1 \end{bmatrix}, \quad \beta = \begin{bmatrix} 0 \\ 1 \\ 1 \end{bmatrix}, \quad \gamma = \begin{bmatrix} 2 \\ 1 \\ -1 \end{bmatrix}, \quad \delta = \begin{bmatrix} 1 \\ -1 \\ 2 \end{bmatrix}$$

in F^3. Decide which of the following sets of vectors in F^3 are linearly dependent. If the vectors are dependent, exhibit a nontrivial linear combination of the vectors which is zero.

(a) $\{\alpha, \beta\}$
(b) $\{\alpha, \beta, \gamma\}$
(c) $\{\alpha, \beta, \delta\}$
(d) $\{\alpha, \beta, \gamma, \delta\}$

2.2 Show that $\mathbb{C}$ is a finite dimensional vector space over $\mathbb{R}$ with $\dim_{\mathbb{R}} \mathbb{C} = 2$.

2.3 Let

$$\alpha = \begin{bmatrix} 1 \\ 1 \\ 2 \\ 1 \end{bmatrix}, \quad \beta = \begin{bmatrix} 2 \\ 1 \\ 3 \\ 1 \end{bmatrix}, \quad \gamma = \begin{bmatrix} 3 \\ 2 \\ 5 \\ x \end{bmatrix}$$

in $\mathbb{R}^4$. For what values of x are these vectors linearly dependent?

2.4 Show that $B = \{1, t, t^2, \ldots, t^n\}$ is a basis of $\mathscr{P}_n$. Compute the coordinate map $[*]_B : \mathscr{P}_n \mapsto \mathbb{R}^{n+1}$.

2.5 Find a basis of $\mathscr{P}_3$ containing $\alpha_1 = 1 + t$ and $\alpha_2 = 1 - t^2$.

2.6 Suppose $A \in M_{m \times n}$ is a matrix in row reduced echelon form. Show $\dim RS(A)$ is the number of nonzero rows of A.

2.7 In Exercise 2.6, show $\dim RS(A) = \dim CS(A)$.

2.8 Suppose α, β, γ are linearly independent in V. Show that $\alpha + \beta$, $\beta + \gamma, \gamma + \alpha$ are also linearly independent in V.

2.9 In the proof of Theorem 2.12, give the argument a second time, i.e., show that after suitably relabeling $\alpha_2, \ldots, \alpha_n$, if need be, $\{\beta_1, \beta_2, \alpha_3, \ldots, \alpha_n\}$ is a basis of V.

2.10 Let $f_1, \ldots, f_n \in \mathbb{R}[t]$ be nonzero polynomials. Suppose the degrees of the f_k are all different. Show that $f_1, \ldots, f_n$ are linearly independent.

2.11 Find a basis for $V = \{A \in M_{m \times n} | A \text{ is upper triangular}\}$. Compute $\dim V$.

2.12 Find a basis for $V = \{A \in M_{n \times n} | A = A^t\}$. Compute $\dim V$.

2.13 Find a basis for $V = \{A \in M_{n \times n} | A = -A^t\}$. Compute $\dim V$.

2.14 Let V be a finite dimensional vector space over F. Let W be a subspace of V. Show there exists a subspace W' of V such that $W + W' = V$, and $W \cap W' = \{o\}$. The subspace W' is called a complement of W.

2.15 In Exercise 14, suppose $V = \mathbb{R}^2$. What do all complements of a one-dimensional subspace W look like? Suppose $V = \mathbb{R}^3$, and $\dim W = 2$. What do all complements of W look like?

2.16 Compute a coordinate map $[*]_B : M_{2 \times 3} \mapsto F^6$ by taking B to be the matrix units of $M_{2 \times 3}$.

2.17 Use your answer from Exercise 2.16 to show the following matrices are linearly dependent

$$A_1 = \begin{bmatrix} 1 & 0 & 0 \\ 1 & 0 & 1 \end{bmatrix}, \quad A_2 = \begin{bmatrix} 2 & -1 & 1 \\ 1 & 2 & 3 \end{bmatrix},$$

$$A_3 = \begin{bmatrix} -1 & -1 & 1 \\ 2 & 0 & 1 \end{bmatrix}, \quad A_4 = \begin{bmatrix} 7 & -2 & 2 \\ 3 & 6 & 10 \end{bmatrix}$$

2.18 Prove (c) of (2.25).

2.19 Use a coordinate map to argue the following. If dim $V = n$, then any $n + 1$ vectors in V are linearly dependent.

2.20 Suppose $V = \mathbb{R}^3$. Set $B = \{\alpha_1, \alpha_2, \alpha_3\}$ where

$$\alpha_1 = \begin{bmatrix} 3 \\ 2 \\ 1 \end{bmatrix}, \quad \alpha_2 = \begin{bmatrix} -1 \\ 1 \\ -3 \end{bmatrix}, \quad \alpha_3 = \begin{bmatrix} 2 \\ 1 \\ 0 \end{bmatrix}.$$

Show B is a basis of V. Compute the coordinate map $[*]_B : \mathbb{R}^3 \mapsto \mathbb{R}^3$.

2.21 Let V be a finite dimensional vector space over F. For any two bases B and C of V, show the change of basis matrix $M(C, B)$ is invertible with inverse $M(B, C)$.

2.22 Let $A \in M_{m \times n}$ and $B \in F^m$. Show the equation $AX = B$ has a solution if and only if dim $CS(A) =$ dim $CS(A \mid B)$.

2.23 Let V be a finite dimensional vector space over F. Suppose B is a basis of V. If W is a subspace of V, show $[W]_B = \{[\gamma]_B \mid \gamma \in W\}$ is a subspace of F^n. Here $n = $ dim V.

2.24 In Exercise 2.23, show dim $W = $ dim$[W]_B$.

3. THE RANK OF A MATRIX

Let $A \in M_{m \times n}$. Recall the column space of A, $CS(A)$, is the subspace of F^m spanned by the columns of A. In particular, $CS(A)$ is a finite dimensional vector space over F. The dimension of $CS(A)$ is called the rank of A.

Definition 3.1 Let $A \in M_{m \times n}$. The dimension of $CS(A)$ is called the rank of A.

In this book, we will let rk(A) denote the rank of A. Thus, rk(A) = dim $CS(A)$. In particular, rk$(A) \geqslant 0$. Since $CS(A) \subseteq F^m$, Theorem 2.21(c) implies rk$(A) \leqslant m$. On the other hand, $CS(A) = L(\text{Col}_1(A), \ldots, \text{Col}_n(A))$.

Therefore, Theorem 2.21(b) implies $\mathrm{rk}(A) \leqslant n$. Thus, for any $m \times n$ matrix A, $0 \leqslant \mathrm{rk}(A) \leqslant \min\{m, n\}$. Notice that $\mathrm{rk}(A) = 0$ if and only if $A = 0$.

Example 3.2 Let

$$A = \begin{bmatrix} 1 & 2 \\ 2 & 4 \\ 3 & 6 \end{bmatrix}, \quad B = \begin{bmatrix} 1 & 1 & 5 \\ 0 & 1 & 3 \\ 1 & 1 & 5 \end{bmatrix}, \quad C = \begin{bmatrix} 2 & -1 & 0 & 4 & 6 \\ 1 & 2 & 1 & 1 & 1 \\ -1 & 0 & 3 & 2 & 8 \end{bmatrix}$$

Since $\mathrm{Col}_2(A) = 2\mathrm{Col}_1(A)$, $\mathrm{rk}(A) = 1$. $\mathrm{Col}_1(B)$ and $\mathrm{Col}_2(B)$ are clearly linearly independent. Since $\mathrm{Col}_3(B) = 2\mathrm{Col}_1(B) + 3\mathrm{Col}_2(B)$, $\dim CS(B) = 2$. Therefore, $\mathrm{rk}(B) = 2$. The first three columns of C form an invertible matrix. Hence, $CS(C) = F^3$. Therefore, $\mathrm{rk}(C) = 3$.

We can also compute the rank of A from the row space, $RS(A)$, of A. Recall that $RS(A)$ is the subspace of $M_{1 \times n}$ spanned by the rows of A. We need the following theorem.

Theorem 3.3 Let $A \in M_{m \times n}$. Then $\dim RS(A) = \dim CS(A)$.

Proof. Let $A = (a_{ij}) \in M_{m \times n}$. Suppose $k = \dim RS(A)$. Since $RS(A) = L(\mathrm{Row}_1(A), \ldots, \mathrm{Row}_m(A)) \subseteq M_{1 \times n}$, Theorem 2.21 implies $k \leqslant \min\{m, n\}$. If $k = 0$, then $A = 0$, and the theorem is obvious. Hence, we can assume $k \geqslant 1$.

Since $\dim RS(A) = k$, there exist k vectors $\alpha_1, \ldots, \alpha_k \in RS(A)$ such that $\{\alpha_1, \ldots, \alpha_k\}$ is a basis of $RS(A)$. Each α_i is a row vector of size n. Let $\alpha_i = (w_{i1}, \ldots, w_{in})$ for $i = 1, \ldots, k$. Then there exist scalars $c_{ij} \in F$ such that

3.4
$$\mathrm{Row}_i(A) = (a_{i1}, \ldots, a_{in}) = c_{i1}\alpha_1 + \cdots + c_{ik}\alpha_k, \quad i = 1, \ldots, m$$

Comparing the pth entries on the left and right in equation (3.4) gives the following system of equations.

3.5
$$\begin{aligned} a_{1p} &= c_{11}w_{1p} + c_{12}w_{2p} + \cdots + c_{1k}w_{kp} \\ a_{2p} &= c_{21}w_{1p} + c_{22}w_{2p} + \cdots + c_{2k}w_{kp} \\ &\vdots \quad\quad \vdots \quad\quad \vdots \quad\quad\quad \vdots \\ a_{mp} &= c_{m1}w_{1p} + c_{m2}w_{2p} + \cdots + c_{mk}w_{kp} \end{aligned} \qquad p = 1, \ldots, n$$

These equations can be rewritten as follows.

3.6

$$\text{Col}_p(A) = w_{1p} \begin{bmatrix} c_{11} \\ c_{21} \\ \vdots \\ c_{m1} \end{bmatrix} + w_{2p} \begin{bmatrix} c_{12} \\ c_{22} \\ \vdots \\ c_{m2} \end{bmatrix} + \cdots + w_{kp} \begin{bmatrix} c_{1k} \\ c_{2k} \\ \vdots \\ c_{mk} \end{bmatrix}$$

Here $p = 1, \ldots, n$. Let $\xi_r = [c_{1r} \cdots c_{mr}]^t \in F^m$ for $r = 1, \ldots, k$. Then equation (3.6) implies $CS(A) \subseteq L(\xi_1, \ldots, \xi_k)$. It now follows from Theorem 2.21 that $\dim CS(A) \leqslant k = \dim RS(A)$.

We have now shown that $\dim CS(A) \leqslant \dim RS(A)$ for any matrix A. To complete the proof of the theorem, we need to show $\dim RS(A) \leqslant \dim CS(A)$. For any $m \times n$ matrix A, suppose we replace A with A^t in the above argument. We get $\dim CS(A^t) \leqslant \dim RS(A^t)$. The columns of A^t are just the rows of A (written as columns). In particular, $\dim CS(A^t) = \dim RS(A)$. Similarly, $\dim RS(A^t) = \dim CS(A)$. Therefore,

$$\dim RS(A) = \dim CS(A^t) \leqslant \dim RS(A^t) = \dim CS(A).$$

Hence, $\dim RS(A) \leqslant \dim CS(A)$, and the proof of Theorem 3.3 is complete.
$\square$

Theorem 3.3 implies $\text{rk}(A) = \dim CS(A) = \dim RS(A)$. Theorem 2.11 implies that $\dim CS(A)$ ($\dim RS(A)$) is the maximum number of columns (rows) of A which are linearly independent. Thus, the rank of A is the maximum number of columns or rows of A which are linearly independent. In our first corollary, we record some of the facts about rank which are now apparent.

Corollary 3.7   Let $A \in M_{m \times n}$.

(a) $0 \leqslant \text{rk}(A) \leqslant \min\{m, n\}$.
(b) $\text{rk}(A) = \text{rk}(A^t)$.
(c) $\text{rk}(A) = \text{rk}(PAQ)$ for any invertible matrices P and Q.

Proof. (a) We had already noted this fact before the proof of Theorem 3.3.
(b) $\text{rk}(A) = \dim RS(A) = \dim CS(A^t) = \text{rk}(A^t)$.
(c) We have seen in Theorem I.5.13 that any invertible matrix P is a

product of elementary matrices. Thus, $A \underset{r}{\sim} PA$. Therefore, $RS(A) = RS(PA)$ by Corollary I.5.20. Theorem 3.3 then implies

$$\text{rk}(A) = \dim RS(A) = \dim RS(PA) = \text{rk}(PA).$$

Since any invertible matrix Q is a product of elementary matrices, $PA \underset{c}{\sim} PAQ$. In particular, $CS(PA) = CS(PAQ)$ by Theorem I.5.22(c). Thus,

$$\text{rk}(PA) = \dim CS(PA) = \dim CS(PAQ) = \text{rk}(PAQ).$$

$\square$

Since the rank of A is the dimension of the row space of A, we have the following method for computing rank.

Corollary 3.8 Let $A \in M_{m \times n}$. Suppose $A \underset{r}{\sim} E$ where E is an $m \times n$ matrix in echelon form. Then $\text{rk}(A)$ is the number of nonzero rows in E.

Proof. Since $A \underset{r}{\sim} E$, there exists an invertible matrix P such that $PA = E$. In particular, Corollary 3.7 implies $\text{rk}(A) = \text{rk}(E)$. By Theorem 3.3, $\text{rk}(E) = \dim RS(E)$. It is clear from the definition of a matrix E in echelon form (Definition I.4.14), that the nonzero rows of E are linearly independent. Thus, the number of nonzero rows in E is precisely the dimension of $RS(E)$. $\square$

Thus, to compute the rank of A, row reduce A to a matrix E in echelon form, and count the number of nonzero rows in E. There are many computer algorithms available which reduce A to a matrix in row reduced echelon form. Any of these programs can be used to compute the rank of A.

Example 3.9 Let

$$A = \begin{bmatrix} 1 & 1 & 1 & 0 & 2 & 3 & 0 & 4 \\ 1 & 1 & 1 & 1 & 2 & 3 & 0 & 5 \\ 2 & 2 & 2 & 0 & 4 & 6 & 1 & 15 \\ 4 & 4 & 4 & 1 & 8 & 12 & 1 & 24 \\ 2 & 2 & 2 & -1 & 4 & 6 & 1 & 14 \end{bmatrix}$$

The reader can easily check that A is row equivalent to the following matrix in row reduced echelon form

$$E = \begin{bmatrix} 1 & 1 & 1 & 0 & 2 & 3 & 0 & 4 \\ 0 & 0 & 0 & 1 & 0 & 0 & 0 & 1 \\ 0 & 0 & 0 & 0 & 0 & 0 & 1 & 7 \\ 0 & 0 & 0 & 0 & 0 & 0 & 0 & 0 \\ 0 & 0 & 0 & 0 & 0 & 0 & 0 & 0 \end{bmatrix}$$

Thus, Corollary 3.8 implies $\mathrm{rk}(A) = 3$.

Recall that two matrices $A, B \in M_{m \times n}$ are said to be equivalent ($A \approx B$) if $PAQ = B$ for some invertible matrices P and Q. We have seen in Section I.5 that this is the same as saying B can be obtained from A by performing a finite number of row and column operations on A. It follows from Corollary 3.7(c) that if A is equivalent to B, then A and B have the same rank. The converse of this statement is also true.

Corollary 3.10   Let $A, B \in M_{m \times n}$. Then $A \approx B$ if and only if $\mathrm{rk}(A) = \mathrm{rk}(B)$.

Proof.   Suppose A and B have the same rank r. If $r = 0$, then $A = B = O$, and certainly $A \approx B$. Hence, we can assume $r > 0$. By Theorem I.5.25, there exist invertible matrices P_1, P_2, Q_1, and Q_2 such that

$$P_1 A Q_1 = \begin{bmatrix} I_s & O \\ \hline O & O \end{bmatrix} \quad \text{and} \quad P_2 B Q_2 = \begin{bmatrix} I_u & O \\ \hline O & O \end{bmatrix}$$

Since $\mathrm{rk}(A) = \mathrm{rk}(B) = r$, Corollary 3.7(c) implies $s = \mathrm{rk}(P_1 A Q_1) = \mathrm{rk}(A) = r = \mathrm{rk}(B) = \mathrm{rk}(P_2 B Q_2) = u$. In particular, $P_1 A Q_1 = P_2 B Q_2$. But then, $(P_2^{-1} P_1) A (Q_1 Q_2^{-1}) = B$. Since the product of invertible matrices is invertible, $A \approx B$. $\square$

Thus, the rank of A is an invariant of the equivalence class containing A. By this, we mean any matrix equivalent to A has the same rank as A.

Our last corollary to Theorem 3.3 gives an upper bound for the rank of a product of two matrices.

Corollary 3.11   Let $A \in M_{m \times n}$ and $B \in M_{n \times p}$. Then

$$\mathrm{rk}(AB) \leqslant \min\{\mathrm{rk}(A), \mathrm{rk}(B)\}$$

Proof. We have seen in (3.22) and (3.23) in Chapter I that $CS(AB) \subseteq CS(A)$, and $RS(AB) \subseteq RS(B)$. Using Theorems 2.21 and 3.3, we have

$$\mathrm{rk}(AB) = \dim CS(AB) \leqslant \dim CS(A) = \mathrm{rk}(A).$$

Also,

$$\text{rk}(AB) = \dim RS(AB) \leqslant \dim RS(B) = \text{rk}(B).$$

Thus, $\text{rk}(AB) \leqslant \min\{\text{rk}(A), \text{rk}(B)\}$. $\qquad\qquad\qquad\qquad\square$

There are some lower bounds for the rank of AB also. See Exercise 3.15.

We can now return to a topic left unfinished in Chapter I. We claimed in Chapter I that any row reduced echelon form of a matrix is unique. This assertion follows easily from our next theorem.

Theorem 3.12 Let $A, B \in M_{m \times n}$. Suppose A and B are both in row reduced echelon form. If $RS(A) = RS(B)$, then $A = B$.

Proof. Let k denote the number of nonzero rows of A. If $k = 0$, then $A = O$. In this case, $(o) = RS(A) = RS(B)$, and, consequently, $B = O$. Hence, we can assume $k \geqslant 1$.

We have seen in Corollary 3.8 that $k = \dim RS(A)$. Thus, $k = \dim RS(B)$, and, consequently, B has precisely k nonzero rows. The rows numbered greater than k (if any) in both A and B are zero. Hence, we can drop these rows from the proof and assume $k = m$.

Partition A and B into rows with the following notation:

3.13
$$A = \begin{bmatrix} A_1 \\ \vdots \\ A_m \end{bmatrix}, \quad B = \begin{bmatrix} B_1 \\ \vdots \\ B_m \end{bmatrix}$$

Since we are assuming $k = m$, $A_1, \ldots, A_m$ are linearly independent in $M_{1 \times n}$. Similarly, $B_1, \ldots, B_m$ are linearly independent in $M_{1 \times n}$. The first nonzero entry in each row of A is a 1. Suppose j_i is the number of the column in which $\text{Row}_i(A)$ begins with a 1. Since A is in row reduced echelon form, $j_1 < j_2 < \cdots < j_m$, and A has the following form:

3.14

$$A = \begin{bmatrix} 0 & \cdots & 0 & 1 & * & \cdots & * & 0 & * & \cdots & * & 0 & * & \cdots \\ 0 & \cdots & 0 & 0 & 0 & \cdots & 0 & 1 & * & \cdots & * & 0 & * & \cdots \\ 0 & \cdots & 0 & 0 & 0 & \cdots & 0 & 0 & 0 & \cdots & 0 & 1 & * & \\ \vdots & & \vdots & \vdots & \vdots & & \vdots & \vdots & \vdots & & \vdots & \vdots & \vdots & \end{bmatrix}$$

The *s in equation (3.14) denote entries of A whose precise values are not relevant to this proof. If the leading 1s of B occur in columns $p_1, \ldots, p_m$, then $p_1 < p_2 < \cdots < p_m$. B has a form similar to that in equation (3.14) with $j_1, \ldots, j_m$ replaced with $p_1, \ldots, p_m$.

Since $RS(A) = RS(B)$, there exist scalars $x_1, \ldots, x_m$ and $y_1, \ldots, y_m$ in F such that the following equations are valid.

3.15   (a) $A_1 = x_1 B_1 + \cdots + x_m B_m.$
       (b) $B_1 = y_1 A_1 + \cdots + y_m A_m.$

We will argue that $x_1 = y_1 = 1$ and $x_2 = y_2 = \cdots = x_m = y_m = 0$.

If $j_1 > p_1$, then (b) of equation (3.15) is clearly impossible. Hence, $j_1 \leqslant p_1$. Similarly, (a) of equation (3.15) implies $p_1 \leqslant j_1$. Thus, $j_1 = p_1$. Since both rows A_1 and B_1 now have a 1 in their j_1th entry, equation (3.15) implies $x_1 = y_1 = 1$. Thus, $j_1 = p_1, x_1 = y_1 = 1$ and the equations in (3.15) have the following form.

3.16   (a) $A_1 = B_1 + x_2 B_2 + \cdots + x_m B_m.$
       (b) $B_1 = A_1 + y_2 A_2 + \cdots + y_m A_m.$

We next argue $x_2 = y_2 = 0$. Suppose $j_2 > p_2$. Then the first two rows of A and B have the following form.

3.17

$$
\begin{array}{llllllllllllll}
 & & & & j_1 & & & & p_2 & & & j_2 & & \\
A_1 = [0 & \cdots & 0 & 1 & * & \cdots & * & a & * & \cdots & * & 0 & * & \cdots] \\
A_2 = [0 & \cdots & 0 & 0 & 0 & \cdots & 0 & 0 & 0 & \cdots & 0 & 1 & * & \cdots] \\
B_1 = [0 & \cdots & 0 & 1 & * & \cdots & * & 0 & * & \cdots & & & &] \\
B_2 = [0 & \cdots & 0 & 0 & 0 & \cdots & 0 & 1 & * & \cdots & & & &]
\end{array}
$$

As the notation indicates, the a in A_1 is the p_2th entry of A_1. Equation (3.16)(b) immediately implies $a = 0$. Since $B_2 \in RS(A)$, there exist scalars $z_1, \ldots, z_m$ such that $B_2 = z_1 A_1 + \cdots + z_m A_m$. Since the p_2th entry of A_1 is zero, $B_2 = z_1 A_1 + \cdots + z_m A_m$ implies the p_2th entry in B_2 is zero. But the p_2th entry of B_2 is 1. This contradiction implies j_2 cannot be larger than p_2. We conclude that $j_2 \leqslant p_2$. Reversing the roles of A and B in this argument, gives $p_2 \leqslant j_2$. Thus, $j_2 = p_2$. The equations in (3.16) now easily imply $x_2 = y_2 = 0$.

We have now shown that $j_1 = p_1$, $j_2 = p_2$, $x_1 = y_1 = 1$, and $x_2 = y_2 = 0$. The equations in (3.16) now have the following form.

3.18   (a) $A_1 = B_1 + x_3 B_3 + \cdots + x_m B_m.$
       (b) $B_1 = A_1 + y_3 A_3 + \cdots + y_m A_m.$

We can now repeat the general argument again. Suppose $j_3 \neq p_3$. Write down the first three rows of A and B and derive a contradiction as before. We get $j_3 = p_3$. (The interested reader should work out the details of this next step. See Exercise 3.11.) The equations in (3.18) now imply $x_3 = y_3 = 0$. Continuing this argument, we get $x_i = y_i = 0$ for all $i = 2,$ $\ldots, m$. So, $A_1 = B_1$.

Now if $A_1 = B_1$ and $RS(A) = RS(B)$, then

$$RS \begin{bmatrix} A_2 \\ \vdots \\ A_m \end{bmatrix} = RS \begin{bmatrix} B_2 \\ \vdots \\ B_m \end{bmatrix}$$

It is now clear that repeated applications of the above argument give $A_i = B_i$ for $i = 2, \ldots, m$. Thus, $A = B$, and the proof of Theorem 3.12 is complete. $\square$

Corollary 3.19 Let $A \in M_{m \times n}$. Then there is precisely one matrix in row reduced echelon form which is row equivalent to A.

Proof. We have seen in Theorem I.4.18 that A can be row reduced to a matrix E in row reduced echelon form. Since, $A \underset{r}{\sim} E$, $RS(A) = RS(E)$ by Corollary I.5.20.

Suppose $A \underset{r}{\sim} E'$ for some $m \times n$ matrix E' in row reduced echelon form. The same argument then shows $RS(A) = RS(E')$. In particular, $RS(E) = RS(E')$. Theorem 3.12 then implies $E = E'$. $\square$

Thus, a matrix A has only one matrix E in row reduced echelon form which is row equivalent to A. This unique matrix E is called *the* row reduced echelon form of A. The row reduced echelon form of A is computed using the Gaussian elimination procedure outlined in the proof of Theorem I.4.18.

Another application of Theorem 3.12, is the converse of Corollary I.5.20.

Corollary 3.20 Let A, $B \in M_{m \times n}$. Then $A \underset{r}{\sim} B$ if and only if $RS(A) = RS(B)$.

Proof. If $A \underset{r}{\sim} B$, then $RS(A) = RS(B)$ by Corollary I.5.20. Suppose $RS(A) = RS(B)$. Let E (E') denote the row reduced echelon forms of A (B). Then $RS(E) = RS(A) = RS(B) = RS(E')$. Theorem 3.12 then implies $E = E'$. Therefore, $A \underset{r}{\sim} E = E' \underset{r}{\sim} B$. In particular, $A \underset{r}{\sim} B$. $\square$

Before stating our next theorem, let us formally introduce a subspace which was studied in section one of this chapter.

Definition 3.21 Let $A \in M_{m \times n}$. The set $NS(A) = \{\xi \in F^n \mid A\xi = O\}$ is called the null space of A.

$NS(A)$ is a subspace of F^n (Example 1.7). The dimension of $NS(A)$ is called the nullity of A. In this book, we will let $v(A)$ denote the nullity of A. Thus, $v(A) = \dim NS(A)$. One of the more important theorems in linear algebra is the assertion that $v(A) + \text{rk}(A) = n$ for any matrix A.

Theorem 3.22 Let $A \in M_{m \times n}$. Then $v(A) + \text{rk}(A) = n$.

Proof. $v(A) = \dim NS(A)$ and $\text{rk}(A) = \dim CS(A)$. Thus, we must argue $\dim NS(A) + \dim CS(A) = n$. There are two trivial cases to take care of first. Suppose $v(A) = 0$. Then $NS(A) = (o)$. This means the homogeneous system of equations $AX = O$ has only the trivial solution $X = O$. In particular, the columns of A are linearly independent. Thus, $\dim CS(A) = n$, and $v(A) + \text{rk}(A) = n$. Hence, we can assume $v(A) > 0$, i.e., $NS(A) \neq (o)$.

Suppose $v(A) = n$. Then $\dim NS(A) = n = \dim F^n$. Therefore, $NS(A) = F^n$. In particular, $A\varepsilon_i = O$ for each vector ε_i in the canonical basis $\underline{\varepsilon} = \{\varepsilon_1, \ldots, \varepsilon_n\}$ of F^n. But $A\varepsilon_i = \text{Col}_i(A)$. Thus, $A = O$, $\text{rk}(A) = 0$, and $v(A) + \text{rk}(A) = n + 0 = n$. Hence, we can assume $1 \leqslant v(A) < n$. Set $r = v(A)$.

Let $\{\gamma_1, \ldots, \gamma_r\}$ be a basis of $NS(A)$. By Theorem 2.9, $\{\gamma_1, \ldots, \gamma_r\}$ can be expanded to a basis $\Gamma = \{\gamma_1, \ldots, \gamma_r, \beta_1, \ldots, \beta_s\}$ of F^n. Clearly, $r + s = n$. Since $v(A) = r$, it suffices to show $s = \dim CS(A)$. By Corollary I.3.12, $CS(A) = \{A\xi \mid \xi \in F^n\}$. We claim that $B = \{A\beta_1, \ldots, A\beta_s\}$ is a basis of $CS(A)$.

Clearly, $B \subseteq CS(A)$. Let $\alpha \in CS(A)$. Then $\alpha = A\xi$ for some $\xi \in F^n$. Since Γ is a basis of F^n, $\xi = x_1\gamma_1 + \cdots + x_r\gamma_r + y_1\beta_1 + \cdots + y_s\beta_s$ for some scalars $x_i, y_j \in F$. Since $\gamma_1, \ldots, \gamma_r \in NS(A)$,

$$\alpha = A\xi = x_1 A\gamma_1 + \cdots + x_r A\gamma_r + y_1 A\beta_1 + \cdots + y_s A\beta_s$$
$$= y_1 A\beta_1 + \cdots + y_s A\beta_s.$$

Thus, $\alpha \in L(A\beta_1, \ldots, A\beta_s)$. It follows that $CS(A) = L(A\beta_1, \ldots, A\beta_s)$.

We next argue the vectors $A\beta_1, \ldots, A\beta_s$ are linearly independent. Suppose $z_1 A\beta_1 + \cdots + z_s A\beta_s = o$. Then $A(z_1\beta_1 + \cdots + z_s\beta_s) = O$. Therefore, $z_1\beta_1 + \cdots + z_s\beta_s \in NS(A) = L(\gamma_1, \ldots, \gamma_r)$. Hence, there exist scalars $w_1, \ldots, w_r$ in F such that $z_1\beta_1 + \cdots + z_s\beta_s = w_1\gamma_1 + \cdots + w_r\gamma_r$. Since Γ is a basis of F^n, we conclude that $w_1 = \cdots = w_r = z_1 = \cdots = z_s = 0$. In particular, the vectors $A\beta_1, \ldots, A\beta_s$ are linearly independent.

We have now shown that $\{A\beta_1, \ldots, A\beta_s\}$ is a basis of $CS(A)$. In particular, $\dim CS(A) = s$. Since $r + s = n$, this completes the proof of the theorem. $\qquad\qquad\square$

Theorem 3.22 has interesting ramifications for nonsingular matrices.

Theorem 3.23   Let $A \in M_{n \times n}$. Then the following statements are equivalent:

(a) A is nonsingular.
(b) $\text{rk}(A) = n$.
(c) $v(A) = 0$.
(d)   The columns of A are linearly independent.
(e)   The rows of A are linearly independent.
(f) $CS(A) = F^n$.
(g) $RS(A) = M_{1 \times n}$.

Proof.   We have seen in Theorem I.5.13 that A is nonsingular if and only if $v(A) = 0$. Thus, (a) and (c) are equivalent. Since $\text{rk}(A) + v(A) = n$, (b) and (c) are equivalent. Since $\dim CS(A) = \dim RS(A)$ by Theorem 3.3, the remaining statements are all equivalent to (b). $\qquad\qquad\square$

We finish this section with another application of the rank function.

Theorem 3.24   Let $A \in M_{m \times n}$ and $B \in F^m$. The linear system of equations $AX = B$ has a solution if and only if $\text{rk}(A) = \text{rk}(A \,|\, B)$.

Proof.   At this stage, this proof is rather obvious. The equation $AX = B$ has a solution $\xi = [x_1 \cdots x_n]^t$ if and only if $x_1 \text{Col}_1(A) + \cdots + x_n \text{Col}_n(A) = B$. Thus, $AX = B$ has a solution if and only if $CS(A) = CS(A \,|\, B)$. Since $CS(A) \subseteq CS(A \,|\, B)$, Theorem 2.21 implies $CS(A) = CS(A \,|\, B)$ if and only if $\dim CS(A) = \dim CS(A \,|\, B)$, i.e., if and only if $\text{rk}(A) = \text{rk}(A \,|\, B)$. Thus, $AX = B$ has a solution if and only if $\text{rk}(A) = \text{rk}(A \,|\, B)$. $\qquad\qquad\square$

Exercises for Section 3

3.1   Find a basis for $CS(A)$, $RS(A)$, and $NS(A)$ when A is the following matrix:

(a) $A = [x_1 \; \cdots \; x_n]$

(b) $A = [x_1 \; \cdots \; x_n]^t$

(c) $A = \begin{bmatrix} 1 & 2 & 3 \\ 0 & 0 & 0 \\ 1 & 2 & 3 \end{bmatrix}$

(d) $A = \begin{bmatrix} 1 & 2 & 0 & 1 & 4 \\ 1 & 1 & 1 & 3 & 1 \end{bmatrix}$

(e) $A = \begin{bmatrix} 1 & 0 & 2 & 1 \\ -1 & 1 & -1 & 1 \\ 5 & -3 & 7 & -1 \end{bmatrix}$

(f) The matrix A in Example 3.9.

3.2   In each problem in Exercise 3.1, verify that $v(A) + rk(A) = n$.

3.3   Let

$$A = \begin{bmatrix} 1 & 1 & 5 & 1 \\ 0 & 1 & 3 & 3 \\ 1 & 1 & 3 & 1 \end{bmatrix}$$

Determine all column vectors B such that $AX = B$ has a solution.

3.4   Let $A \in M_{m \times n}$ and $B \in F^m$. Suppose $A\xi = B$. Show that the set $\{\xi + \eta \mid \eta \in NS(A)\}$ is the complete set of solutions to $AX = B$.

3.5   What are the various possibilities for the number of linearly independent solutions to $AX = O$ when A is a 4×7 matrix?

3.6   Let $A \in M_{m \times n}$. Show that $\dim CS(A^t) = \dim RS(A)$, and $\dim RS(A^t) = \dim CS(A)$ as claimed in the proof of Theorem 3.3.

3.7   Is Corollary 3.10 true with $\approx$ replaced with $\underset{r}{\sim}$?

3.8   Suppose $A, B \in M_{m \times n}$ such that $rk(A) = rk(B)$. Does it follow that $RS(A) = RS(B)$?

3.9   Give an example of two matrices A and B for which $rk(AB) < \min\{rk(A), rk(B)\}$.

3.10   Compute the rank and nullity of the following two matrices:

(a)

$$A = \begin{bmatrix} 2 & 1 & 0 & 4 \\ 1 & -1 & 1 & 2 \\ 3 & 1 & 2 & -1 \\ 12 & 6 & 5 & 3 \end{bmatrix}$$

(b)

$$A = \begin{bmatrix} 0 & 1 & 2 & 1 & -1 & 3 & 2 \\ 1 & 1 & -1 & 2 & 1 & 1 & 0 \\ 1 & 2 & 1 & 3 & 2 & 1 & 1 \\ -1 & 2 & 7 & 1 & -2 & 5 & 5 \end{bmatrix},$$

3.11 Repeat the argument one more time in the proof of Theorem 3.12. Thus, show $j_3 = p_3$ and $x_3 = y_3 = 0$ in the equations in (3.18). Then give the general inductive step in the argument.

3.12 Let $A, B \in M_{m \times n}$ and assume A and B are both in column reduced echelon form. If $CS(A) = CS(B)$, show $A = B$.

3.13 Use Exercise 3.12 to show the column reduced echelon form of a matrix is unique.

3.14 Let $A, B \in M_{m \times n}$. Prove the following inequalities:
(a) $\mathrm{rk}(A + B) \leqslant \mathrm{rk}(A) + \mathrm{rk}(B)$.
(b) $v(A + B) \geqslant v(A) + v(B) - n$.

3.15 Let $A \in M_{m \times n}$ and $B \in M_{n \times p}$. Show $\mathrm{rk}(A) + \mathrm{rk}(B) - n \leqslant \mathrm{rk}(AB)$.

3.16 Let $A \in M_{n \times n}$ such that $A^k = O$, but $A^{k-1} \neq O$. Assume $1 \leqslant k \leqslant n$. Suppose $\xi \in F^n$ such that $A^{k-1}\xi \neq o$. Show the vectors $\xi, A\xi, \ldots,$ $A^{k-1}\xi$ are linearly independent. Thus, $\dim L(\xi, A\xi, \ldots, A^{k-1}\xi) = k$.

3.17 Let $A \in M_{n \times n}$. If $\mathrm{rk}(A) = 1$, show $A^2 = xA$ for some scalar $x \neq 0$.

4. LINEAR TRANSFORMATIONS

In this section, we study the maps between two vector spaces over F. We are only interested in those functions which preserve vector addition and scalar multiplication. Such functions are called linear transformations. Throughout this section, V and W will denote two vector spaces over the same set F. As usual, F is either $\mathbb{R}$ or $\mathbb{C}$.

Definition 4.1 A function $T: V \mapsto W$ is called a linear transformation if $T(x\alpha + y\beta) = xT(\alpha) + yT(\beta)$ for all $\alpha, \beta \in V$ and all $x, y \in F$.

The notation $T: V \mapsto W$ is meant to imply the domain of the function T is all of V. The range of T is some subset of W, possibly all of W. In the equation, $T(x\alpha + y\beta) = xT(\alpha) + yT(\beta)$, the sum $x\alpha + y\beta$ is a linear combination of two vectors, α and β, in V. The sum $xT(\alpha) + yT(\beta)$ is a linear combination of $T(\alpha)$ and $T(\beta)$ in W. T is a linear transformation from V to W if T sends the linear combination $x\alpha + y\beta$ in V to the corresponding linear combination $xT(\alpha) + yT(\beta)$ in W. A function with this property is said to preserve addition and scalar multiplication. Thus, linear transformations are precisely those functions (from one vector space to another) which preserve vector addition and scalar multiplication.

It is easy to see directly from the definition that a linear transformation $T: V \mapsto W$ satisfies the following properties.

4.2 (a) $T(\alpha + \beta) = T(\alpha) + T(\beta)$ for all $\alpha, \beta \in V$.

    (b) $T(x\alpha) = xT(\alpha)$ for all $\alpha \in V$ and all $x \in F$.

    (c) $T(o) = o$.

    (d) $T(x_1\alpha_1 + \cdots + x_n\alpha_n) = x_1 T(\alpha_1) + \cdots + x_n T(\alpha_n)$ for all $\alpha_i \in V$, $x_i \in F$.

The two statements in (a) and (b) are clearly equivalent to the definition. The zero on the left in (c) is the zero vector of V. The zero on the right in (c) is the zero vector in W. (c) follows immediately from (b) by setting $x = 0$. The equation in (d) follows from (a) and (b) by induction.

We now consider some examples of linear transformations.

Example 4.3 Let $T: V \mapsto W$ be given by $T(\alpha) = o$ for all $\alpha \in V$. If $\alpha, \beta \in V$ and $x, y \in F$, then $xT(\alpha) + yT(\beta) = xo + yo = o = T(x\alpha + y\beta)$. Thus, T is a linear transformation from V to W. This map is called the zero map from V to W. In the sequel, we will let 0 denote the zero map.

Example 4.4 Suppose $V = W$. The function $T: V \mapsto V$ given by $T(\alpha) = \alpha$ for all $\alpha \in V$ is clearly a linear transformation from V to V. This map is called the identity map on V. Henceforth, we will let I_V denote the identity map on V.

Example 4.5 Let V be a finite dimensional vector space over F. Suppose $n = \dim V$. Set $W = F^n$. If $B = \{\alpha_1, \ldots, \alpha_n\}$ is any (ordered) basis of V, then the coordinate map $[*]_B : V \mapsto F^n$ is a linear transformation. This follows from (b) and (c) of (2.25).

There are several interesting linear transformations on matrices.

Example 4.6 Let $V = M_{m \times n}$ and $W = M_{n \times m}$. The map $T: V \mapsto W$ given by $T(A) = A^t$ is a linear transformation. We have seen in Theorem I.2.35 that $T(A + B) = (A + B)^t = A^t + B^t = T(A) + T(B)$. If $x \in F$, then, clearly $T(xA) = (xA)^t = xA^t = xT(A)$. Thus, applying the transpose is a linear transformation from $M_{m \times n}$ to $M_{n \times m}$.

Example 4.7 Let $A \in M_{m \times n}$. Set $V = M_{n \times p}$ and $W = M_{m \times p}$. Multiplication by A (necessarily on the left) induces a linear transformation $T: M_{n \times p} \mapsto M_{m \times p}$ given by $T(B) = AB$. We have seen in (c) and (f) of Theorem I.2.24 that T is a linear transformation.

One special case of Example 4.7 is worth noting here. Any $m \times n$ matrix A determines a linear transformation $T: F^n \mapsto F^m$ given by $T(\xi) = A\xi$ for any $\xi \in F^n$.

In Example 4.7, if $V = M_{p \times m}$ and $W = M_{p \times n}$, then multiplication on the right with A determines a linear transformation $S: V \mapsto W$ given by $S(B) = BA$. Left and right multiplication by fixed matrices can be combined to produce a third type of linear transformation.

Example 4.8 Let $A \in M_{m \times n}$ and $B \in M_{p \times q}$. Then the map $T: M_{n \times p} \mapsto M_{m \times q}$ given by $T(C) = ACB$ is a linear transformation.

There are many interesting examples of linear transformations in the calculus. Let us examine some of these.

Example 4.9 Let I be a nonempty subset of $\mathbb{R}$. Set $V = \mathbb{R}^I$ (Example 1.8). Let $a \in I$. Then the evaluation map $E_a: \mathbb{R}^I \mapsto \mathbb{R}$ given by $E_a(f(x)) = f(a)$ is a linear transformation. We have

$$E_a(f + g) = (f + g)(a) = f(a) + g(a) = E_a(f) + E_a(g).$$

Similarly,

$$E_a(cf) = (cf)(a) = cf(a) = cE_a(f).$$

Example 4.10 Let I denote an open interval (of positive length) in $\mathbb{R}$. Set $V = C^1(I)$ (Example 1.9) and $W = \mathbb{R}^I$. Then ordinary differentiation determines a linear transformation $D: V \mapsto W$. D is given by $D(f) = df/dt$. The formulas $D(f + g) = D(f) + D(g)$ and $D(cf) = cD(f)$ are familiar from the calculus.

Example 4.11 Let $I = [a, b]$ be a closed interval in $\mathbb{R}$. Set $V = \mathbb{R}(I)$ (Example 1.10). Let $W = \mathbb{R}$. Then integration provides a linear transformation $T: V \mapsto W$ given by $T(f) = \int_a^b f(t)\, dt$.

We can also manufacture new linear transformations from old ones by taking composites. Suppose V, W, and Y are vector spaces over the same set F. If $T: V \mapsto W$ and $S: W \mapsto Y$ are linear transformations, then we can consider the composite map $ST: V \mapsto Y$. Recall that a composite map ST is defined as follows: $ST(\alpha) = S(T(\alpha))$ for all $\alpha \in V$. It is easy to check that ST is a linear transformation from V to Y. If α, $\beta \in V$ and x, $y \in F$, then $ST(x\alpha + y\beta) = S(T(x\alpha + y\beta)) = S(xT(\alpha) + yT(\beta)) = xS(T(\alpha)) + yS(T(\beta)) = x(ST)(\alpha) + y(ST)(\beta)$.

Example 4.12   Let $V = C^1(I)$ as in Example 4.10. Set $W = \mathbb{R}$. Let $a \in I$. The map $S: C^1(I) \mapsto \mathbb{R}$ given by $S(f) = (df/dt)(a)$ is a linear transformation. This follows from the fact that S is clearly the composite of E_a and D given in Examples 4.9 and 4.10. Thus, $S = E_a D$.

Definition 4.13   The set of all linear transformations from V to W will be denoted by $\mathrm{Hom}(V, W)$.

The letters Hom in Definition 4.13 are an abbreviation for the word homomorphism. Algebraists often call linear transformations homomorphisms. We will see that the set $\mathrm{Hom}(V, W)$ is itself a vector space over F. For now, we observe that $\mathrm{Hom}(V, W)$ is always a nonempty set since $0 \in \mathrm{Hom}(V, W)$ by Example 4.3.

There are several definitions which pertain to a given linear transformation $T \in \mathrm{Hom}(V, W)$. We present these definitions next.

Definition 4.14   Let $T \in \mathrm{Hom}(V, W)$.

(a) $\mathrm{Ker}(T) = \{\alpha \in V \mid T(\alpha) = o\}$.
(b) $\mathrm{Im}(T) = \{T(\alpha) \in W \mid \alpha \in V\}$.
(d) T is injective (monomorphism, 1-1) if $\mathrm{Ker}(T) = (o)$.
(d) T is surjective (epimorphism, onto) if $\mathrm{Im}(T) = W$.
(e) T is an isomorphism if T is both injective and surjective.

The set $\mathrm{Ker}(T)$ is called the kernel of T (or sometimes the null space of T). The set $\mathrm{Im}(T)$ is called the image of T (or sometimes the range of T). Both of these sets are subspaces.

Lemma 4.15   Let $T \in \mathrm{Hom}(V, W)$.

(a) $\mathrm{Ker}(T)$ is a subspace of V.
(b) $\mathrm{Im}(T)$ is a subspace of W.

Proof.   (a) Let α, $\beta \in \mathrm{Ker}(T)$ and x, $y \in F$. By Theorem 1.13, it suffices to show $x\alpha + y\beta \in \mathrm{Ker}(T)$. Since α and β are vectors in $\mathrm{Ker}(T)$, $T(\alpha) = T(\beta) = o$. Therefore, $T(x\alpha + y\beta) = xT(\alpha) + yT(\beta) = xo + yo = o$. Thus, $x\alpha + y\beta \in \mathrm{Ker}(T)$.

(b) Let $\gamma, \delta \in \text{Im}(T)$. Then there exist vectors $\alpha, \beta \in V$ such that $T(\alpha) = \gamma$ and $T(\beta) = \delta$. If $x, y \in F$, then $T(x\alpha + y\beta) = xT(\alpha) + yT(\beta) = x\gamma + y\delta$. Therefore, $x\gamma + y\delta \in \text{Im}(T)$. Theorem 1.13 implies $\text{Im}(T)$ is a subspace of W. $\qquad\square$

If $T \in \text{Hom}(V, W)$ and $\alpha, \beta \in V$, then we have the following equivalent statements: $T(\alpha) = T(\beta) \Leftrightarrow T(\alpha - \beta) = o \Leftrightarrow \alpha - \beta \in \text{Ker}(T)$. In particular, T is injective (i.e., $\text{Ker}(T) = (o)$) if and only if T is a 1-1 function from V to W. The reader will recall that a function $f : V \mapsto W$ is said to be 1-1 if f satisfies the following property: $f(\alpha) = f(\beta)$ if and only if $\alpha = \beta$ for all $\alpha, \beta \in V$. If T is injective, and $T(\alpha) = T(\beta)$ for two vectors $\alpha, \beta \in V$, then $\alpha - \beta \in \text{Ker}(T) = (o)$. Therefore, $\alpha = \beta$. Hence, T is a 1-1 map from V to W. Conversely, if T is 1-1 and $\alpha \in \text{Ker}(T)$, then $o = T(\alpha) = T(o)$. Thus, $\alpha = o$ and T is injective.

The symbols $T(V)$ are used in function theory to denote the image of T. Thus, $T(V) = \text{Im}(T)$. We will occasionally use this notation for $\text{Im}(T)$. If $T \in \text{Hom}(V, W)$ is injective and $T(V) = W$, then T is an isomorphism. In this case, V and W are isomorphic. Let us introduce this definition formally.

Definition 4.16 Let V and W be two vectors spaces over F. We say V and W are isomorphic if there exists a linear transformation $T : V \mapsto W$ which is an isomorphism.

If V and W are isomorphic, we will write $V \cong W$.

Suppose $V \cong W$. Then $W \cong V$. In other words, the expression $V \cong W$ is symmetric in V and W. To see this, suppose $T : V \mapsto W$ is an isomorphism. Then T is both injective, i.e., 1-1, and surjective, i.e., onto. In particular, the function T has a well-defined inverse function $T^{-1} : W \mapsto V$. Recall that the inverse function T^{-1} is defined by the following equation: $T^{-1}(\beta) = \alpha$ if and only if $T(\alpha) = \beta$. Since T is a linear transformation, T^{-1} is also a linear transformation. For suppose $\beta_1, \beta_2 \in W$ and $x, y \in F$. Since T is onto, there exist $\alpha_1, \alpha_2 \in V$ such that $T(\alpha_1) = \beta_1$ and $T(\alpha_2) = \beta_2$. Since T is a linear map,

$$T(x\alpha_1 + y\alpha_2) = xT(\alpha_1) + yT(\alpha_2) = x\beta_1 + y\beta_2.$$

Using the definition of T^{-1}, we have

$$T^{-1}(x\beta_1 + y\beta_2) = x\alpha_1 + y\alpha_2 = xT^{-1}(\beta_1) + yT^{-1}(\beta_2).$$

Thus, $T^{-1} : W \mapsto V$ is a linear transformation. Since T^{-1} has inverse T, clearly T^{-1} is 1-1 and onto. In particular, T^{-1} is an isomorphism, and $W \cong V$. We have now proven the following theorem.

Theorem 4.17 Let $T \in \text{Hom}(V, W)$. If T is an isomorphism, then $T^{-1}: W \mapsto V$ is a linear transformation and an isomorphism.

Theorem 4.17 implies $V \cong W$ if and only if $W \cong V$. From a linear algebra point of view, two vector spaces which are isomorphic are structurally identical. Only the names of the vectors are being changed by the isomorphism. For example, if $T: V \mapsto W$ is an isomorphism, and α_1, $\ldots, \alpha_r \in V$, then $\alpha_1, \ldots, \alpha_r$ are linearly independent in V if and only if $T(\alpha_1)$, $\ldots, T(\alpha_r)$ are linearly independent in W. Also, $\beta \in L(\alpha_1, \ldots, \alpha_r)$ if and only if $T(\beta) \in L(T(\alpha_1), \ldots, T(\alpha_r))$. $\dim L(\alpha_1, \ldots, \alpha_r) = s$ if and only if $\dim L(T(\alpha_1), \ldots, T(\alpha_r)) = s$. These assertions are left as exercises at the end of this section.

Let us consider some of the isomorphisms that have appeared in our examples.

Example 4.18 The identity map $I_V: V \mapsto V$ is certainly injective and surjective. Consequently, I_V is an isomorphism.

Example 4.19 Let $T: M_{m \times n} \mapsto M_{n \times m}$ be the transpose map given in Example 4.6. Thus, $T(A) = A^t$. Clearly, T is both injective and surjective. Hence, T is an isomorphism. In particular, $M_{m \times n} \cong M_{n \times m}$.

Example 4.20 Let $A \in M_{n \times n}$ be an invertible matrix. The map $T: M_{n \times p} \mapsto M_{n \times p}$ given by $T(B) = AB$ is injective and surjective. Thus, T is an isomorphism.

A special case of Example 4.20 is worth mentioning. If A is an invertible $n \times n$ matrix, then the linear transformation $T: F^n \mapsto F^n$ given by $T(\xi) = A\xi$, is an isomorphism.

Coordinate maps are also isomorphisms. We prove this in our next theorem.

Theorem 4.21 Let V be a finite dimensional vector space over F of dimension n. Let $B = \{\alpha_1, \ldots, \alpha_n\}$ be an (ordered) basis of V. Then the coordinate map $[*]_B: V \mapsto F^n$ is an isomorphism.

Proof. We have already observed in the equations in (2.25) that $[*]_B$ is a linear transformation from V to F^n. If $\gamma \in \text{Ker}([*]_B)$, then

$[\gamma]_B = [x_1 \cdots x_n]^t = O$ in F^n. Therefore, $x_1 = \cdots = x_n = 0$. But, $\gamma = x_1\alpha_1 + \cdots + x_n\alpha_n$. Therefore, $\gamma = o$. Thus, $[*]_B$ is injective.

Suppose $[y_1 \cdots y_n]^t \in F^n$. Then $[\gamma]_B = [y_1 \cdots y_n]^t$ where $\gamma = y_1\alpha_1 + \cdots + y_n\alpha_n$. In particular, $[*]_B$ is a surjective map. Hence, $[*]_B$ is an isomorphism. ☐

Thus, $[*]_B : V \cong F^n$. We have already commented on the fact that isomorphic vector spaces are structurally identical. Thus, Theorem 4.21 says that, up to isomorphism, there is only one finite dimensional vector space of dimension n, namely F^n. In particular, $M_{m \times n} \cong M_{n \times m} \cong F^{mn}$ and $\mathscr{P}_n \cong F^{n+1}$. Also $CS(A) \cong F^{\mathrm{rk}(A)}$ and $RS(A) \cong F^{\mathrm{rk}(A)}$ for any $m \times n$ matrix A.

In our next two theorems, we explore how linear transformations and bases are related. Our first theorem says a linear transformation is totally determined by its action on a basis of V.

Theorem 4.22 Let V be a finite dimensional vector space over F with basis $B = \{\alpha_1, \ldots, \alpha_n\}$. Let W be any vector space over F.

(a) If $S, T \in \mathrm{Hom}(V, W)$ and $S(\alpha_i) = T(\alpha_i)$ for all $i = 1, \ldots, n$, then $S = T$.
(b) Let $\beta_1, \ldots, \beta_n$ be n vectors (not necessarily distinct) in W. Then there exists a unique $T \in \mathrm{Hom}(V, W)$ such that $T(\alpha_i) = \beta_i$ for all $i = 1, \ldots, n$.

Proof. (a) Let $\gamma \in V$. Since B is a basis of V, there exist scalars $x_1, \ldots, x_n \in F$ such that $\gamma = x_1\alpha_1 + \cdots + x_n\alpha_n$. Since S and T are linear transformations, we have $T(\gamma) = x_1 T(\alpha_1) + \cdots + x_n T(\alpha_n) = x_1 S(\alpha_1) + \cdots + x_n S(\alpha_n) = S(\gamma)$. Since γ is an arbitrary vector in V, we conclude $T = S$.

(b) Define a map $P : F^n \mapsto W$ by the formula

$$P([x_1 \cdots x_n]^t) = x_1\beta_1 + \cdots + x_n\beta_n.$$

The map P is a linear transformation. To see this, let $\xi_1 = [x_1 \cdots x_n]^t$ and $\xi_2 = [y_1 \cdots y_n]^t$ be two column vectors in F^n. Then for any scalars $x, y \in F$, we have

$$
\begin{aligned}
P(x\xi_1 + y\xi_2) &= P([xx_1 + yy_1 \cdots xx_n + yy_n]^t) \\
&= (xx_1 + yy_1)\beta_1 + \cdots + (xx_n + yy_n)\beta_n \\
&= x(x_1\beta_1 + \cdots + x_n\beta_n) + y(y_1\beta_1 + \cdots + y_n\beta_n) \\
&= xP(\xi_1) + yP(\xi_2). \text{ Thus, } P \in \mathrm{Hom}(F^n, W).
\end{aligned}
$$

Consider the composite linear transformation $T = P[*]_B$. Since $[*]_B : V \mapsto F^n$ and $P : F^n \mapsto W$, $T \in \mathrm{Hom}(V, W)$. We remind the reader that $[\alpha_i]_B = \varepsilon_i$, the ith vector in the canonical basis of F^n. Thus, for all $i = 1, \ldots, n$, we have $T(\alpha_i) = P[\alpha_i]_B = P\varepsilon_i = P([0 \cdots 0 \ 1 \ 0 \cdots 0]^t) = \beta_i$.

Thus, T is a linear transformation from V to W such that $T(\alpha_i) = \beta_i$ for all $i = 1, \ldots, n$. The fact that T is the only linear transformation from V to W with this property follows from (a).   □

Suppose V is a finite dimensional vector space over F with dim $V = n$. Let W be an arbitrary vector space over F. If $\beta_1 \ldots, \beta_n$ are n arbitrary vectors in W, and $\alpha_1 \ldots, \alpha_n$ are any n linearly independent vectors in V, then Theorem 4.22 implies there exists a unique linear transformation $T \in \text{Hom}(V, W)$ such that $T(\alpha_i) = \beta_i$ for all $i = 1, \ldots, n$. The effect of T on all other vectors in V is given by (d) of equation (4.2).

Theorem 4.23   Let V be a finite dimensional vector space over F. Let $T \in \text{Hom}(V, W)$.

(a)   If T is surjective, then W is finite dimensional over F and dim $V \geqslant$ dim W.
(b)   Suppose dim $V =$ dim W. If T is injective, then T is an isomorphism.
(c)   Suppose dim $V =$ dim W. If T is surjective, then T is an isomorphism.
(d)   dim $\text{Ker}(T)$ + dim $\text{Im}(T)$ = dim V.

Proof. (a) Let $B = \{\alpha_1, \ldots, \alpha_n\}$ be a basis of V. Then $W = L(T(\alpha_1), \ldots, T(\alpha_n))$. To see this, let $\gamma \in W$. Since T is surjective, there exists a vector $\delta \in V$ such that $T(\delta) = \gamma$. Since B is a basis of V, $\delta = x_1\alpha_1 + \cdots + x_n\alpha_n$. Thus, $\gamma = T(\delta) = x_1 T(\alpha_1) + \cdots + x_n T(\alpha_n) \in L(T(\alpha_1), \ldots, T(\alpha_n))$. Since γ is an arbitrary vector in W, we conclude $W = L(T(\alpha_1), \ldots, T(\alpha_n))$. In particular, W is finite dimensional. Theorem 2.11 implies that the set $\{T(\alpha_1), \ldots, T(\alpha_n)\}$ contains a basis of W. Therefore, dim $W \leqslant$ dim V.

(b) Again let $B = \{\alpha_1, \ldots, \alpha_n\}$ be a basis of V. Since T is injective, the vectors $T(\alpha_1), \ldots, T(\alpha_n)$ are linearly independent in W. For suppose, $x_1 T(\alpha_1) + \cdots + x_n T(\alpha_n) = o$. Then $T(x_1\alpha_1 + \cdots + x_n\alpha_n) = o$. Therefore, $x_1\alpha_1 + \cdots + x_n\alpha_n \in \text{Ker}(T)$. Since T is injective, $\text{Ker}(T) = (o)$. Thus, $x_1\alpha_1 + \cdots + x_n\alpha_n = o$. Since B is a basis of V, $x_1 = \cdots = x_n = 0$. Thus, $T(\alpha_1), \ldots, T(\alpha_n)$ are linearly independent.

Theorem 2.9 implies the set $B' = \{T(\alpha_1), \ldots, T(\alpha_n)\}$ can be expanded to a basis of W. But $n =$ dim $V =$ dim W. Consequently, B' is a basis of W. We can now argue that T is surjective, and, hence, an isomorphism. Let $\gamma \in W$. Since B' is a basis of W, $\gamma = x_1 T(\alpha_1) + \cdots + x_n T(\alpha_n)$ for some $x_i \in F$. But then $\gamma = T(x_1\alpha_1 + \cdots + x_n\alpha_n) \in \text{Im}(T)$. Thus, $W = \text{Im}(T)$, and the proof of (b) is complete.

(c) Again let $B = \{\alpha_1, \ldots, \alpha_n\}$ be a basis of V. We have observed in the proof of (a) that the set $B' = \{T(\alpha_1), \ldots, T(\alpha_n)\}$ contains a basis of W when

T is surjective. Since dim $W = n$, B' is a basis of W. We can now argue that T is injective, and hence, an isomorphism.

Suppose $\gamma \in \text{Ker}(T)$. Write $\gamma = x_1\alpha_1 + \cdots + x_n\alpha_n$ for some $x_i \in F$. Then $o = T(\gamma) = x_1 T(\alpha_1) + \cdots + x_n T(\alpha_n)$. Since B' is a basis of W, $x_1 = \cdots = x_n = 0$. In particular, $\gamma = o$, and T is injective.

(d) This proof is exactly the same as the proof of Theorem 3.22. Let $\{\gamma_1, \ldots, \gamma_r\}$ be a basis (possibly empty) of $\text{Ker}(T)$. Expand this set to a basis $\{\gamma_1, \ldots, \gamma_r, \beta_1, \ldots, \beta_s\}$ of V. Then $r = \dim \text{Ker}(T)$ and $r + s = \dim V$. We can then argue that $\{T(\beta_1), \ldots, T(\beta_s)\}$ is a basis of $\text{Im}(T)$. We leave the details of this argument to the exercises at the end of this section. $\quad\square$

Theorem 4.23(a) has one important consequence which is worth mentioning. Suppose V and W are vector spaces over F. We do not assume either vector space is finite dimensional. Let $T \in \text{Hom}(V, W)$. Suppose Z is a finite dimensional subspace of V. Restricting T to Z, we get a linear transformation $T : Z \mapsto W$. Let us denote the image of the restriction of T to Z by $T(Z)$. Then $T(Z)$ is a subspace of W and $T : Z \mapsto T(Z)$ is a surjective linear transformation. Theorem 4.23(a) implies $\dim Z \geqslant \dim T(Z)$. Thus, the map $Z \mapsto T(Z)$ which sends (finite dimensional) subspaces of V to subspaces in W can never increase the dimension.

Theorem 4.23(b) and (c) are particularly useful when dealing with an endomorphism of V, i.e., a linear transformation $T \in \text{Hom}(V, V)$. Suppose $\dim V < \infty$, and we wish to argue that T is an isomorphism. By Theorem 4.23(b) or (c), it is enough to show T is injective (or T is surjective). Consider the following example.

Example 4.24 Let $V = M_{n \times n}$. Suppose P is an invertible matrix in V. Define an endomorphism $T : V \mapsto V$ by $T(A) = PAP^{-1}$. Since P is invertible, it is easy to check that $\text{Ker}(T) = (o)$. Thus, T is injective. Theorem 4.23(b) implies T is an isomorphism.

W finish this section with a few observations about the set $\text{Hom}(V, W)$. Suppose $S, T \in \text{Hom}(V, W)$. Define the sum, $S + T$, of S and T to be the function from V to W given by $(S + T)(\alpha) = S(\alpha) + T(\alpha)$ for all $\alpha \in V$. The map $S + T$ is a linear transformation. To see this, let $\alpha, \beta \in V$ and $x, y \in F$. Then

$$
\begin{aligned}
(S + T)(x\alpha + y\beta) &= S(x\alpha + y\beta) + T(x\alpha + y\beta) \\
&= xS(\alpha) + yS(\beta) + xT(\alpha) + yT(\beta) \\
&= x(S(\alpha) + T(\alpha)) + y(S(\beta) + T(\beta)) \\
&= x(S + T)(\alpha) + y(S + T)(\beta).
\end{aligned}
$$

Hence, $S + T \in \mathrm{Hom}(V, W)$. Thus, the sum of any two linear transformations from V to W is another linear transformation from V to W.

Define scalar multiplication cT as follows. If $T \in \mathrm{Hom}(V, W)$, and $c \in F$, then cT is the function from V to W given by $(cT)(\alpha) = cT(\alpha)$. The scalar product cT is also a linear transformation from V to W. We have $(cT)(x\alpha + y\beta) = cT(x\alpha + y\beta) = xcT(\alpha) + ycT(\beta) = x(cT)(\alpha) + y(cT)(\beta)$. Thus, $cT \in \mathrm{Hom}(V, W)$.

The reader can easily check that these definitions of addition and scalar multiplication satisfy axioms V1 through V8 in Definition 1.1. The zero vector in $\mathrm{Hom}(V, W)$ is the zero map 0. The negative of a linear transformation T is $(-1)T$. Hence, we have the following theorem.

Theorem 4.25 $\mathrm{Hom}(V, W)$ is a vector space over F.

Whenever we refer to $\mathrm{Hom}(V, W)$ as a vector space, we will mean addition is $S + T$ and scalar multiplication is cT as defined above.

If V and W are finite dimensional vector spaces over F, then $\mathrm{Hom}(V, W)$ is also finite dimensional. To see this, suppose $B = \{\alpha_1, \ldots, \alpha_n\}$ is a basis of V and $C = \{\beta_1, \ldots, \beta_m\}$ is a basis of W. For each $i = 1, \ldots, m$ and $j = 1, \ldots, n$, let S_{ij} be the unique linear transformation from V to W defined by the following equations.

4.26
$$S_{ij}(\alpha_p) = \begin{cases} o & \text{if } p \neq j \\ \beta_i & \text{if } p = j, \end{cases} \quad p = 1, \ldots, n$$

Theorem 4.22(b) implies that such a linear transformation $S_{ij} \in \mathrm{Hom}(V, W)$ exists for each i and j.

We claim that $D = \{S_{ij} \mid i = 1, \ldots, m; \, j = 1, \ldots, n\}$ is a basis for $\mathrm{Hom}(V, W)$. We first argue that the S_{ij} are linearly independent in $\mathrm{Hom}(V, W)$. Suppose $\sum_{i=1}^{m} \sum_{j=1}^{n} x_{ij} S_{ij} = 0$ in $\mathrm{Hom}(V, W)$. Let p be a fixed integer with $1 \leqslant p \leqslant n$. Then using (4.26), we have

$$o = \left(\sum_{i=1}^{m} \sum_{j=1}^{n} x_{ij} S_{ij} \right)(\alpha_p)$$
$$= \sum_{i=1}^{m} \sum_{j=1}^{n} x_{ij} S_{ij}(\alpha_p)$$
$$= \sum_{i=1}^{m} x_{ip} S_{ip}(\alpha_p) = \sum_{i=1}^{m} x_{ip} \beta_i.$$

Since C is a basis of W, we conclude $x_{1p} = \cdots = x_{mp} = 0$. Since p is arbitrary, $x_{ij} = 0$ for all i and j. Thus, the linear transformations in D are linearly independent vectors in $\text{Hom}(V, W)$.

We next show that $L(D) = \text{Hom}(V, W)$. Let $T \in \text{Hom}(V, W)$. For each $j = 1, \ldots, n$, $T(\alpha_j)$ is a vector in $W = L(\beta_1, \ldots, \beta_m)$. Hence, for each $j = 1, \ldots, n$, there exist scalars $a_{ij} \in F$ such that

$$T(\alpha_j) = a_{1j}\beta_1 + a_{2j}\beta_2 + \cdots + a_{mj}\beta_m.$$

Consider the linear transformation

$$\sum_{i=1}^{m} \sum_{j=1}^{n} a_{ij}S_{ij} \in L(D).$$

For any $p = 1, \ldots, n$, we have

$$\left(\sum_{i=1}^{m} \sum_{j=1}^{n} a_{ij}S_{ij} \right)(\alpha_p) = \sum_{i=1}^{m} \sum_{j=1}^{n} a_{ij}S_{ij}(\alpha_p)$$

$$= \sum_{i=1}^{m} a_{ip}\beta_i = T(\alpha_p)$$

It now follows from Theorem 4.22(a) that $T = \sum_{i=1}^{m} \sum_{j=1}^{n} a_{ij}S_{ij}$. Thus, $T \in L(D)$. Since T is arbitrary, we conclude that $L(D) = \text{Hom}(V, W)$. Thus, the set D is a basis of the vector space $\text{Hom}(V, W)$. Since the number of vectors in D is mn, we have proven the following theorem.

Theorem 4.27 Let V and W be finite dimensional vector spaces over F. Then $\text{Hom}(V, W)$ is also a finite dimensional vector space over F. In this case, $\dim \text{Hom}(V, W) = (\dim V)(\dim W)$.

Exercises for Section 4

4.1 Let $V = W = \mathscr{P}_3$. Which of the following functions listed below are linear transformations from V to W?
   (a) $T(p(t)) = t \, dp/dt$.
   (b) $T(a_0 + a_1 t + a_2 t^2 + a_3 t^3) = a_0 + a_1 t$.
   (c) $T(a_0 + a_1 t + a_2 t^2 + a_3 t^3) = |a_0| + |a_1|t + a_2 t^2 + a_3 t^3$.
   (d) $T(p(t)) = p(t) - 1$.
   (e) $T(p(t)) = p(t - 1)$.

4.2 Let $V = M_{2 \times 3}(\mathbb{R})$. Which of the following functions listed below are endomorphism of V, i.e., linear transformations from V to V?

   (a) $T\left(\begin{bmatrix} a_1 & a_2 & a_3 \\ b_1 & b_2 & b_3 \end{bmatrix} \right) = \begin{bmatrix} b_1 & b_2 & b_3 \\ a_1 & a_2 & a_3 \end{bmatrix}$

(b) $T\left(\begin{bmatrix} a_1 & a_2 & a_3 \\ b_1 & b_2 & b_3 \end{bmatrix}\right) = \begin{bmatrix} a_1 & a_2 & 0 \\ b_1 & b_2 & 0 \end{bmatrix}$

(c) $T\left(\begin{bmatrix} a_1 & a_2 & a_3 \\ b_1 & b_2 & b_3 \end{bmatrix}\right) = \begin{bmatrix} |a_1| & |a_2| & |a_3| \\ |b_1| & |b_2| & |b_3| \end{bmatrix}$

(d) $T\left(\begin{bmatrix} a_1 & a_2 & a_3 \\ b_1 & b_2 & b_3 \end{bmatrix}\right) = \begin{bmatrix} a_1^2 & a_1 a_2 & a_1 a_3 \\ b_1 & b_2 & b_3 \end{bmatrix}$

4.3 $\mathbb{C}$ is a vector space over $\mathbb{R}$ and over $\mathbb{C}$. Show that complex conjugation is an $\mathbb{R}$-linear transformation from $\mathbb{C}$ to $\mathbb{C}$, but not a $\mathbb{C}$-linear transformation.

4.4   Is the map $A \mapsto A^{-1}$ a linear transformation on the set of invertible $n \times n$ matrices?

4.5   Is the map $T: M_{m \times n}(\mathbb{C}) \mapsto M_{n \times m}(\mathbb{C})$ given by $A \mapsto A^*$ a linear transformation?

4.6   Let W be a subspace of a vector space V. Show that the inclusion map $W \hookrightarrow V$ is a linear transformation.

4.7   Suppose $T: \mathscr{P}_2 \mapsto \mathscr{P}_4$ is a linear transformation. If $T(1) = t$, $T(1 - t) = t^3 + t^4$, and $T(t^2 + t + 1) = t^4 - t + 1$, compute $T(p)$ for any $p \in \mathscr{P}_2$.

4.8   Let $T: V \mapsto W$ be an isomorphism. Show the following statements are true:

(a) $\alpha_1, \ldots, \alpha_n$ are linearly independent in V if and only if $T(\alpha_1), \ldots, T(\alpha_n)$ are linearly independent in W.

(b) $\beta \in L(\alpha_1, \ldots, \alpha_n)$ if and only if $T(\beta) \in L(T(\alpha_1), \ldots, T(\alpha_n))$.

(c) $\dim L(\alpha_1, \ldots, \alpha_n) = \dim L(T(\alpha_1), \ldots, T(\alpha_n))$.

(d) $V_1 \cap V_2 = V_3$ for subspaces V_i in V if and only if $T(V_1) \cap T(V_2) = T(V_3)$ in W.

(e) $V_1 + V_2 = V_3$ in V if and only if $T(V_1) + T(V_2) = T(V_3)$ in W.

4.9   Show that the map S in Example 4.20 is both injective and surjective.

4.10   Let $V = \mathbb{R}[t]$. Show that $B = \{1, t, t^2, t^3, \ldots\}$ is a basis for this infinite dimensional vector space. Construct linear transformations from V to V which show that (b) and (c) in Theorem 4.23 are false for infinite dimensional vector spaces.

4.11   Give a careful proof of Theorem 4.23(d).

4.12   Let $T: \mathscr{P}_4 \mapsto M_{2 \times 3}$ be a linear transformation. What are the various possibilities for $\dim \mathrm{Ker}(T)$ and $\dim \mathrm{Im}(T)$?

4.13   Let $T \in \mathrm{Hom}(V, W)$ and suppose V and W are both finite dimensional. If $\dim V > \dim W$, show that T cannot be injective. Is T necessarily surjective?

4.14 Let $T \in \text{Hom}(V, W)$ and $S \in \text{Hom}(W, Z)$. Show the following statements are true:
 (a) If S and T are injective, so is ST.
 (b) If S and T are surjective, so is ST.
 (c) If S and T are isomorphisms, so is ST.

4.15 Let $S, T \in \text{Hom}(V, W)$. If S and T are injective, what can you say about $S + T$?

4.16 Let

$$A = \begin{bmatrix} 1 & 2 & 0 & 1 \\ 1 & 4 & 1 & 0 \\ -1 & 0 & -1 & 2 \end{bmatrix} \in M_{3 \times 4}(\mathbb{R})$$

Consider the map $T: \mathbb{R}^4 \mapsto \mathbb{R}^3$ given by $T(\xi) = A\xi$.
 (a) Show that T is a linear transformation.
 (b) Find a basis for $\text{Ker}(T)$.
 (c) Find a basis for $\text{Im}(T)$.
 (d) Verify Theorem 4.23(d) in this case.

4.17 Show that the function $T: \mathbb{R}^2 \mapsto \mathbb{R}^2$ given by

$$T\left(\begin{bmatrix} x \\ y \end{bmatrix}\right) = \begin{bmatrix} x \cos(\theta) - y \sin(\theta) \\ x \sin(\theta) + y \cos(\theta) \end{bmatrix}$$

is an isomorphism. This is a well known function in the calculus. Draw a picture of what T does to a vector in $\mathbb{R}^2$.

4.18 Let V be a finite dimensional vector space over F, and let S, $T \in \text{Hom}(V, V)$. If $ST = I_V$, show S is an isomorphism with $S^{-1} = T$.

4.19 Exercise 4.18 is not true for infinite dimensional vector spaces. Let $V = \mathbb{R}[t]$. Define $J, D \in \text{Hom}(V, V)$ by the following formulas: $D(p) = dp/dt$ and $J(p) = \int_0^t p(u)\, du$.
 (a) Show D and J are linear transformations.
 (b) Show $\text{Ker}(D) \neq (o)$. Thus, D is not injective.
 (c) Show that $DJ = I_V$, but $JD \neq I_V$.
 In Exercises 4.20 through 4.23, assume V is finite dimensional.

4.20 Let $T \in \text{Hom}(V, V)$. Suppose T is not injective. Show there exists a nonzero $S \in \text{Hom}(V, V)$ such that $TS = O$.

4.21 Let $T \in \text{Hom}(V, V)$. Suppose T is not surjective. Show there exists a nonzero $S \in \text{Hom}(V, V)$ such that $ST = O$.

4.22 Let $S, T \in \text{Hom}(V, F)$. Suppose $T(\alpha) = o$ whenever $S(\alpha) = o$. Show that $T = xS$ for some $x \in F$.

4.23 Suppose $S, T \in \text{Hom}(V, W)$ with $\dim W > 1$. If $T(\alpha) = o$ whenever $S(\alpha) = o$, does it follow that $T = xS$ for some $x \in F$?

5.   MATRIX REPRESENTATIONS OF LINEAR TRANSFORMATIONS

In this section, we will discuss the relationships between matrices and linear transformations. Throughout this section, V and W will denote finite dimensional vector spaces over F ($=\mathbb{R}$ or $\mathbb{C}$). We assume dim $V = n$ and dim $W = m$. Let $T \in \text{Hom}(V, W)$. We have seen in Theorem 4.22 that T is totally determined by its action on a basis of V. This suggests a simple way to associate an $m \times n$ matrix with the linear transformation T.

Definition 5.1   Let $B = \{\alpha_1, \ldots, \alpha_n\}$ be an ordered basis of V and let $C = \{\beta_1, \ldots, \beta_m\}$ be an ordered basis of W. Then $\Gamma(B, C)(T)$ is the $m \times n$ matrix whose columns are given by

$$\Gamma(B, C)(T) = [[T(\alpha_1)]_C \,|\, [T(\alpha_2)]_C \,|\, \cdots \,|\, [T(\alpha_n)]_C]$$

In Definition 5.1, $[*]_C : W \mapsto F^m$ is the coordinate map on W determined by the basis C. Since each $T(\alpha_i) \in W$, $[T(\alpha_i)]_C$ is a column vector of size m. Thus, $\Gamma(B, C)(T)$ is an $m \times n$ matrix with entries in F.

It is clear that the matrix $\Gamma(B, C)(T)$ depends not only on the two bases B and C, but also on the order of the vectors in B and C. If we permute the vectors in B (C), then the columns (rows) of $\Gamma(B, C)(T)$ are permuted. As with coordinate maps, this point will cause no real confusion in the sequel. We will always pick fixed bases B and C (of V and W, respectively) and a fixed ordering of the vectors in B and C. If we change the ordering of the vectors in B or C for some reason, then we will change the names of the reordered bases to different symbols.

The $m \times n$ matrix $\Gamma(B, C)(T)$ is called the matrix representation of T with respect to the (ordered) bases B and C. The columns of $\Gamma(B, C)(T)$ are a complete description of the action of T on B.

Example 5.2   Let $V = \mathscr{P}_3$ and $W = \mathscr{P}_2$. Suppose $D : V \mapsto W$ is ordinary differentiation. We have seen (Example 4.10) that D is a linear transformation. To construct a matrix representation of D, we must choose bases of V and W. Let $B = \{1, t, t^2, t^3\}$ and $C = \{1, t, t^2\}$. B is a basis of $\mathscr{P}_3$ and C is a basis of $\mathscr{P}_2$. Then we have the following equations:

$$[D(1)]_C = [0]_C = \begin{bmatrix} 0 \\ 0 \\ 0 \end{bmatrix},$$

$$[D(t)]_C = [1]_C = \begin{bmatrix} 1 \\ 0 \\ 0 \end{bmatrix}$$

$$[D(t^2)]_C = [2t]_C = \begin{bmatrix} 0 \\ 2 \\ 0 \end{bmatrix},$$

$$[D(t^3)]_C = [3t^2]_C = \begin{bmatrix} 0 \\ 0 \\ 3 \end{bmatrix}$$

Therefore

$$\Gamma(B, C)(D) = \begin{bmatrix} 0 & 1 & 0 & 0 \\ 0 & 0 & 2 & 0 \\ 0 & 0 & 0 & 3 \end{bmatrix}$$

is the matrix representation of D with respect to the bases B and C.

If we change B and/or C in Example 5.2, then we of course get a new 3×4 matrix representing D.

Example 5.3 In the previous example, suppose we let $B_1 = \{1 + t, 1 - t, t + t^2, 1 + 3t + t^3\}$ and $C_1 = \{2 + t, 1 + t, t^2\}$. The reader can easily check that B_1 is a basis of $\mathscr{P}_3$ and C_1 is a basis of $\mathscr{P}_2$. We now have the following equations:

$$[D(1 + t)]_{C_1} = [1]_{C_1} = \begin{bmatrix} 1 \\ -1 \\ 0 \end{bmatrix}$$

since

$$1 = (2 + t) - (1 + t) + 0t^2.$$

$$[D(1 - t)]_{C_1} = [-1]_{C_1} = \begin{bmatrix} -1 \\ 1 \\ 0 \end{bmatrix}$$

since

$$-1 = (-1)(2 + t) + (1)(1 + t) + 0t^2.$$

$$[D(t + t^2)]_{C_1} = [1 + 2t]_{C_1} = \begin{bmatrix} -1 \\ 3 \\ 0 \end{bmatrix}$$

since

$$1 + 2t = (-1)(2 + t) + 3(1 + t) + 0t^2.$$

$$[D(1 + 3t + t^3)]_{C_1} = [3 + 3t^2]_{C_1} = \begin{bmatrix} 3 \\ -3 \\ 3 \end{bmatrix}$$

since

$$3 + 3t^2 = 3(2 + t) - 3(1 + t) + 3t^2.$$

These equations imply the matrix representation of D with respect to B_1 and C_1 is the following 3×4 matrix

$$\Gamma(B_1, C_1)(D) = \begin{bmatrix} 1 & -1 & -1 & 3 \\ -1 & 1 & 3 & -3 \\ 0 & 0 & 0 & 3 \end{bmatrix}$$

Let us consider one more example before continuing.

Example 5.4 Suppose $V = W = \mathbb{R}^3$. Let T be the function which rotates a vector in $\mathbb{R}^3$ through 90° in the counterclockwise direction around the z-axis. Let S be the function which rotates a vector in $\mathbb{R}^3$ through 90° in the counterclockwise direction around the x-axis. It is easy to check that S, $T \in \text{Hom}(\mathbb{R}^3, \mathbb{R}^3)$. Suppose we wish to find a matrix representation of the composite linear transformation ST.

A basis for $\mathbb{R}^3$ is the canonical basis $\underline{\varepsilon} = \{\varepsilon_1, \varepsilon_2, \varepsilon_3\}$. From our descriptions of the two maps, we have the following equations:

5.5
$$\begin{aligned} T(\varepsilon_1) &= \varepsilon_2, & S(\varepsilon_1) &= \varepsilon_1 \\ T(\varepsilon_2) &= -\varepsilon_1, & S(\varepsilon_2) &= \varepsilon_3 \\ T(\varepsilon_3) &= \varepsilon_3, & S(\varepsilon_3) &= -\varepsilon_2 \end{aligned}$$

and, therefore

$$ST(\varepsilon_1) = \varepsilon_3$$
$$ST(\varepsilon_2) = -\varepsilon_1$$
$$ST(\varepsilon_3) = -\varepsilon_2$$

Hence

$$\Gamma(\underline{\varepsilon}, \underline{\varepsilon})(ST) = \begin{bmatrix} 0 & -1 & 0 \\ 0 & 0 & -1 \\ 1 & 0 & 0 \end{bmatrix}$$

is a matrix representation of ST.

Suppose we keep the bases B and C fixed and vary the linear transformation T. As T ranges over $\mathrm{Hom}(V, W)$, the matrix $\Gamma(B, C)(T)$ ranges over $M_{m \times n}$. Thus, $\Gamma(B, C)(*): \mathrm{Hom}(V, W) \mapsto M_{m \times n}$ is a well-defined function. Our first theorem in this section says this map is an isomorphism.

Theorem 5.6 Let V and W be finite dimensional vector spaces of dimensions n and m, respectively. Let B be a basis of V and C a basis of W. Then the function $\Gamma(B, C)(*): \mathrm{Hom}(V, W) \mapsto M_{m \times n}$ is an isomorphism of vector spaces.

Proof. We first argue that $\Gamma(B, C)(*): \mathrm{Hom}(V, W) \mapsto M_{m \times n}$ is a linear transformation. Thus, if $T, S \in \mathrm{Hom}(V, W)$ and $x \in F$, we must verify the following equations.

5.7 (a) $\Gamma(B, C)(S + T) = \Gamma(B, C)(S) + \Gamma(B, C)(T)$.
(b) $\Gamma(B, C)(xT) = x\Gamma(B, C)(T)$.

Suppose $B = \{\alpha_1, \ldots, \alpha_n\}$. Then $(S + T)(\alpha_i) = S(\alpha_i) + T(\alpha_i)$ for all i. Since $[*]_C: W \mapsto F^m$ is a linear transformation, we have

$$\begin{aligned}
\Gamma(B, C)(S + T) &= [[(S + T)(\alpha_1)]_C | \cdots | [(S + T)(\alpha_n)]_C] \\
&= [[S(\alpha_1) + T(\alpha_1)]_C | \cdots | [S(\alpha_n) + T(\alpha_n)]_C] \\
&= [[S(\alpha_1)]_C + [T(\alpha_1)]_C | \cdots | [S(\alpha_n)]_C + [T(\alpha_n)]_C] \\
&= [[S(\alpha_1)]_C | \cdots | [S(\alpha_n)]_C] \\
&\quad + [[T(\alpha_1)]_C | \cdots | [T(\alpha_n)]_C] \\
&= \Gamma(B, C)(S) + \Gamma(B, C)(T)
\end{aligned}$$

We have proven (a) of (5.7).

As for (b), we have

$$\Gamma(B,\ C)(xT) = [[(xT)(\alpha_1)]_C | \cdots | [(xT)(\alpha_n)]_C]$$
$$= [[xT(\alpha_1)]_C | \cdots | [xT(\alpha_n)]_C]$$
$$= [x[T(\alpha_1)]_C | \cdots | x[T(\alpha_n)]_C]$$
$$= x[[T(\alpha_1)]_C | \cdots | [T(\alpha_n)]_C]$$
$$= x\Gamma(B,\ C)(T).$$

We have now shown that $\Gamma(B,\ C)(*)$ is a linear transformation from $\mathrm{Hom}(V, W)$ to $M_{m \times n}$.

We next argue that $\Gamma(B, C)(*)$ is injective. Suppose $\Gamma(B, C)(T) = O$ for some $T \in \mathrm{Hom}(V, W)$. Then $[T(\alpha_i)]_C = O$ for all $i = 1, \ldots, n$. Since $[*]_C$ is an isomorphism, $T(\alpha_i) = o$ for all i. Since B is a basis of V, equation (4.2) implies $T = O$. Thus, $\Gamma(B, C)(*)$ is injective.

We have seen in Example 2.18 that $\dim M_{m \times n} = mn$. Also, $\dim \mathrm{Hom}(V, W) = (\dim V)(\dim W) = mn$ by Theorem 4.27. It now follows that $\Gamma(B, C)(*)$ is an isomorphism by Theorem 4.23(b). $\square$

Thus, $\mathrm{Hom}(V, W) \cong M_{m \times n}$ via the linear transformation $\Gamma(B, C)(*)$. One important consequence of the equations in (5.7) is the fact that a matrix representation of a linear combination, $x_1 T_1 + \cdots + x_n T_n$, of linear transformations $T_1, \cdots T_n$ is the corresponding linear combination of the matrix representations of the T_i. Thus

$$\Gamma(B,\ C)\left(\sum_{i=1}^{n} x_i T_i\right) = \sum_{i=1}^{n} x_i \Gamma(B,\ C)(T_i)$$

Example 5.8   Find a matrix representation for the linear transformation $2S - 5T$ in Example 5.4. The equations in (5.5) imply

$$\Gamma(\varepsilon,\ \varepsilon)(T) = \begin{bmatrix} 0 & -1 & 0 \\ 1 & 0 & 0 \\ 0 & 0 & 1 \end{bmatrix}$$

and

$$\Gamma(\varepsilon,\ \varepsilon)(S) = \begin{bmatrix} 1 & 0 & 0 \\ 0 & 0 & -1 \\ 0 & 1 & 0 \end{bmatrix}$$

Therefore, Theorem 5.6 implies

$$\Gamma(\underline{\varepsilon},\, \underline{\varepsilon})(2S - 5T) = 2\Gamma(\underline{\varepsilon},\, \underline{\varepsilon})(S) - 5\Gamma(\underline{\varepsilon},\, \underline{\varepsilon})(T)$$

$$= 2 \begin{bmatrix} 1 & 0 & 0 \\ 0 & 0 & -1 \\ 0 & 1 & 0 \end{bmatrix} - 5 \begin{bmatrix} 0 & -1 & 0 \\ 1 & 0 & 0 \\ 0 & 0 & 1 \end{bmatrix} = \begin{bmatrix} 2 & 5 & 0 \\ -5 & 0 & -2 \\ 0 & 2 & -5 \end{bmatrix}$$

Before presenting our next result, let us recall a special case of Example 4.7. Suppose $A \in M_{m \times n}$. Then A induces a linear transformation $T_A : F^n \mapsto F^m$ given by $T_A(\xi) = A\xi$. The map T_A is just left multiplication by the matrix A. In the rest of this section, we will shorten our notation when dealing with this map, and write A for T_A. Thus, the symbols $A : F^n \mapsto F^m$ stand for the linear transformation which sends ξ to $A\xi$.

Now suppose B and C are bases of V and W, respectively. For each $T \in \operatorname{Hom}(V, W)$, we have the following diagram.

5.9

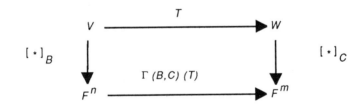

In the diagram in (5.9), $n = \dim V$ and $m = \dim W$. In particular, the two vertical maps $[*]_B$ and $[*]_C$ are isomorphisms. $\Gamma(B, C)(T)$ is the matrix representation of T with respect to B and C. The lower horizontal arrow, labeled $\Gamma(B, C)(T)$, is left multiplication by the $m \times n$ matrix $\Gamma(B, C)(T)$. Our next theorem is the assertion that diagram (5.9) is commutative. By this, we mean the two composite maps from V to F^m in (5.9) are the same.

Theorem 5.10   Let V and W be finite dimensional vector spaces over F of dimensions n and m, respectively. Let B be a basis of V and C a basis of W. Then for any $T \in \operatorname{Hom}(V, W)$ and for any $\gamma \in V$, $[T(\gamma)]_C = \Gamma(B, C)(T) [\gamma]_B$.

Proof.   Fix $T \in \operatorname{Hom}(V, W)$. In terms of composites of linear transformations, the theorem claims $[*]_C T = \Gamma(B, C)(T)[*]_B$ on V. Both $[*]_C T$

and $\Gamma(B, C)(T)[*]_B$ are linear transformations from V to F^m. These two maps are the same provided they agree on B (Theorem 4.22).

Let $B = \{\alpha_1, \ldots, \alpha_n\}$ and $C = \{\beta_1, \ldots, \beta_m\}$. Then $\Gamma(B, C)(T)$ is the $m \times n$ matrix given in Definition 5.1. As usual, let ε_i be the ith vector of the canonical basis $\underline{\varepsilon}$ of F^n. Then $[\alpha_i]_B = \varepsilon_i$ for $i = 1, \ldots, n$. Thus,

$$\Gamma(B, C)(T)[\alpha_i]_B = \Gamma(B, C)(T)\varepsilon_i = \text{Col}_i(\Gamma(B, C)(T))$$

$$= [T(\alpha_i)]_C.$$

In particular, the two linear transformations $\Gamma(B, C)(T)[*]_B$ and $[*]_C T$ agree on the basis B. This completes the proof of Theorem 5.10. $\qquad\square$

Theorem 5.10 implies that if we start in V (diagram (5.9)) and go around the diagram to F^m in the clockwise direction (i.e., $[*]_C T$), or if we start in V and go around the diagram to F^m in the counterclockwise direction (i.e., $\Gamma(B, C)(T)[*]_B$), we get the same result. This is what we mean when we say the diagram is commutative.

The end result of Theorem 5.10 is that the theory of linear transformations $T: V \mapsto W$ is completely equivalent to left multiplication by $A = \Gamma(B, C)(T)$ from F^n to F^m. The vertical maps in (5.9) are isomorphisms, and, consequently, questions about T can be rephrased as matrix arithmetic questions about A. For example, we have the following corollary to Theorem 5.10.

Corollary 5.11 With the same notation as in Theorem 5.10, set $A = \Gamma(B, C)(T)$. Then

(a) $\text{Ker}(T) \cong NS(A)$.
(b) $\text{Im}(T) \cong CS(A)$.

Proof.   (a) We claim $[*]_B: \text{Ker}(T) \mapsto NS(A)$ is an isomorphism. We have seen that $[*]_B: V \mapsto F^n$ is an isomorphism. Restrict $[*]_B$ to the subspace $\text{Ker}(T)$. We then have an injective linear transformation (which we continue to call $[*]_B$) on $\text{Ker}(T)$ to F^n. Therefore, it suffices to show $[*]_B$ maps $\text{Ker}(T)$ onto $NS(A)$.

Suppose $\gamma \in \text{Ker}(T)$. Then $[T(\gamma)]_C = [o]_C = O$. Since the diagram in (5.9) is commutative, $A[\gamma]_B = [T(\gamma)]_C = O$. Therefore, $[\gamma]_B \in NS(A)$. We have now shown that $[*]_B(\text{Ker}(T)) \subseteq NS(A)$.

Suppose $\xi \in NS(A)$. Then $A\xi = O$. Since $[*]_B: V \mapsto F^n$ is an isomorphism, there exists a vector $\gamma \in V$ such that $[\gamma]_B = \xi$. Thus, $A[\gamma]_B = A\xi = O$. Since (5.9) is commutative, $[T(\gamma)]_C = A[\gamma]_B = O$. Since $[*]_C$ is an isomorphism, we conclude that $T(\gamma) = o$. Thus, $\gamma \in \text{Ker}(T)$. We have now shown that $[*]_B : \mathrm{Ker}(T) \mapsto NS(A)$ is surjective, and consequently an isomorphism.

(b) We claim that $[*]_C : \mathrm{Im}(T) \mapsto CS(A)$ is an isomorphism. As in (a), we need only argue the restriction of $[*]_C$ to $\mathrm{Im}(T)$ maps $\mathrm{Im}(T)$ onto $CS(A)$. First, let $\xi \in \mathrm{Im}(T)$. Then $\xi = T(\gamma)$ for some $\gamma \in V$. Since (5.9) is commutative, $[\xi]_C = [T(\gamma)]_C = A[\gamma]_B \in CS(A)$. Hence, $[*]_C(\mathrm{Im}(T)) \subseteq CS(A)$.

Suppose $\lambda \in CS(A)$. Then $\lambda = A\delta$ for some $\delta \in F^n$. Since $[*]_B$ is an isomorphism, there exists a vector $\mu \in V$ such that $[\mu]_B = \delta$. Since (5.9) is commutative, $\lambda = A\delta = A[\mu]_B = [T(\mu)]_C$. Since $T(\mu) \in \mathrm{Im}(T)$, we conclude $[*]_C : \mathrm{Im}(T) \mapsto CS(A)$ is surjective. This completes the proof of (b).

$\square$

We can extend the definitions of rank and nullity to linear transformations in the obvious way.

Definition 5.12 Let $T \in \mathrm{Hom}(V, W)$. The rank of T is the dimension of $\mathrm{Im}(T)$. The nullity of T is the dimension of $\mathrm{Ker}(T)$.

We will let $\mathrm{rk}(T)$ denote the rank of T. Therefore, $\mathrm{rk}(T) = \dim \mathrm{Im}(T)$. We will let $v(T)$ denote the nullity of T. Hence, $v(T) = \dim \mathrm{Ker}(T)$. Corollary 5.11 implies that $\mathrm{rk}(T) = \mathrm{rk}(A)$ and $v(T) = v(A)$ for any matrix representation A of T.

Notice that Theorem 4.23(d) is now a trivial consequence of Theorem 3.22. For suppose $T \in \mathrm{Hom}(V, W)$ and let A be some matrix representation of T. Thus, $A = \Gamma(B, C)(T)$ for two bases B and C of V and W, respectively. Then A is an $m \times n$ matrix, and $\mathrm{rk}(A) + v(A) = n$ by Theorem 3.22. Using Corollary 5.11, we have

$$\dim \mathrm{Im}(T) + \dim \mathrm{Ker}(T) = \mathrm{rk}(T) + v(T) = \mathrm{rk}(A) + v(A)$$
$$= n = \dim V.$$

Consider the following simple example.

Example 5.13 Compute the rank and nullity of the linear transformation $D : \mathscr{P}_3 \mapsto \mathscr{P}_2$ given by $D(p) = dp/dt$.

We had seen in Example 5.2 that

$$A = \begin{bmatrix} 0 & 1 & 0 & 0 \\ 0 & 0 & 2 & 0 \\ 0 & 0 & 0 & 3 \end{bmatrix}$$

is a matrix representation of D. Therefore, $\mathrm{rk}(D) = \mathrm{rk}(A) = 3$ and $v(D) = v(A) = 1$.

Before stating our next theorem, let us point out one more interesting feature of the diagram in (5.9). Theorem 5.10 implies $\Gamma(B, C)(T)$ is an $m \times n$ matrix for which the diagram in (5.9) commutes. In fact, $\Gamma(B, C)(T)$ is the only $m \times n$ matrix for which (5.9) commutes. To be more specific, suppose $E \in M_{m \times n}$ and $E[\gamma]_B = [T(\gamma)]_C$ for all $\gamma \in V$. Then $E = \Gamma(B, C)(T)$. To see this, use the same notation as in the proof of Theorem 5.10. For any $i = 1, \ldots, n$,

$$\text{Col}_i(\Gamma(B, C)(T)) = [T(\alpha_i)]_C = E[\alpha_i]_B = E\varepsilon_i = \text{Col}_i(E).$$

Since i is any integer between 1 and n, $E = \Gamma(B, C)(T)$. Thus, for fixed bases B and C, the matrix representation, $\Gamma(B, C)(T)$, is the unique $m \times n$ matrix for which the diagram in (5.9) is commutative. We can use this idea to compute a matrix representation of a composite of two linear transformations.

Theorem 5.14 Let V, W, and Z denote finite dimensional vector spaces over F. Let n, m, and p be the dimensions of V, W, and Z respectively. Suppose B, C, and D are bases of V, W, and Z, respectively. Let $T \in \text{Hom}(V, W)$ and $S \in \text{Hom}(W, Z)$. Then

$$\Gamma(B, D)(ST) = \Gamma(C, D)(S)\Gamma(B, C)(T).$$

Proof.   Consider the following diagram:

5.15

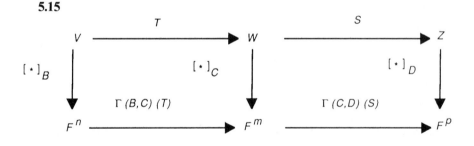

By Theorem 5.10, the left- and right-hand squares in (5.15) commute. Thus, for any vector $\gamma \in V$,

$$[(ST)(\gamma)]_D = [S(T(\gamma))]_D = \Gamma(C, D)(S)[T(\gamma)]_C$$
$$= \Gamma(C, D)(S)\Gamma(B, C)(T)[\gamma]_B.$$

We have pointed out in the paragraph before this proof, that the $p \times n$

matrix $\Gamma(B, D)(ST)$ is the only matrix for which

$$[(ST)(\gamma)]_D = \Gamma(B, D)(ST)[\gamma]_B.$$

Thus,

$$\Gamma(B, D)(ST) = \Gamma(C, D)(S)\Gamma(B, C)(T). \qquad \square$$

Hence, a matrix representation of ST is given by the product of the corresponding matrix representations of S and T. Let us return to Example 5.4 for an illustration of these ideas.

Example 5.16 Suppose S and T are the rotations discussed in Example 5.4. The equations in (5.5) imply

$$\Gamma(\underline{\varepsilon}, \underline{\varepsilon})(T) = \begin{bmatrix} 0 & -1 & 0 \\ 1 & 0 & 0 \\ 0 & 0 & 1 \end{bmatrix}$$

and

$$\Gamma(\underline{\varepsilon}, \underline{\varepsilon})(S) = \begin{bmatrix} 1 & 0 & 0 \\ 0 & 0 & -1 \\ 0 & 1 & 0 \end{bmatrix}$$

Theorem 5.14 implies

$$\Gamma(\underline{\varepsilon}, \underline{\varepsilon})(ST) = \begin{bmatrix} 1 & 0 & 0 \\ 0 & 0 & -1 \\ 0 & 1 & 0 \end{bmatrix}\begin{bmatrix} 0 & -1 & 0 \\ 1 & 0 & 0 \\ 0 & 0 & 1 \end{bmatrix}$$

$$= \begin{bmatrix} 0 & -1 & 0 \\ 0 & 0 & -1 \\ 1 & 0 & 0 \end{bmatrix}$$

We can use Theorem 5.14 to compute a matrix representation of T^{-1} when T is an isomorphism.

Theorem 5.17 Suppose V and W are finite dimensional vector spaces over F of the same dimension n. Let B and C be bases of V and W, respectively. Suppose $T \in \text{Hom}(V, W)$ is an isomorphism. Then $\Gamma(B, C)(T)$ is invertible and $\Gamma(C, B)(T^{-1}) = \Gamma(B, C)(T)^{-1}$.

Proof. We have the following commutative diagram:

5.18

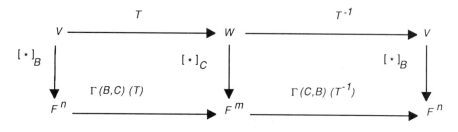

From Theorem 5.14,

$$\Gamma(B, B)(T^{-1}T) = \Gamma(C, B)(T^{-1})\Gamma(B, C)(T).$$

But $T^{-1}T = I_V$ and, hence, $\Gamma(B, B)(T^{-1}T) = \Gamma(B, B)(I_V) = I_n$. Thus, $I_n = \Gamma(C, B)(T^{-1})\Gamma(B, C)(T)$. It now follows from Theorem I.5.13 that $\Gamma(B, C)(T)$ is invertible with inverse $\Gamma(C, B)(T^{-1})$.   □

Thus, if T is an isomorphism, a matrix representation of T^{-1} is the inverse of the matrix representation of T.

Example 5.19 Find a matrix representation of T^{-1} in Example 5.4. We know from the equations in (5.5) that

$$\Gamma(\varepsilon, \varepsilon)(T) = \begin{bmatrix} 0 & -1 & 0 \\ 1 & 0 & 0 \\ 0 & 0 & 1 \end{bmatrix}$$

It is easy to check that

$$\begin{bmatrix} 0 & -1 & 0 \\ 1 & 0 & 0 \\ 0 & 0 & 1 \end{bmatrix}^{-1} = \begin{bmatrix} 0 & 1 & 0 \\ -1 & 0 & 0 \\ 0 & 0 & 1 \end{bmatrix}$$

(Use the algorithm in (5.14) of Chapter I). Theorem 5.17 implies

$$\Gamma(\varepsilon, \varepsilon)(T^{-1}) = \begin{bmatrix} 0 & 1 & 0 \\ -1 & 0 & 0 \\ 0 & 0 & 1 \end{bmatrix}$$

The matrix representation $\Gamma(B, C)(T)$ of a linear transformation $T: V \mapsto W$ of course depends on the two bases B and C of V and W, respectively. If we change B and C, we get a new matrix representation of T. It is an easy matter to keep track of how $\Gamma(B, C)(T)$ changes with B and C. Before stating the result, let us recall a few facts about change of bases matrices.

Suppose $B_1 = \{\alpha_1, \ldots, \alpha_n\}$ and $B_2 = \{\alpha'_1, \ldots, \alpha'_n\}$ are two bases of V. Then each basis determines a coordinate map $[*]_{B_i}: V \mapsto F^n$. The change of bases matrix $M(B_1, B_2)$ is the $n \times n$ matrix whose columns are defined by the following equation.

5.20 $$M(B_1, B_2) = [[\alpha_1]_{B_2} | \cdots | [\alpha_n]_{B_2}]$$

Theorem 2.28 implies $M(B_1, B_2)[\gamma]_{B_1} = [\gamma]_{B_2}$ for all $\gamma \in V$. In terms of a diagram, Theorem 2.28 implies the following is commutative.

5.21

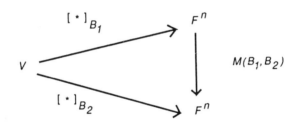

The vertical map $M(B_1, B_2): F^n \mapsto F^n$ in (5.21) means left multiplication by $M(B_1, B_2)$. We have observed in Exercise 2.21 that $M(B_1, B_2)$ is an invertible matrix with inverse $M(B_2, B_1)$. In particular, (5.21) is a commutative diagram of isomorphisms.

We can now state the change of basis theorem.

Theorem 5.22 Suppose V and W are finite dimensional vector spaces over F of dimensions n and m, respectively. Let $T \in \text{Hom}(V, W)$. Let B_1 and B_2 be bases of V, and let C_1 and C_2 be bases of W. Then

$$\Gamma(B_2, C_2)(T) = M(C_1, C_2)\Gamma(B_1, C_1)(T)M(B_1, B_2)^{-1}.$$

Proof. The proof we will give here is called a diagram chase. We put (5.9) together with (5.21) and chase around the figure. Therefore, consider the following diagram of linear transformations.

5.23

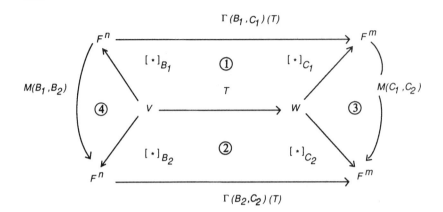

We have divided the diagram in (5.23) into four parts, labeled ①, ②, ③, and ④. ③ and ④ are commutative diagrams of isomorphisms as in (5.21). In particular, we can back up along the arrows in these diagrams by taking the inverse map. Diagrams ① and ② are commutative by Theorem 5.10. We now have the following equations:

5.24

$$M(C_1, C_2)\Gamma(B_1, C_1)(T) = M(C_1, C_2)[*]_{C_1}T[*]_{B_1}^{-1} \quad ①$$
$$= [*]_{C_2}T[*]_{B_1}^{-1} \quad ③$$
$$= \Gamma(B_2, C_2)(T)[*]_{B_2}[*]_{B_1}^{-1} \quad ②$$
$$= \Gamma(B_2, C_2)(T)M(B_1, B_2) \quad ④$$

The numbers out on the right in (5.24) indicate which commutative diagram in (5.23) is giving the equality in question. Since $M(B_1, B_2)$ is invertible, we can rewrite the beginning and the end of (5.24) as follows:

$$M(C_1, C_2)\Gamma(B_1, C_1)(T)M(B_1, B_2)^{-1} = \Gamma(B_2, C_2)(T).$$

This completes the proof of Theorem 5.22.   □

Recall that two $m \times n$ matrices A_1 and A_2 are said to be equivalent if there exist invertible matrices P and Q such that $PA_1Q = A_2$. Since change of bases matrices are invertible, we have the following corollary to Theorem 5.22.

Corollary 5.25 Any two matrix representations of the same linear transformation are equivalent.

Theorem 5.22 can be used to find the simplest matrix representation of a linear transformation T. By this we mean the matrix representation of T which contains the maximum number of zeros as entries. Assume as usual that V has dimension n and W has dimension m. Let B_1 be a basis of V and let C_1 be a basis of W. Set $A_1 = \Gamma(B_1, C_1)(T)$. Theorem I.5.25 implies there exist invertible matrices $P \in M_{m \times m}$ and $Q \in M_{n \times n}$ such that

$$PA_1Q = \left[\begin{array}{c|c} I_k & O \\ \hline O & O \end{array}\right]$$

Here $k = \text{rk}(A_1) = \text{rk}(T)$.

Suppose $C_1 = \{\beta_1, \ldots, \beta_m\}$. Let $P^{-1} = (p_{ij}) \in M_{m \times m}$. For each $j = 1, \ldots, m$, set $\beta_j' = \sum_{i=1}^{m} p_{ij}\beta_i$. Since P^{-1} is invertible, an easy argument shows $C_2 = \{\beta_1', \ldots, \beta_m'\}$ is a basis of W. Also,

$$M(C_2, C_1) = [[\beta_1']_{C_1} | \cdots | [\beta_m']_{C_1}] = P^{-1}.$$

In particular, $P = M(C_2, C_1)^{-1} = M(C_1, C_2)$.

Suppose $B_1 = \{\alpha_1, \ldots, \alpha_n\}$. Let $Q = (q_{rs}) \in M_{n \times n}$. For each $s = 1, \ldots, n$, let $\alpha_s' = \sum_{r=1}^{n} q_{rs}\alpha_r$. Again since Q is invertible, $B_2 = \{\alpha_1' \ldots, \alpha_n'\}$ is a basis of V. Also,

$$M(B_1, B_2)^{-1} = M(B_2, B_1) = [[\alpha_1']_{B_1} | \cdots | [\alpha_n']_{B_1}] = Q.$$

We are now in a position to apply Theorem 5.22.

$$\Gamma(B_2, C_2)(T) = M(C_1, C_2)\Gamma(B_1, C_1)(T)M(B_1, B_2)^{-1}$$

$$= PA_1Q = \left[\begin{array}{c|c} I_k & O \\ \hline O & O \end{array}\right]$$

we have now proven the following theorem.

Theorem 5.26 Suppose V and W are finite dimensional vector spaces over F. Let $T \in \text{Hom}(V, W)$. Suppose $\text{rk}(T) = k$. Then there exist bases B and C of V and W, respectively such that

$$\Gamma(B, C)(T) = \left[\begin{array}{c|c} I_k & O \\ \hline O & O \end{array}\right]$$

We finish this section with a special case of Theorem 5.22. Suppose V is a finite dimensional vector space over F of dimension n. Let $T \in \text{Hom}(V, V)$. If B_1 is a basis of V, then $\Gamma(B_1, B_1)(T)$ is the matrix representation of T with respect to B_1. Suppose B_2 is a second basis of V.

Then the relation between $\Gamma(B_2, B_2)(T)$ and $\Gamma(B_1, B_1)(T)$ is easily determined from Theorem 5.22. We have

5.27 $\Gamma(B_2, B_2)(T) = M(B_1, B_2)\Gamma(B_1, B_1)(T)M(B_1, B_2)^{-1}$

Recall that two $n \times n$ matrices A_1 and A_2 are said to be similar if $PA_1P^{-1} = A_2$ for some invertible matrix $P \in M_{n \times n}$. Thus, (5.27) says that $\Gamma(B_1, B_1)(T)$ and $\Gamma(B_2, B_2)(T)$ are similar.

The question: What is the simplest matrix which is similar to a given matrix A? is more difficult than the corresponding question for equivalence. The interested reader could consult Ref. 2, 3, or 5 to see how this question is answered.

Exercises for Section 5

5.1   Consider $\mathbb{C}$ as a vector space over $\mathbb{R}$. Find a matrix representation for complex conjugation $z \mapsto \bar{z}$.

5.2   Consider the maps S and T in Example 5.4.
 (a)   Show that S and T are both linear transformations.
 (b)   Find a matrix representation for the following linear transformations: (i) TS; (ii) $2T - 3S$; (iii) $TS^2 - ST + T$; (iv) $2T - S + I$ where I is the identity on $\mathbb{R}^3$.

5.3   Let $\int : \mathscr{P}_n \mapsto \mathscr{P}_{n+1}$ be the function defined as follows:

$$\int(a_0 + a_1 t + \cdots + a_n t^n) = a_0 t + (a_1/2)t^2 + \cdots + (a_n/n + 1)t^{n+1}$$

 (a)   Show that $\int$ is a linear transformation.
 (b)   Compute a matrix representation of $\int$.
 (c)   Compute the rank and nullity of $\int$.

5.4   Let

$$\alpha_1 = \begin{bmatrix} 1 \\ 1 \\ 1 \end{bmatrix}, \quad \alpha_2 = \begin{bmatrix} 1 \\ -1 \\ 0 \end{bmatrix}, \quad \alpha_3 = \begin{bmatrix} 2 \\ 1 \\ 1 \end{bmatrix}$$

in $\mathbb{R}^3$. Show that $B = \{\alpha_1, \alpha_2, \alpha_3\}$ is a basis of $\mathbb{R}^3$. Compute $\Gamma(B, B)(T)$ and $\Gamma(B, B)(S)$ for T and S as in Example 4.5.

5.5   Let $T : M_{2 \times 3} \to M_{2 \times 2}$ be the function

$$T\left(\begin{bmatrix} a & b & c \\ e & f & g \end{bmatrix}\right) = \begin{bmatrix} a & b \\ e & f \end{bmatrix}$$

Show that T is a linear transformation, and compute the matrix representation of T with respect to the matrix units in $M_{2 \times 3}$ and $M_{2 \times 2}$. What is the rank and nullity of T?

5.6 Let $T: M_{2 \times 3} \mapsto M_{3 \times 2}$ be given by $T(A) = A^t$. Compute the matrix representation of T with respect to the matrix units in $M_{2 \times 3}$ and $M_{3 \times 2}$. Now do this problem for $M_{m \times n}$.

5.7 Let $T: \mathscr{P}_n \mapsto \mathbb{R}$ be given by $T(p) = p(2)$. Compute a matrix representation of T.

5.8 Let $A \in M_{m \times n}$. Define $T: F^n \mapsto F^m$ by $T(\xi) = A\xi$. Let $\underline{\varepsilon}$ and $\underline{\varepsilon}'$ denote the canonical bases in F^n and F^m, respectively. Compute $\Gamma(\underline{\varepsilon}, \underline{\varepsilon}')(T)$.

5.9 Let

$$A = \begin{bmatrix} 1 & 0 & 2 \\ 1 & 1 & 1 \\ 1 & -1 & 3 \end{bmatrix}$$

Define $T: \mathbb{R}^3 \mapsto \mathbb{R}^3$ by $T(\xi) = A\xi$. Let

$$B = \left\{ \begin{bmatrix} 1 \\ 0 \\ 1 \end{bmatrix}, \begin{bmatrix} 1 \\ 1 \\ 1 \end{bmatrix}, \begin{bmatrix} -1 \\ 1 \\ -3 \end{bmatrix} \right\} \quad \text{and} \quad C = \left\{ \begin{bmatrix} 1 \\ 2 \\ 1 \end{bmatrix}, \begin{bmatrix} 0 \\ 1 \\ 1 \end{bmatrix}, \begin{bmatrix} -2 \\ 1 \\ 2 \end{bmatrix} \right\}$$

(a) Verify that B and C are bases of $\mathbb{R}^3$.
(b) Compute $\Gamma(B, B)(T)$.
(c) Compute $\Gamma(C, C)(T)$.
(d) Compute $\Gamma(B, C)(T)$.

5.10 In Exercise 5.9, find an invertible matrix P such that $PAP^{-1} = \Gamma(B, B)(T)$.

5.11 Suppose $B = \{\alpha_1, \ldots, \alpha_n\}$ is a basis of a vector space V. Let $\alpha'_j = \sum_{i=1}^{n} p_{ij}\alpha_i$ for $j = 1, \ldots, n$. Show that $B' = \{\alpha'_1, \ldots, \alpha'_n\}$ is a basis of V if and only if $P = (p_{ij})$ is invertible.

5.12 Show that $\Gamma(B, B)(I_V) = I_n$ for any basis B of V. Here $n = \dim V$.

5.13 Let $T: M_{1 \times 2} \mapsto M_{1 \times 4}$ be given by the formula

$$T(x_1, x_2) = (x_1, x_1 - x_2, 2x_1 + 3x_2, x_1 + x_2)$$

(a) Show that T is a linear transformation.
(b) Compute $\mathrm{rk}(T)$ and $\nu(T)$.
(c) Show that the transposes of the canonical bases of F^2 and F^4 are bases for $M_{1 \times 2}$ and $M_{1 \times 4}$, respectively.
(d) Compute the matrix representation of T with respect to the bases in (c).

5.14 In Exercise 5.13, find bases B and C of $M_{1 \times 2}$ and $M_{1 \times 4}$, respectively, such that

$$\Gamma(B, C)(T) = \left[\begin{array}{c|c} I_k & O \\ \hline O & O \end{array}\right]$$

with $k = \mathrm{rk}(T)$.

5.15 Let $A, B \in M_{m \times n}$. Show that A and B are equivalent if and only if A and B are both matrix representations of the same linear transformation T.

5.16 Let $T : \mathscr{P}_3 \mapsto \mathbb{R}^3$ be given by

$$T(p) = \begin{bmatrix} p(0) \\ p'(0) + 2p(0) \\ p''(0) - p'(1) \end{bmatrix}$$

Here p' and p'' denote the first and second derivatives of p.

(a) Show that T is a linear transformation.

(b) Compute the matrix representation of T with respect to $B = \{1, t, t^2, t^3\}$ and $C = \underline{\varepsilon} = \{\varepsilon_1, \varepsilon_2, \varepsilon_3\}$.

5.17 Let $T : \mathbb{R}^2 \mapsto \mathbb{R}^2$ be a linear transformation such that

$$T(\varepsilon_1) = \begin{bmatrix} 5 \\ 6 \end{bmatrix} \quad \text{and} \quad T(\varepsilon_2) = \begin{bmatrix} -2 \\ -2 \end{bmatrix}$$

Find a basis B of $\mathbb{R}^2$ such that $\Gamma(B, B)(T)$ is a diagonal matrix.

5.18 Let $T : M_{2 \times 2}(\mathbb{R}) \mapsto \mathscr{P}_2$ be given by the formula

$$T\left(\begin{bmatrix} a & b \\ c & d \end{bmatrix}\right) = (a + d) + (c - 2b)t + (b - 3c)t^2$$

(a) Show that T is a linear transformation.

(b) Compute a matrix representation of T.

III
Determinants

1. PERMUTATIONS AND THE DEFINITION OF THE DETERMINANT

There are several ways to present the basic theory of the determinant. In this book, we will study permutations first. Many of the sign changes in theorems dealing with determinants are a direct consequence of basic facts about permutations. Also, permutations occur all over mathematics, and consequently, every student should know the basic facts about these creatures. Hence, we will discuss permutations first and then introduce the definition of the determinant at the end of this section.

We have already encountered examples of permutations in both Chapters I and II. A permutation of a finite set S is a rearrangement of the elements of S. To be more precise, let $\Delta(n)$ denote the set of natural numbers from 1 to n. Thus, $\Delta(n) = \{1, 2, 3, \ldots, n\}$. A permutation of size n is defined as follows.

Definition 1.1   A permutation (of size n) is a bijective function from $\Delta(n)$ to $\Delta(n)$.

Recall that a function $\sigma : \Delta(n) \mapsto \Delta(n)$ is bijective if σ is both 1-1 and onto. σ is 1-1 if whenever $x, y \in \Delta(n)$, and $x \neq y$, then $\sigma(x) \neq \sigma(y)$. σ is onto if every $y \in \Delta(n)$ is the image, $y = \sigma(x)$, of some $x \in \Delta(n)$.

We will let S_n denote the set of all permutations (of size n). Thus, $S_n = \{\sigma : \Delta(n) \mapsto \Delta(n) \mid \sigma$ is bijective$\}$. A function $\sigma \in S_n$ will be called a permutation of $\Delta(n)$. It is a simple matter to count how many functions are members of S_n. Suppose a bijective map $\sigma : \Delta(n) \mapsto \Delta(n)$ is to be constructed. Since $\Delta(n)$ is a finite set, it suffices to construct a function σ which is 1-1 from $\Delta(n)$ to $\Delta(n)$. σ will automatically be onto. The value of σ at 1, i.e. $\sigma(1)$, can be any one of n choices from the set $\Delta(n)$. Since σ must be 1-1, there are now $n - 1$ choices left for $\sigma(2)$ ($\sigma(1) \neq \sigma(2)$). Having chosen values for $\sigma(1)$ and $\sigma(2)$, there are now $n - 2$ choices left for $\sigma(3)$, since $\sigma(3) \neq \sigma(1)$ or $\sigma(2)$. Continuing with this analysis, there are $n(n - 1)(n - 2) \cdots (3)(2)(1) = n!$ different possibilities for σ. Thus, S_n is a finite set containing $n!$ elements. Equivalently, there are precisely $n!$ distinct permutations of $\Delta(n)$.

There is a convenient way to represent permutations of $\Delta(n)$. Let $\sigma \in S_n$. Suppose $\sigma(1) = j_1$, $\sigma(2) = j_2$, ..., $\sigma(n) = j_n$. Since σ is bijective, $\Delta(n) = \{j_1, \ldots, j_n\}$. We can represent the action of σ on $\Delta(n)$ with the following $2 \times n$ matrix of integers.

1.2
$$\sigma = \begin{bmatrix} 1 & 2 & 3 & \cdots & n \\ j_1 & j_2 & j_3 & \cdots & j_n \end{bmatrix}$$

Consider the following example.

Example 1.3 Let $n = 3$. Then the permutations in S_3 can be listed as follows:

$$\sigma_1 = \begin{bmatrix} 1 & 2 & 3 \\ 1 & 2 & 3 \end{bmatrix}, \quad \sigma_2 = \begin{bmatrix} 1 & 2 & 3 \\ 1 & 3 & 2 \end{bmatrix}, \quad \sigma_3 = \begin{bmatrix} 1 & 2 & 3 \\ 2 & 1 & 3 \end{bmatrix}$$

$$\sigma_4 = \begin{bmatrix} 1 & 2 & 3 \\ 2 & 3 & 1 \end{bmatrix}, \quad \sigma_5 = \begin{bmatrix} 1 & 2 & 3 \\ 3 & 1 & 2 \end{bmatrix}, \quad \sigma_6 = \begin{bmatrix} 1 & 2 & 3 \\ 3 & 2 & 1 \end{bmatrix}$$

For instance, σ_5 is the bijection of $\Delta(3)$ given by $\sigma_5(1) = 3$, $\alpha_5(2) = 1$, and $\sigma_5(3) = 2$.

Example 1.4 Let $n = 4$. Then S_4 contains 24 permutations. Using the notation in (1.2), these permutations are as follows:

$$\sigma_1 = \begin{bmatrix} 1 & 2 & 3 & 4 \\ 1 & 2 & 3 & 4 \end{bmatrix}, \quad \sigma_2 = \begin{bmatrix} 1 & 2 & 3 & 4 \\ 1 & 2 & 4 & 3 \end{bmatrix}, \quad \sigma_3 = \begin{bmatrix} 1 & 2 & 3 & 4 \\ 1 & 3 & 2 & 4 \end{bmatrix}$$

$$\sigma_4 = \begin{bmatrix} 1 & 2 & 3 & 4 \\ 1 & 3 & 4 & 2 \end{bmatrix}, \quad \sigma_5 = \begin{bmatrix} 1 & 2 & 3 & 4 \\ 1 & 4 & 2 & 3 \end{bmatrix}, \quad \sigma_6 = \begin{bmatrix} 1 & 2 & 3 & 4 \\ 1 & 4 & 3 & 2 \end{bmatrix}$$

$$\sigma_7 = \begin{bmatrix} 1 & 2 & 3 & 4 \\ 2 & 1 & 3 & 4 \end{bmatrix}, \quad \sigma_8 = \begin{bmatrix} 1 & 2 & 3 & 4 \\ 2 & 1 & 4 & 3 \end{bmatrix}, \quad \sigma_9 = \begin{bmatrix} 1 & 2 & 3 & 4 \\ 2 & 3 & 1 & 4 \end{bmatrix}$$

$$\sigma_{10} = \begin{bmatrix} 1 & 2 & 3 & 4 \\ 2 & 3 & 4 & 1 \end{bmatrix}, \quad \sigma_{11} = \begin{bmatrix} 1 & 2 & 3 & 4 \\ 2 & 4 & 1 & 3 \end{bmatrix}, \quad \sigma_{12} = \begin{bmatrix} 1 & 2 & 3 & 4 \\ 2 & 4 & 3 & 1 \end{bmatrix}$$

$$\sigma_{13} = \begin{bmatrix} 1 & 2 & 3 & 4 \\ 3 & 1 & 2 & 4 \end{bmatrix}, \quad \sigma_{14} = \begin{bmatrix} 1 & 2 & 3 & 4 \\ 3 & 1 & 4 & 2 \end{bmatrix}, \quad \sigma_{15} = \begin{bmatrix} 1 & 2 & 3 & 4 \\ 3 & 2 & 1 & 4 \end{bmatrix}$$

$$\sigma_{16} = \begin{bmatrix} 1 & 2 & 3 & 4 \\ 3 & 2 & 4 & 1 \end{bmatrix}, \quad \sigma_{17} = \begin{bmatrix} 1 & 2 & 3 & 4 \\ 3 & 4 & 1 & 2 \end{bmatrix}, \quad \sigma_{18} = \begin{bmatrix} 1 & 2 & 3 & 4 \\ 3 & 4 & 2 & 1 \end{bmatrix}$$

$$\sigma_{19} = \begin{bmatrix} 1 & 2 & 3 & 4 \\ 4 & 1 & 2 & 3 \end{bmatrix}, \quad \sigma_{20} = \begin{bmatrix} 1 & 2 & 3 & 4 \\ 4 & 1 & 3 & 2 \end{bmatrix}, \quad \sigma_{21} = \begin{bmatrix} 1 & 2 & 3 & 4 \\ 4 & 2 & 1 & 3 \end{bmatrix}$$

$$\sigma_{22} = \begin{bmatrix} 1 & 2 & 3 & 4 \\ 4 & 2 & 3 & 1 \end{bmatrix}, \quad \sigma_{23} = \begin{bmatrix} 1 & 2 & 3 & 4 \\ 4 & 3 & 1 & 2 \end{bmatrix}, \quad \sigma_{24} = \begin{bmatrix} 1 & 2 & 3 & 4 \\ 4 & 3 & 2 & 1 \end{bmatrix}$$

For instance, σ_{20} is the bijection of $\Delta(4)$ given by $\sigma_{20}(1) = 4$, $\sigma_{20}(2) = 1$, $\sigma_{20}(3) = 3$, and $\sigma_{20}(4) = 2$.

Permutations (of size n) induce bijective maps on any finite set of n distinct objects. Suppose $T = \{A_1, \ldots, A_n\}$ where $A_1, \ldots, A_n$ are n distinct objects. If $\sigma \in S_n$, then σ induces a bijective map (which we continue to call σ) from T to T given by $\sigma(A_i) = A_{\sigma(i)}$ for all $i = 1, \ldots, n$. This idea is used in many diverse contexts in mathematics. Consider the following example.

Example 1.5   Let

$$I_3 = \begin{bmatrix} 1 & 0 & 0 \\ 0 & 1 & 0 \\ 0 & 0 & 1 \end{bmatrix}$$

denote the 3×3 identity matrix. Compute the six permutation matrices obtained from I_3 by taking all possible permutations of the columns of I_3.

To do this in a systematic way, partition I_3 into columns: $I_3 = [\varepsilon_1 \,|\, \varepsilon_2 \,|\, \varepsilon_3]$. Here $\underline{\varepsilon} = \{\varepsilon_1, \varepsilon_2, \varepsilon_3\}$ is the canonical basis of F^3. Let S_3 act on $\underline{\varepsilon}$ via $\sigma(\varepsilon_i) = \varepsilon_{\sigma(i)}$. Each $\sigma \in S_3$ determines a new matrix $I(\sigma) = [\varepsilon_{\sigma(1)} \,|\, \varepsilon_{\sigma(2)} \,|\, \varepsilon_{\sigma(3)}]$. Using the same notation as in Example 1.3, we have

$$I(\sigma_1) = \begin{bmatrix} 1 & 0 & 0 \\ 0 & 1 & 0 \\ 0 & 0 & 1 \end{bmatrix}, \quad I(\sigma_2) = \begin{bmatrix} 1 & 0 & 0 \\ 0 & 0 & 1 \\ 0 & 1 & 0 \end{bmatrix},$$

$$I(\sigma_3) = \begin{bmatrix} 0 & 1 & 0 \\ 1 & 0 & 0 \\ 0 & 0 & 1 \end{bmatrix}, \quad I(\sigma_4) = \begin{bmatrix} 0 & 0 & 1 \\ 1 & 0 & 0 \\ 0 & 1 & 0 \end{bmatrix}$$

$$I(\sigma_5) = \begin{bmatrix} 0 & 1 & 0 \\ 0 & 0 & 1 \\ 1 & 0 & 0 \end{bmatrix}, \quad I(\sigma_6) = \begin{bmatrix} 0 & 0 & 1 \\ 0 & 1 & 0 \\ 1 & 0 & 0 \end{bmatrix}$$

These matrices are the complete set of 3×3 permutation matrices.

Several of our previous results in this book can be written more succinctly using permutations. For example, the observation that the order of the vectors $\alpha_1, \ldots, \alpha_n$ in the definition of a basis of V is not important can be phrased as follows: $\{\alpha_1, \ldots, \alpha_n\}$ is a basis of V if and only if $\{\alpha_{\sigma(1)}, \ldots, \alpha_{\sigma(n)}\}$ is a basis of V for any $\sigma \in S_n$.

Since permutations (of size n) are functions from $\Delta(n)$ to $\Delta(n)$, they can be composed. The composition or product of any two bijective maps from $\Delta(n)$ to $\Delta(n)$ is again a bijective map. Hence, the product $\sigma\tau$ of any two permutations $\sigma, \tau \in S_n$ is another permutation in S_n. The representation of $\sigma\tau$ is easily determined from equation (1.2). To compute $\sigma\tau(j)$ for any $j \in \Delta(n)$, compute $\tau(j)$ from the representation of τ in (1.2), and then compute $\sigma(\tau(j))$ from the representation of σ.

Example 1.6 Suppose

$$\tau = \begin{bmatrix} 1 & 2 & 3 & 4 & 5 \\ 5 & 4 & 3 & 1 & 2 \end{bmatrix} \quad \text{and} \quad \sigma = \begin{bmatrix} 1 & 2 & 3 & 4 & 5 \\ 2 & 1 & 4 & 3 & 5 \end{bmatrix}$$

are two permutations in S_5. Using equation (1.2)

$$\sigma\tau = \begin{bmatrix} 1 & 2 & 3 & 4 & 5 \\ 5 & 3 & 4 & 2 & 1 \end{bmatrix} \quad \text{and} \quad \tau\sigma = \begin{bmatrix} 1 & 2 & 3 & 4 & 5 \\ 4 & 5 & 1 & 3 & 2 \end{bmatrix}$$

Notice that the product of permutations is not in general commutative. In Example 1.6, $\sigma\tau \neq \tau\sigma$. The product of two permutations $\sigma, \tau \in S_n$ is easily computed from the matrix representations of σ and τ. However, notice that the matrix representation of $\sigma\tau$ is not the product of the matrix representations of σ and τ. The composition of permutations in S_n satisfies the following properties.

1.7 (a) $\sigma(\tau\gamma) = (\sigma\tau)\gamma$ for all $\sigma, \tau, \gamma \in S_n$.

(b) There exists a unique permutation $I \in S_n$ such that $\sigma I = I\sigma = \sigma$ for all $\sigma \in S_n$.

(c) For any $\sigma \in S_n$, there exists a unique $\tau \in S_n$ such that $\sigma\tau = \tau\sigma = I$.

The first property in (1.7) is just the familiar fact that the composition of functions is associative. The unique permutation I in (b) of (1.7) is the identity map from $\Delta(n)$ to $\Delta(n)$. Thus

$$I = \begin{bmatrix} 1 & 2 & 3 & \cdots & n \\ 1 & 2 & 3 & \cdots & n \end{bmatrix}$$

We will always denote the identity map from $\Delta(n)$ to $\Delta(n)$ by I. Given $\sigma \in S_n$, the unique permutation τ described in (c) of (1.7) is just the inverse σ^{-1} of σ. Since σ is bijective, σ has a well-defined inverse map $\sigma^{-1}: \Delta(n) \mapsto \Delta(n)$ given by $\sigma^{-1}(i) = j$ if and only if $\sigma(j) = i$. Clearly, σ^{-1} is a permutation on $\Delta(n)$ and $\sigma\sigma^{-1} = \sigma^{-1}\sigma = I$.

In algebra, a nonempty set S together with a binary operation $(\alpha, \beta) \mapsto \alpha\beta$ (from $S \times S$ to S) which satisfies the properties listed in (1.7) is called a group. Thus, S_n is a finite group with the binary operation being composition of functions. S_n is usually called the symmetric group (on n letters).

Every permutation has a sign associated with it. In order to understand what the sign of a permutation is, we need to consider cycles and transpositions.

Definition 1.8 Let $\sigma \in S_n$. σ is called a cycle of length r if there exist r distinct integers $i_1, i_2, \ldots, i_r \in \Delta(n)$ such that

(a) $\sigma(i_1) = i_2, \sigma(i_2) = i_3, \ldots, \sigma(i_{r-1}) = i_r$, and $\sigma(i_r) = i_1$.

(b) $\sigma(j) = j$ for any $j \in \Delta(n) - \{i_1, \ldots, i_r\}$.

If σ is a cycle of length one, then $\sigma = I$. This case is not very interesting. Suppose σ is a cycle of length $r > 1$. Then σ cyclically permutes r integers $i_1, \ldots, i_r$ and leaves all other elements of $\Delta(n)$ invariant.

Example 1.9

$$\sigma_1 = \begin{bmatrix} 1 & 2 & 3 & 4 & 5 & 6 \\ 3 & 1 & 4 & 2 & 5 & 6 \end{bmatrix} \quad \text{is a cycle of length 4 in } S_6$$

$$\sigma_2 = \begin{bmatrix} 1 & 2 & 3 & 4 & 5 & 6 \\ 2 & 3 & 4 & 5 & 6 & 1 \end{bmatrix} \quad \text{is a cycle of length 6 in } S_6$$

$$\sigma_3 = \begin{bmatrix} 1 & 2 & 3 & 4 & 5 & 6 \\ 2 & 3 & 1 & 6 & 4 & 5 \end{bmatrix} \quad \text{is a product of two 3-cycles in } S_6$$

Suppose σ is a cycle of length r in S_n. We can suppose $r > 1$, and

$$\sigma(i_1) = i_2, \ldots, \sigma(i_{r-1}) = i_r \quad \text{and} \quad \sigma(i_r) = i_1.$$

In this case, we will abbreviate our matrix representation of σ given in (1.2), and simply write $\sigma = (i_1, i_2, \ldots, i_r)$. Thus, in Example 1.9, $\sigma_1 = (1, 3, 4, 2)$, $\sigma_2 = (1, 2, 3, 4, 5, 6)$, and $\sigma_3 = (1, 2, 3)(4, 6, 5)$. Notice that the notation $(i_1, \ldots, i_r)$ for a cycle of length r is not quite unique. Clearly, $(1, 3, 4, 2) = (3, 4, 2, 1) = (4, 2, 1, 3)$ etc.

Two cycles $\sigma = (i_1, i_2, \ldots, i_r)$ and $\tau = (j_1, j_2, \ldots, j_s)$ are said to be disjoint if $\{i_1, \ldots, i_r\} \cap \{j_1, \ldots, j_s\} = \varnothing$. Thus, σ and τ are disjoint if their representations $(i_1, \ldots, i_r)$ and $(j_1, \ldots, j_s)$ have no integer in common. In Example 1.9, σ_3 is a product of two disjoint cycles $(1, 2, 3)$ and $(4, 6, 5)$. This is always the case. Following the paths of various elements of $\Delta(n)$ under a fixed permutation, we get the general result.

1.10   Every permutation is a unique product of disjoint cycles.

An example here is as good as a proof.

Example 1.11   Let

$$\sigma = \begin{bmatrix} 1 & 2 & 3 & 4 & 5 & 6 & 7 & 8 & 9 & 10 & 11 \\ 4 & 3 & 1 & 2 & 6 & 7 & 8 & 5 & 10 & 9 & 11 \end{bmatrix} \in S_{11}$$

Then $\sigma = (1, 4, 2, 3)(5, 6, 7, 8)(9, 10)$ is the unique product of disjoint cycles giving σ.

A cycle of length two is called a transposition (2-cycle). A simple computation establishes the following identity.

1.12

$$(i_1, i_2, \ldots, i_r) = (i_1, i_r)(i_1, i_{r-1})(i_1, i_{r-2}) \cdots (i_1, i_2)$$

Thus, every cycle is a product of transpositions. Combining this statement with (1.10), yields the following theorem.

Theorem 1.13 Every permutation in S_n is a product of transpositions.

The identity map $I : \Delta(n) \mapsto \Delta(n)$ is a product of any transposition with itself, i.e., $I = (a, b)(a, b)$. In particular, the representation of a permutation as a product of transpositions is not unique. Here is a less trivial example.

Example 1.14 Let $(1, 2, 4, 3)$ be a cycle in S_4. Then $(1, 2, 4, 3) = (1, 3)(1, 4)(1, 2)$. But, also $(1, 2, 4, 3) = (4, 3, 1, 2)$. Therefore, $(1, 2, 4, 3) = (4, 2)(4, 1)(4, 3)$.

Although the factorization of a permutation into transpositions is not unique, we do have the following important theorem.

Theorem 1.15 Let $\sigma \in S_n$. If σ is a product of an even number of transpositions, then any factorization of σ into a product of transpositions must contain an even number of terms. Similarly, if σ is a product of an odd number of transpositions, then any factorization of σ into a product of transpositions must contain an odd number of terms.

Proof. Let $X_1, \ldots, X_n$ be indeterminates over the field $\mathbb{R}$. Let $P(X_1, \ldots, X_n)$ denote the following polynomial in the variables $X_1, \ldots, X_n$.

1.16
$$P(X_1, \ldots, X_n) = \prod_{i<j} (X_i - X_j)$$

The product in equation (1.16) is taken over all ordered pairs (i, j) such that $1 \leqslant i < j \leqslant n$. For any $\sigma \in S_n$, define

$$\sigma(P) = P(X_{\sigma(1)}, \ldots, X_{\sigma(n)}) = \prod_{i<j} (X_{\sigma(i)} - X_{\sigma(j)})$$

Thus, $\sigma(P)$ is just P with the variables $X_1, \ldots, X_n$ permuted by σ. It is easy to see that $\sigma(P) = \pm P$ for any $\sigma \in S_n$.

Now suppose $\sigma = (p, q)$ is a transposition in S_n. We can assume $1 \leqslant p < q \leqslant n$. We claim $\sigma(P) = -P$. Note that $\sigma(P)$ is the polynomial obtained from P by interchanging X_p and X_q. Thus, the only factors of P (in (1.16)) which are affected by σ are those which contain either X_p or X_q. Let us list these factors as follows.

1.17

$$X_1 - X_p, X_2 - X_p, \ldots, X_{p-1} - X_p | X_p - X_{p+1}, \ldots,$$
$$X_p - X_{q-1} | X_p - X_{q+1}, \ldots, X_p - X_n$$
$$X_1 - X_q, X_2 - X_q, \ldots, X_{p-1} - X_q | X_{p+1} - X_q, \ldots,$$
$$X_{q-1} - X_q | X_q - X_{q+1}, \ldots, X_q - X_n$$
$$X_p - X_q$$

The vertical bars in (1.17) enclose $2(q - p - 1)$ factors whose signs we will want to change (in order to make the subscripts strictly ascend as in P) when X_p and X_q are interchanged. Thus, $\sigma(P) = (-1)^{2(q-p-1)+1}P = -P$.

Thus, a single transposition applied to P changes the sign of P. It easily follows from this remark that $\sigma(P) = P$ if σ is a product of an even number of transpositions, and $\sigma(P) = -P$ if σ is a product of an odd number of transpositions. For a fixed $\sigma \in S_n$, $\sigma(P)$ is either P or $-P$ but not both. Thus, if σ is a product of an even number of transpositions, then $\sigma(P) = P$, and any factorization of σ into transpositions must contain an even number of terms. If σ is a product of an odd number of transpositions, then $\sigma(P) = -P$, and any factorization of σ into transpositions must contain an odd number of terms. □

We can now define the parity of a permutation.

Definition 1.18 A permutation $\sigma \in S_n$ is said to be even if σ is a product of an even number of transpositions. σ is said to be odd if σ is a product of an odd number of transpositions.

Theorems 1.13 and 1.15 guarantee that every permutation is either even or odd, but not both. Notice that the definitions imply that I is even since $I = (a, b)(a, b)$ for any $a, b \in \Delta(n)$. On the other hand, any transposition (a, b) is odd. We have the following familiar sounding rules for the parity of products.

1.19 (a) The product of two even permutations is even.
     (b) The product of two odd permutations is even.
     (c) The product of an even permutation and an odd permutation (or the product of an odd and even permutation) is odd.

The sign of a permutation is defined next.

Definition 1.20 Let $\sigma \in S_n$. The sign of σ, written sgn(σ), is defined as follows:

$$\text{sgn}(\sigma) = \begin{cases} 1 & \text{if } \sigma \text{ is even} \\ -1 & \text{if } \sigma \text{ is odd} \end{cases}$$

As σ varies over S_n, the sgn(σ) defines a function sgn($*$): $S_n \mapsto \{-1, 1\}$. Notice that sgn(I) = 1, and sgn((a, b)) = -1 for any $a \neq b$ in $\Delta(n)$.

Let us return to Examples 1.3 and 1.4 and compute the signs of these permutations.

Example 1.21 In S_3, each permutation factors as follows.

$\sigma_1 = I$; therefore, sgn(σ_1) = 1.
$\sigma_2 = (2, 3)$; therefore, sgn(σ_2) = -1.
$\sigma_3 = (1, 2)$; therefore, sgn(σ_3) = -1.
$\sigma_4 = (1, 2, 3) = (1, 3)(1, 2)$; therefore, sgn($\sigma_4$) = 1.
$\sigma_5 = (1, 3, 2) = (1, 2)(1, 3)$; therefore, sgn($\sigma_5$) = 1.
$\sigma_6 = (1, 3)$; therefore, sgn(σ_6) = -1.

Example 1.22 For S_4, factor each permutation given in Example 1.4 into transpositions and count the number of terms. The resulting signs are as follows:

sgn(σ_1) = 1,	sgn(σ_9) = 1,	sgn(σ_{17}) = 1
sgn(σ_2) = -1,	sgn(σ_{10}) = -1,	sgn(σ_{18}) = -1
sgn(σ_3) = -1,	sgn(σ_{11}) = -1,	sgn(σ_{19}) = -1
sgn(σ_4) = 1,	sgn(σ_{12}) = 1,	sgn(σ_{20}) = 1
sgn(σ_5) = 1,	sgn(σ_{13}) = 1,	sgn(σ_{21}) = 1
sgn(σ_6) = -1,	sgn(σ_{14}) = -1,	sgn(σ_{22}) = -1
sgn(σ_7) = -1,	sgn(σ_{15}) = -1,	sgn(σ_{23}) = -1
sgn(σ_8) = 1,	sgn(σ_{16}) = 1,	sgn(σ_{24}) = 1

We can now introduce the definition of the determinant of a square matrix A.

Definition 1.23 Let $A = (a_{ij}) \in M_{n \times n}$. The determinant of A, written det(A), is the following number:

$$\det(A) = \sum_{\sigma \in S_n} \text{sgn}(\sigma) a_{1\sigma(1)} a_{2\sigma(2)} \cdots a_{n\sigma(n)}$$

The symbols $\sum_{\sigma \in S_n}$ mean add the products $\text{sgn}(\sigma)a_{1\sigma(1)} \cdots a_{n\sigma(n)}$ as σ ranges over all of S_n. The set S_n contains $n!$ distinct permutations. Hence, there are $n!$ summands to be added together when computing $\det(A)$. As A varies over $M_{n \times n}$, $\det(A)$ determines a function $\det(*): M_{n \times n} \mapsto F$ ($= \mathbb{R}$ or $\mathbb{C}$). Notice that $\det(A)$ is only defined when A is a square matrix. Nonsquare matrices do not have determinants.

We will present the basic computational theorems about the determinant in the next section. We close this section with specific formulas for $\det(A)$ when $n = 1$, 2, or 3.

Example 1.24 Suppose $A = (a) \in M_{1 \times 1}$. S_1 consists of the single permutation $I: \Delta(1) \mapsto \Delta(1)$. $\text{sgn}(I) = 1$. Definition 1.23 then implies $\det(A) = a$.

Example 1.25 Suppose

$$A = \begin{bmatrix} a_{11} & a_{12} \\ a_{21} & a_{22} \end{bmatrix} \in M_{2 \times 2}$$

S_2 consists of two permutations, I and the transposition $\sigma = (1, 2)$. The signs of these permutations are $\text{sgn}(I) = 1$ and $\text{sgn}(\sigma) = -1$. Definition 1.23 then implies

$$\det(A) = \text{sgn}(I)a_{11}a_{22} + (\text{sgn}(\sigma))a_{1\sigma(1)}a_{2\sigma(2)}$$
$$= a_{11}a_{22} - a_{12}a_{21}.$$

The reader will notice that the formula obtained for $\det(A)$ in Example 1.25 is the familiar one used in the exercises in Chapter I.

Example 1.26 Suppose

$$A = \begin{bmatrix} a_{11} & a_{12} & a_{13} \\ a_{21} & a_{22} & a_{23} \\ a_{31} & a_{32} & a_{33} \end{bmatrix} \in M_{3 \times 3}$$

S_3 consists of the six permutations $\sigma_1 = I, \sigma_2, \ldots, \sigma_6$ listed in Example 1.3. The signs of these permutations were computed in Example 1.21.

Definition 1.23 then implies

$$\det(A) = \text{sgn}(I)a_{11}a_{22}a_{33} + \text{sgn}(\sigma_2)a_{11}a_{23}a_{32}$$
$$+ \text{sgn}(\sigma_3)a_{12}a_{21}a_{33} + \text{sgn}(\sigma_4)a_{12}a_{23}a_{31}$$
$$+ \text{sgn}(\sigma_5)a_{13}a_{21}a_{32} + \text{sgn}(\sigma_6)a_{13}a_{22}a_{31}$$
$$= a_{11}a_{22}a_{33} - a_{11}a_{23}a_{32} - a_{12}a_{21}a_{33}$$
$$+ a_{12}a_{23}a_{31} + a_{13}a_{21}a_{32} - a_{13}a_{22}a_{31}$$

Again notice that the formula obtained for $\det(A)$ in Example 1.26 is the familiar one used in the calculus.

Exercises for Section 1

1.1 Let $\sigma: \Delta(n) \mapsto \Delta(n)$ be a function. Show that σ is a permutation if either σ is 1-1 or σ is onto.
1.2 Show that all 3×3 permutation matrices are listed in Example 1.5.
1.3 Suppose we form all 3×3 matrices that can be obtained from I_3 by permuting the rows of I_3. Show we get the same set of matrices as those in Example 1.5.
1.4 Is Exercise 1.3 true for an arbitrary $3 = 3$ matrix, i.e., if we form all matrices from A by permuting columns, and all matrices from A by permuting rows, do we get the same set of matrices?
1.5 Show that S_n is a group, i.e., prove the assertions in (1.7).
1.6 Prove the statement in (1.10).
1.7 Verify the transposition equation given in (1.12).
1.8 Give an example of a cycle in S_n which has two factorizations (into transpositions) which contain a different number of terms.
1.9 Express the permutation $(1, 2, 3)(4, 5)(1, 6, 7, 8, 9)(1, 5)$ as a product of disjoint cycles.
1.10 Let $\sigma = (1, 2, 3, \ldots, n) \in S_n$. Show that $\sigma^{-1} = (n, n-1, n-2, \ldots, 1)$.
1.11 Let $\sigma \in S_n$. Show that $\text{sgn}(\sigma) = \text{sgn}(\sigma^{-1})$.
1.12 In the proof of Theorem 1.15, show that $\sigma(P) = \pm P$.
1.13 Show there are precisely $n!/2$ even permutations in S_n.
1.14 Which of the following permutations are even?
  (a) $(1, 2, 3)(1, 2)$
  (b) $(1, 2, 3)(4, 5)(1, 6, 7, 8, 9)(1, 5)$
  (c) $(1, 3, 5)(4, 3, 2)(1, 4, 5)$

1.15 Factor every permutation in S_4 into transpositions and verify the signs given in Example 1.22.

1.16 Is $\det(*): M_{n \times n} \mapsto F$ a linear transformation?

1.17 Repeat the computations in Example 1.26 for an arbitrary 4×4 matrix A.

1.18 Compute $\det(A)$ when A is each of the following matrices:

(a) $\begin{bmatrix} 1 & 2 & 0 \\ 1 & 0 & 1 \\ 1 & 3 & -4 \end{bmatrix}$

(b) $\begin{bmatrix} 0 & 1 & 2 \\ 1 & 1 & 1 \\ -1 & 2 & 1 \end{bmatrix}$

(c) $\begin{bmatrix} 1 & 2 & 3 & 4 \\ 1 & 0 & 1 & 2 \\ 1 & 4 & 0 & -1 \\ -1 & -1 & -1 & 2 \end{bmatrix}$

2. BASIC THEOREMS ON DETERMINANTS

In this section, we will present some of the basic theorems about determinants. The definition of the determinant of A when $A = (a_{ij}) \in M_{n \times n}$ is as follows.

2.1
$$\det(A) = \sum_{\sigma \in S_n} \operatorname{sgn}(\sigma) a_{1\sigma(1)} a_{2\sigma(2)} \cdots a_{n\sigma(n)}$$

Notice that the determinant of A is the sum (with various signs) of all possible products of entries of A, one entry taken from each row and column of A. From a computational point of view, the determinant is a difficult number to evaluate even when the size of A is small. If $n = 4$, $\det(A)$ has $4! = 24$ summands to be computed. If $n = 5$, then $\det(A)$ has $5! = 120$ summands to be computed. We will soon see, there are better, i.e., faster, ways to compute the determinant than directly from the definition. Corollary 2.17 in this section is perhaps the best technique for computing the determinant when n is large. At any rate, we will present several theorems in this section which allow us to evaluate $\det(A)$ without extensive computations. For instance, we have the following theorem whose proof is a simple consequence of the definition.

Theorem 2.2 Let $A \in M_{n \times n}$.

(a) If A has a zero row (or column), then $\det(A) = 0$.
(b) If A is a lower (or upper) triangular matrix, then $\det(A)$ is the product of the diagonal entries of A.

Proof. (a) We had observed that, up to signs, $\det(A)$ is a sum of all possible products of entries of A, one entry taken from each row and column of A. Thus, if some row (column) of A is zero, then every summand in (2.1) is zero. Hence, $\det(A) = 0$.

(b) We will prove the assertion in (b) for a lower triangular matrix A. The corresponding argument for upper triangular matrices is left to the exercises at the end of this section. Suppose A is a lower triangular matrix. Then A has the following form.

2.3

$$A = \begin{bmatrix} a_{11} & 0 & 0 & \cdots & 0 \\ a_{21} & a_{22} & 0 & \cdots & 0 \\ \vdots & \vdots & & & \vdots \\ a_{n1} & a_{n2} & \cdots & \cdots & a_{nn} \end{bmatrix}$$

We claim $\det(A) = a_{11} a_{22} \cdots a_{nn}$. This is a simple consequence of the definition. Suppose $\sigma \in S_n$. If $\sigma(1) \neq 1$, then $a_{1\sigma(1)} = 0$ from (2.3). Thus, the only summands of (2.1) which can possibly be nonzero are those in which $\sigma(1) = 1$. Now if $\sigma(1) = 1$, then $\sigma(2) \neq 1$. If $\sigma(2) > 2$, then again $a_{2\sigma(2)} = 0$ from (2.3). Thus, the only summands of (2.1) that can possibly be nonzero are those in which $\sigma(1) = 1$ and $\sigma(2) = 2$. Continuing this argument, we see there is only one summand in (2.1) which can possibly be nonzero, namely the summand corresponding to I. Therefore, $\det(A) = a_{11} a_{22} \cdots a_{nn}$ □

There is one special case of Theorem 2.2(b) which is worth emphasizing here. Recall that a square matrix $A = (a_{ij})$ is diagonal if $a_{ij} = 0$ whenever $i \neq j$. Diagonal matrices will become increasingly more important as we develop the material in the rest of this chapter. Consequently, we introduce some special notation for diagonal matrices. We will let $\text{Diag}(d_1, \ldots, d_n)$ denote an $n \times n$ diagonal matrix whose (i, i)th entry is d_i for all $i = 1, \ldots, n$. Thus

2.4

$$\begin{bmatrix} d_1 & 0 & 0 & \cdots & 0 \\ 0 & d_2 & 0 & \cdots & 0 \\ \vdots & \vdots & \vdots & & \vdots \\ 0 & 0 & 0 & \cdots & d_n \end{bmatrix} = \text{Diag}(d_1, \ldots, d_n)$$

Theorem 2.2(b) implies $\det(\mathrm{Diag}(d_1,\ldots,d_n)) = d_1 d_2 \cdots d_n$. For example, $\det(I_n) = 1$ for every $n \geqslant 1$.

Our next result says $\det(A)$ is a multilinear function of the rows of A. Let $A = (a_{ij}) \in M_{n \times n}$. Suppose A is partitioned into rows.

2.5

$$A = \begin{bmatrix} R_1 \\ \hline R_2 \\ \hline \vdots \\ \hline R_n \end{bmatrix}$$

Here $R_i = [a_{i1}, a_{i2}, \ldots, a_{in}] \in M_{1 \times n}$ for all $i = 1, \ldots, n$. For the rest of this chapter, it will be notationally convenient to write the row partition of A given in equation (2.5) on one line as follows: $A = (R_1; R_2; \ldots; R_n)$. We put the semicolons between the R_i in this notation to remind the reader that $R_1, \ldots, R_n$ are the rows of A. We will always let R_1 denote the first row of A, R_2 the second row of A, etc. With this notation, the row partition in equation (2.5) can be rewritten as follows.

2.5′

$$A = \begin{bmatrix} R_1 \\ \hline R_2 \\ \hline \vdots \\ \hline R_n \end{bmatrix} = (R_1; R_2; \ldots; R_n)$$

We can now state the following important result about the determinant.

Theorem 2.6 Let $A = (a_{ij}) = (R_1; R_2; \ldots; R_n) \in M_{n \times n}$.

(a) For any $x \in F$, and for any $i = 1, \ldots, n$,

$$\det((R_1; \ldots; R_{i-1}; xR_i; R_{i+1}; \ldots; R_n)) = x \det(A)$$

(b) Suppose for some $i = 1, \ldots, n$, $R_i = B + C$ in $M_{1 \times n}$. Then

$$\det(A) = \det((R_1; \ldots; R_{i-1}; B; R_{i+1}; \ldots; R_n))$$
$$+ \det((R_1; \ldots; R_{i-1}; C; R_{i+1}; \ldots; R_n))$$

Proof. Both of these formulas follow directly from the definition in (2.1).

(a)

$$(R_1; \ldots; R_{i-1}; xR_i; R_{i+1}; \ldots; R_n) = \begin{bmatrix} a_{11} & a_{12} & \cdots & a_{1n} \\ \vdots & \vdots & & \vdots \\ xa_{i1} & xa_{i2} & \cdots & xa_{in} \\ \vdots & \vdots & & \vdots \\ a_{n1} & a_{n2} & \cdots & a_{nn} \end{bmatrix} i$$

Thus

$$\det((R_1; \ldots; R_{i-1}; xR_i; R_{i+1}; \ldots; R_n))$$

$$= \sum_{\sigma \in S_n} \text{sgn}(\sigma) a_{1\sigma(1)} \cdots a_{(i-1)\sigma(i-1)} (x a_{i\sigma(i)}) a_{(i+1)\sigma(i+1)} \cdots a_{n\sigma(n)}$$

$$= x \sum_{\sigma \in S_n} \text{sgn}(\sigma) a_{1\sigma(1)} a_{2\sigma(2)} \cdots a_{n\sigma(n)} = x \det(A).$$

(b) Suppose some row of A, say R_i, is a sum of two row vectors B and C in $M_{1 \times n}$. Let $B = [b_1, \ldots, b_n]$ and $C = [c_1, \ldots, c_n]$. Then

$$R_i = [a_{i1}, \ldots, a_{in}] = B + C = [b_1 + c_1, \ldots, b_n + c_n],$$

and

$$A = \begin{bmatrix} a_{11} & a_{12} & \cdots & a_{1n} \\ \vdots & \vdots & & \vdots \\ b_1 + c_1 & b_2 + c_2 & \cdots & b_n + c_n \\ \vdots & \vdots & & \vdots \\ a_{n1} & a_{n2} & \cdots & a_{nn} \end{bmatrix} i$$

Thus

$$\det(A) = \sum_{\sigma \in S_n} \text{sgn}(\sigma) a_{1\sigma(1)} a_{2\sigma(2)} \cdots a_{n\sigma(n)}$$

$$= \sum_{\sigma \in S_n} \text{sgn}(\sigma) a_{1\sigma(1)} \cdots a_{(i-1)\sigma(i-1)} (b_{\sigma(i)} + c_{\sigma(i)}) a_{(i+1)\sigma(i+1)} \cdots a_{n\sigma(n)}$$

$$= \sum_{\sigma \in S_n} \text{sgn}(\sigma) a_{1\sigma(1)} \cdots a_{(i-1)\sigma(i-1)} b_{\sigma(i)} a_{(i+1)\sigma(i+1)} \cdots a_{n\sigma(n)}$$

$$+ \sum_{\sigma \in S_n} \text{sgn}(\sigma) a_{1\sigma(1)} \cdots a_{(i-1)\sigma(i-1)} c_{\sigma(i)} a_{(i+1)\sigma(i+1)} \cdots a_{n\sigma(n)}$$

$$= \det((R_1; \ldots; R_{i-1}; B; R_{i+1}; \ldots; R_n))$$

$$+ \det((R_1; \ldots; R_{i-1}; C; R_{i+1}; \ldots; R_n)) \qquad \square$$

Theorem 2.6 implies (keeping all rows of A fixed except for row i), the determinant is a linear function of the ith row of A. Functions of n variables which are linear functions of each variable separately (when the other variables are held fixed) are called multilinear functions. Thus, $\det((R_1; \ldots; R_n))$ is a typical example of multilinear function of the rows R_1, $\ldots$, R_n. See Exercise 2.5 for a more precise definition of a multilinear function. Multilinear functions are studied in some detail in [2] and [7].

Example 2.7 Let

$$A = \begin{bmatrix} 1 & 1 \\ -1 & 10 \end{bmatrix}.$$

Then $\det(A) = 11$. On the other hand

$$\begin{bmatrix} 1 & 1 \\ -1 & 10 \end{bmatrix} = \begin{bmatrix} 1 & 1 \\ 2(1) + 3(-1) & 2(2) + 3(2) \end{bmatrix}$$

Therefore, Theorem 2.6 implies

$$11 = \det(A) = \det\begin{bmatrix} 1 & 1 \\ 2 & 4 \end{bmatrix} + \det\begin{bmatrix} 1 & 1 \\ -3 & 6 \end{bmatrix}$$

$$= 2\det\begin{bmatrix} 1 & 1 \\ 1 & 2 \end{bmatrix} + (-3)\det\begin{bmatrix} 1 & 1 \\ 1 & -2 \end{bmatrix} = 2(1) + (-3)(-3)$$

Care must be taken to interpret Theorem 2.6 correctly. The theorem *does not* say the function $\det(*): M_{n \times n} \mapsto F$ is a linear transformation. In fact, $\det(*)$ preserves neither addition nor scalar multiplication in $M_{n \times n}$.

Our next result describes how the determinant of A changes when the rows of A are permuted.

Theorem 2.8 Let $A = (a_{ij}) = (R_1; \ldots; R_n) \in M_{n \times n}$. Suppose $\sigma \in S_n$, and set $B = (R_{\sigma(1)}; R_{\sigma(2)}; \ldots; R_{\sigma(n)})$. Then $\det(B) = \text{sgn}(\sigma)\det(A)$.

Before giving a proof of Theorem 2.8, we need a preliminary result. The matrix B in Theorem 2.8 is obtained from A by permuting (via σ) the rows of A. Thus, B has the following form:

$$B = \begin{bmatrix} a_{\sigma(1)1} & a_{\sigma(1)2} & \cdots & a_{\sigma(1)n} \\ \vdots & \vdots & & \vdots \\ a_{\sigma(n)1} & a_{\sigma(n)2} & \cdots & a_{\sigma(n)n} \end{bmatrix}$$

The proof of Theorem 2.8 is a simple consequence of the following lemma which is of interest in its own right.

Lemma 2.9 Suppose $A = (a_{ij}) = (R_1; \ldots; R_n) \in M_{n \times n}$. If $R_i = R_j$ for some $i \neq j$, then $\det(A) = 0$.

Proof. Suppose $R_i = R_j$ for some $i \neq j$. We can assume $1 \leq i < j \leq n$. The determinant of A has the following form.

2.10
$$\det(A) = \sum_{\sigma \in S_n} \operatorname{sgn}(\sigma) a_{1\sigma(1)} \cdots a_{i\sigma(i)} \cdots a_{j\sigma(j)} \cdots a_{n\sigma(n)}$$

We want to show that the sum in equation (2.10) is zero. We will argue this by showing that each summand in (2.10) is paired with another summand in (2.10) which has the same absolute value, but a different sign.

Fix $\sigma \in S_n$, and let τ be the transposition $\tau = (\sigma(i), \sigma(j))$. Then $\tau\sigma \in S_n$. Consider the two summands of (2.10) given below.

2.11
$$\operatorname{sgn}(\sigma) a_{1\sigma(1)} \cdots a_{i\sigma(i)} \cdots a_{j\sigma(j)} \cdots a_{n\sigma(n)}$$
$$\operatorname{sgn}(\tau\sigma) a_{1\tau\sigma(1)} \cdots a_{i\tau\sigma(i)} \cdots a_{j\tau\sigma(j)} \cdots a_{n\tau\sigma(n)}$$

Since $R_i = R_j$, $a_{i\tau\sigma(i)} = a_{i\sigma(j)} = a_{j\sigma(j)}$, and $a_{j\tau\sigma(j)} = a_{j\sigma(i)} = a_{i\sigma(i)}$. Thus, the cross factors in (2.11) (indicated by the arrows) are equal. If $p \in \Delta(n) - \{i,j\}$, then $\tau\sigma(p) = \sigma(p)$ since τ interchanges only $\sigma(i)$ and $\sigma(j)$. In particular, $a_{p\sigma(p)} = a_{p\tau\sigma(p)}$. We conclude that

$$a_{1\sigma(1)} \cdots a_{i\sigma(i)} \cdots a_{j\sigma(j)} \cdots a_{n\sigma(n)} = a_{1\tau\sigma(1)} \cdots a_{i\tau\sigma(i)} \cdots a_{j\tau\sigma(j)} \cdots a_{n\tau\sigma(n)}.$$

Since τ is a transposition, we have $\operatorname{sgn}(\tau\sigma) = -\operatorname{sgn}(\sigma)$. Therefore, the two summands of equation (2.10) listed in (2.11) have sum zero.

Suppose we had started this argument with the summand

$$\operatorname{sgn}(\tau\sigma) a_{1\tau\sigma(1)} \cdots a_{i\tau\sigma(i)} \cdots a_{j\tau\sigma(j)} \cdots a_{n\tau\sigma(n)}$$
$$= \operatorname{sgn}(\tau\sigma) a_{1\sigma(1)} \cdots a_{i\sigma(j)} \cdots a_{j\sigma(i)} \cdots a_{n\sigma(n)}.$$

Then this summand would be paired with

$$\operatorname{sgn}(\delta\tau\sigma) a_{1\delta\tau\sigma(1)} \cdots a_{i\delta\tau\sigma(i)} \cdots a_{j\delta\tau\sigma(j)} \cdots a_{n\delta\tau\sigma(n)}$$

where δ is the transposition $(\sigma(j), \sigma(i))$. But, $\delta = (\sigma(j), \sigma(i)) = (\sigma(i), \sigma(j)) = \tau$, and $\tau^2 = I$. Thus

$$\operatorname{sgn}(\delta\tau\sigma) a_{1\delta\tau\sigma(1)} \cdots a_{i\delta\tau\sigma(i)} \cdots a_{j\delta\tau\sigma(j)} \cdots a_{n\delta\tau\sigma(n)}$$
$$= \operatorname{sgn}(\sigma) a_{1\sigma(1)} \cdots a_{i\sigma(i)} \cdots a_{j\sigma(j)} \cdots a_{n\sigma(n)}.$$

Hence, the summand

$$\text{sgn}(\tau\sigma)a_{1\tau\sigma(1)} \cdots a_{i\tau\sigma(i)} \cdots a_{j\tau\sigma(j)} \cdots a_{n\tau\sigma(n)}$$

is paired with

$$\text{sgn}(\sigma)a_{1\sigma(1)} \cdots a_{i\sigma(i)} \cdots a_{j\sigma(j)} \cdots a_{n\sigma(n)}.$$

Finally, suppose ξ is a permutation in $S_n - \{\sigma, \tau\sigma\}$. It is easy to see that the indices in the summands of equation (2.10) corresponding to ξ and $(\xi(i), \xi(j))\xi$ are different from the indices of the two summands listed in equation (2.11). Thus, when computing the sum in equation (2.10), the terms corresponding to σ and $(\sigma(i), \sigma(j))\sigma$ cancel, then the terms corresponding to ξ and $(\xi(i), \xi(j))\xi$ cancel. We can continue this argument until there are no terms left in (2.10). Thus, $\det(A) = 0$. $\square$

We can now prove Theorem 2.8.

Proof of 2.8.   Theorem 1.13 implies every permutation is a product of transpositions. Thus, it suffices to prove Theorem 2.8 for a single transposition $\sigma = (i, j)$. We can assume $1 \leqslant i < j \leqslant n$.

Consider the following $n \times n$ matrix C

$$C = (R_1; \ldots; R_{i-1}; R_i + R_j; R_{i+1}; \ldots; R_{j-1}; R_i + R_j; R_{j+1}; \ldots; R_n)$$

Since $\text{Row}_i(C) = R_i + R_j = \text{Row}_j(C)$, Lemma 2.9 implies $\det(C) = 0$. Expand the determinant of C using Theorem 2.6(b).

$$0 = \det((R_1; \ldots; R_{i-1}; R_i + R_j; R_{i+1}; \ldots; R_{j-1}; R_i + R_j; R_{j+1}; \ldots; R_n))$$

$$= \det((R_1; \ldots; R_{i-1}; R_i; R_{i+1}; \ldots; R_{j-1}; R_j; R_{j+1}; \ldots; R_n))$$

$$+ \det((R_1; \ldots; R_{i-1}; R_j; R_{i+1}; \ldots; R_{j-1}; R_i; R_{j+1}; \ldots; R_n))$$

$$+ \det((R_1; \ldots; R_{i-1}; R_i; R_{i+1}; \ldots; R_{j-1}; R_i; R_{j+1}; \ldots; R_n))$$

$$+ \det((R_1; \ldots; R_{i-1}; R_j; R_{i+1}; \ldots; R_{j-1}; R_j; R_{j+1}; \ldots; R_n))$$

$$= \det((R_1; \ldots; R_{i-1}; R_i; R_{i+1}; \ldots; R_{j-1}; R_j; R_{j+1}; \ldots; R_n))$$

$$+ \det((R_1; \ldots; R_{i-1}; R_j; R_{i+1}; \ldots; R_{j-1}; R_i; R_{j+1}; \ldots; R_n))$$

$$= \det((R_1; \ldots; R_n)) + \det((R_{\sigma(1)}; \ldots; R_{\sigma(n)}))$$

Therefore

$$\det((R_{\sigma(1)}; \ldots; R_{\sigma(n)})) = -\det((R_1; \ldots; R_n))$$

Since σ is a transposition, $\text{sgn}(\sigma) = -1$. Thus,

$$\det((R_{\sigma(1)}; \ldots; R_{\sigma(n)})) = \text{sgn}(\sigma) \det((R_1; \ldots; R_n)).$$

This completes the proof of Theorem 2.8. $\qquad\qquad\qquad\qquad\qquad\quad\square$

We should point out here that the properties described in Theorem 2.6 and Lemma 2.9 together with the fact that $\det(I_n) = 1$ uniquely characterize the determinant. By this we mean if $f : M_{n \times n} \mapsto F$ is a function satisfying the assertions in Theorem 2.6 and Lemma 2.9, and $f(I_n) = 1$, then $f(A) = \det(A)$ for all $A \in M_{n \times n}$. We leave this point as an exercise at the end of this section.

We can now prove the most important algebraic property of the determinant. Namely, the determinant preserves products.

Theorem 2.12   Let $A, B \in M_{n \times n}$. Then $\det(AB) = \det(A)\det(B)$.

Proof.   Let $A = (a_{ij})$, and suppose $B = (R_1; \ldots; R_n)$ is the row partition of B. Then for each $i = 1, \ldots, n$

2.13
$$\text{Row}_i(AB) = \sum_{j=1}^{n} a_{ij} R_j$$

Equation (2.13) was established in Chapter I, equation (3.18). By repeated use of Theorem 2.6, we have the following expansion.

2.14
$$\det(AB) = \det\left(\left(\sum_{j=1}^{n} a_{1j} R_j; \sum_{j=1}^{n} a_{2j} R_j; \ldots; \sum_{j=1}^{n} a_{nj} R_j\right)\right)$$

$$= \sum_{i_1, i_2, \ldots, i_n = 1}^{n} a_{1i_1} a_{2i_2} \cdots a_{ni_n} \det((R_{i_1}; R_{i_2}; \ldots; R_{i_n}))$$

In equation (2.14), the indices $i_1, \ldots, i_n$ range independently from 1 to n. However, Lemma 2.9 implies that $\det((R_{i_1}; R_{i_2}; \ldots; R_{i_n})) = 0$ if any $i_p = i_q$ for $p \neq q$. Thus, the only summands in equation (2.14) which are possibly nonzero are those for which $i_1, \ldots, i_n$ are all distinct, i.e., those for which

$$\begin{bmatrix} 1 & 2 & 3 & \cdots & n \\ i_1 & i_2 & i_3 & \cdots & i_n \end{bmatrix}$$

is a permutation in S_n. If

$$\sigma = \begin{bmatrix} 1 & 2 & 3 & \cdots & n \\ i_1 & i_2 & i_3 & & i_n \end{bmatrix}$$

then Theorem 2.8 implies

$$\det((R_{i_1}; R_{i_2}; \ldots; R_{i_n})) = \text{sgn}(\sigma)\det((R_1; R_2; \ldots; R_n))$$
$$= \text{sgn}(\sigma)\det(B).$$

Thus, equation (2.14) can be rewritten as follows.

2.15
$$\det(AB) = \sum_{i_1,i_2,\ldots,i_n=1}^{n} a_{1i_1} \cdots a_{ni_n} \det((R_{i_1}; R_{i_2}; \ldots; R_{i_n}))$$

$$= \sum_{\sigma \in S_n} a_{1\sigma(1)} a_{2\sigma(2)} \cdots a_{n\sigma(n)} \, \text{sgn}(\sigma)\det(B)$$

$$= \det(B) \sum_{\sigma \in S_n} \text{sgn}(\sigma) a_{1\sigma(1)} a_{2\sigma(2)} \cdots a_{n\sigma(n)}$$

$$= \det(B)\det(A). \qquad \square$$

There are three types of elementary matrices: E_{ij}, $E_{ij}(c)$, and $E_i(c)$ corollary describes how the determinant changes when elementary row operations are performed on the matrix. We have seen in Theorem I.5.8 that an elementary row operation is performed on a matrix A by multiplying A on the left by a suitable elementary matrix.

There are three types of elementary matrices E_{ij}, $E_{ij}(c)$, and $E_i(c)$ (Definitions I.5.2, I.5.4, and I.5.6). The matrix E_{ij} is obtained from I_n by interchanging rows i and j of I_n. Hence, if $i \neq j$, Theorem 2.8 implies $\det(E_{ij}) = -1$. The matrix $E_{ij}(c)$ is an upper or lower triangular matrix depending on the relative sizes of i and j. Consequently, Theorem 2.2 implies $\det(E_{ij}(c)) = 1$ if $i \neq j$. The matrix $E_i(c)$ is a diagonal matrix. Hence, $\det(E_i(c)) = c$ by Theorem 2.2. We can now state our first corollary to Theorem 2.12.

Corollary 2.16 Let $A, B \in M_{n \times n}$.

(a) If B is obtained from A by interchanging two distinct rows of A, then $\det(B) = -\det(A)$.
(b) If B is obtained from A by replacing a row of A with itself plus c times another row of A, then $\det(B) = \det(A)$.
(c) If B is obtained from A by multiplying a row of A with c, then $\det(B) = c\det(A)$.

Proof. The proofs of these three assertions follow directly from Theorem I.5.8 and Theorem 2.12. Suppose B is obtained from A by interchanging two distinct rows of A. Then $B = E_{ij}A$ for some $i \neq j$. Thus,

$\det(B) = \det(E_{ij}A) = \det(E_{ij})\det(A) = -\det(A)$. The other two proofs are similar. □

We can use Corollary 2.16 to compute the determinant of any square matrix A. Row reduce A to an upper triangular matrix U keeping track of how the determinant is changing as the necessary row operations are performed. Then compute $\det(U)$ using Theorem 2.2. To be more specific, there exists a permutation matrix $P \in M_{n \times n}$ such that PA has an LU-factorization, $PA = LU$ (Corollary I.6.32). Here L is a lower triangular $n \times n$ matrix with 1's on its diagonal and U is an upper triangular $n \times n$ matrix. A permutation matrix is a product of transpositions E_{ij}. Thus, $P = (\varepsilon_{\sigma(1)}^t; \varepsilon_{\sigma(2)}^t; \dots; \varepsilon_{\sigma(n)}^t)$ for some $\sigma \in S_n$. Here $I_n = (\varepsilon_1^t; \varepsilon_2^t; \dots; \varepsilon_n^t)$ where as usual $\{\varepsilon_1, \dots, \varepsilon_n\}$ is the canonical basis of F^n. In particular, $\det(P) = \operatorname{sgn}(\sigma)$ by Theorem 2.8. If U has diagonal entries $u_{11}, \dots, u_{nn}$, then $\det(L) = 1$ and $\det(U) = u_{11}u_{22} \cdots u_{nn}$. Theorem 2.12 implies $\det(P)\det(A) = \det(L)\det(U)$, or equivalently,

$$\det(A) = \operatorname{sgn}(\sigma)\left(\prod_{i=1}^{n} u_{ii}\right).$$

We have now proven the following corollary.

Corollary 2.17 Let $A \in M_{n \times n}$. Suppose $P = (\varepsilon_{\sigma(1)}^t; \varepsilon_{\sigma(2)}^t; \dots; \varepsilon_{\sigma(n)}^t)$ is a permutation matrix such that PA has LU-factorization, $PA = LU$. If the diagonal entries of U are $u_{11}, \dots, u_{nn}$, then then

$$\det(A) = \operatorname{sgn}(\sigma)\left(\prod_{i=1}^{n} u_{ii}\right).$$

Of course this corollary works best when A itself has an LU-factorization. Consider the following example.

Example 2.18 Let

$$A = \begin{bmatrix} -1 & 2 & 1 & 0 \\ -1 & 4 & 3 & 0 \\ -1 & 4 & 9 & 3 \\ 1 & -3 & 0 & 13 \end{bmatrix}$$

A has the following LU-factorization:

$$A = \begin{bmatrix} 1 & 0 & 0 & 0 \\ 1 & 1 & 0 & 0 \\ 1 & 1 & 1 & 0 \\ -1 & -\frac{1}{2} & \frac{1}{3} & 1 \end{bmatrix} \begin{bmatrix} -1 & 2 & 1 & 0 \\ 0 & 2 & 2 & 0 \\ 0 & 0 & 6 & 3 \\ 0 & 0 & 0 & 12 \end{bmatrix}$$

We derive this factorization using the algorithm given before Theorem I.6.21. Corollary 2.17 implies

$$\det(A) = (-1)(2)(6)(12) = -144.$$

Our next result is also a simple corollary to Theorem 2.12. Because of this result's importance in linear algebra, we list it as a theorem.

Theorem 2.19 Let $A \in M_{n \times n}$. A is nonsingular if and only if $\det(A) \neq 0$.

Proof. If A is nonsingular, then there exists an $n \times n$ matrix B such that $AB = I_n$. By Theorem 2.12, $1 = \det(I_n) = \det(AB) = \det(A)\det(B)$. In particular, $\det(A) \neq 0$.

Suppose $\det(A) \neq 0$. By Theorem I.4.18, there exists an invertible matrix P such that $E = PA$ is the row reduced echelon form of A. $\det(P) \neq 0$ by the first part of this argument. Hence, $\det(E) = \det(P)\det(A) \neq 0$.

E is an upper triangular $n \times n$ matrix. Thus, Theorem 2.2 implies $0 \neq \det(E) = e_{11}e_{22} \cdots e_{nn}$. Here $e_{11}, \ldots, e_{nn}$ are the diagonal entries of E. In particular, no e_{ii} is zero. Since E is in row reduced echelon form, $e_{11} = \cdots = e_{nn} = 1$. But then $\text{rk}(A) = \text{rk}(E) = n$. Thus, A is nonsingular.
                                                                              $\square$

Notice if A is invertible, then

$$1 = \det(I_n) = \det(AA^{-1}) = \det(A)\det(A^{-1}).$$

In particular, $\det(A^{-1}) = (\det(A))^{-1}$.

So far, we have been concentrating on the relationships between the determinant of A and the rows of A. Theorems about rows are equally true about columns. To see this, we need the following result.

Theorem 2.20 Let $A \in M_{n \times n}$. Then $\det(A) = \det(A^t)$.

Proof. Suppose $A = (a_{ij})$. Then $A^t = (b_{ij})$ where $b_{ij} = a_{ji}$ for all $i, j = 1, \ldots, n$. Thus,

2.21
$$\det(A^t) = \sum_{\sigma \in S_n} \operatorname{sgn}(\sigma) b_{1\sigma(1)} b_{2\sigma(2)} \cdots b_{n\sigma(n)}$$

$$= \sum_{\sigma \in S_n} \operatorname{sgn}(\sigma) a_{\sigma(1)1} a_{\sigma(2)2} \cdots a_{\sigma(n)n}$$

Fix $\sigma \in S_n$, and consider the summand $\operatorname{sgn}(\sigma) a_{\sigma(1)1} a_{\sigma(2)2} \cdots a_{\sigma(n)n}$ in equation (2.21). Let $\tau = \sigma^{-1} : \Delta(n) \mapsto \Delta(n)$ be the inverse of σ. Then $\tau \in S_n$ and $\sigma\tau = \tau\sigma = I$. Let $i \in \Delta(n)$ and suppose $\sigma(i) = j$. Then $\tau(j) = i$. In particular $a_{\sigma(i)i} = a_{j\tau(j)}$. It follows that

$$a_{\sigma(1)1} a_{\sigma(2)2} \cdots a_{\sigma(n)n} = a_{1\tau(1)} a_{2\tau(2)} \cdots a_{n\tau(n)}.$$

(We may have to rearrange some factors here.)

If $\sigma = t_1 t_2 \cdots t_p$ where each t_i is a transposition, then clearly $\tau = t_p^{-1} t_{p-1}^{-1} \cdots t_1^{-1} = t_p t_{p-1} \cdots t_1$. In particular, $\operatorname{sgn}(\sigma) = \operatorname{sgn}(\tau)$. Therefore,

$$\operatorname{sgn}(\sigma) a_{\sigma(1)1} a_{\sigma(2)2} \cdots a_{\sigma(n)n} = \operatorname{sgn}(\tau) a_{1\tau(1)} a_{2\tau(2)} \cdots a_{n\tau(n)}$$

where $\tau = \sigma^{-1}$. Finally, the map $\sigma \mapsto \sigma^{-1}$ is clearly a bijective map from S_n to S_n. Using these ideas in equation (2.21), we have the following result.

2.22
$$\det(A^t) = \sum_{\sigma \in S_n} \operatorname{sgn}(\sigma) a_{\sigma(1)1} a_{\sigma(2)2} \cdots a_{\sigma(n)n}$$

$$= \sum_{\tau \in S_n} \operatorname{sgn}(\tau) a_{1\tau(1)} a_{2\tau(2)} \cdots a_{n\tau(n)}$$

$$= \det(A) \qquad \square$$

Since the columns of A are the rows of A^t, Theorem 2.20 allows us to translate Theorems 2.6 and 2.8 into the corresponding statements about columns.

Corollary 2.23 Let $A \in M_{n \times n}$. Partition A into columns $A = [A_1 | \cdots | A_n]$.

(a) For any $i = 1, \ldots, n$ and any $x \in F$,

$$\det([A_1 | \cdots | xA_i | \cdots | A_n]) = x \det([A_1 | \cdots | A_n]).$$

(b) If $A_i = B + C$ in F^n, then

$$\det([A_1 | \cdots | A_{i-1} | B + C | A_{i+1} | \cdots | A_n])$$
$$= \det([A_1 | \cdots | A_{i-1} | B | A_{i+1} | \cdots | A_n])$$
$$+ \det([A_1 | \cdots | A_{i-1} | C | A_{i+1} | \cdots | A_n])$$

(c) For any $\sigma \in S_n$, $\det([A_{\sigma(1)} | \cdots | A_{\sigma(n)}]) = \operatorname{sgn}(\sigma) \det([A_1 | \cdots | A_n])$.

Thus the determinant is a multilinear function of the columns of A as well as the rows of A. Multilinear functions which satisfy Corollary 2.23(c) are called alternating, multilinear functions. The determinant then is an alternating, multilinear function on the rows or columns of A.

We have seen in Section 5 of Chapter I that column operations are performed on a matrix A by multiplying A on the right with elementary matrices. In particular, Theorem 2.12 implies the following analog of Corollary 2.16 for columns.

Corollary 2.24 Let $A, B \in M_{n \times n}$.

(a) If B is obtained from A by interchanging two distinct columns of A, then $\det(B) = -\det(A)$.

(b) If B is obtained from A by replacing a column of A with itself plus c times another column of A, then $\det(B) = \det(A)$.

(c) If B is obtained from A by multiplying a column of A by a constant c, then $\det(B) = c \det(A)$.

We leave the simple proofs of Corollaries 2.23 and 2.24 to the exercises.

Exercises for Section 2

2.1 Find the determinant of the following matrices by using Corollary 2.17:

(a) $\begin{bmatrix} 2 & 1 \\ 1 & 2 \end{bmatrix}$

(b) $\begin{bmatrix} 2 & 1 & 3 \\ 2 & 2 & 5 \\ 0 & -1 & -1 \end{bmatrix}$

(c) $\begin{bmatrix} 2 & 4 & 1 \\ 4 & 6 & -1 \\ 2 & 10 & 2 \end{bmatrix}$

(d) $\begin{bmatrix} 1 & 0 & 1 \\ 2 & 0 & 1 \\ -1 & 1 & 2 \end{bmatrix}$

$$(e) \begin{bmatrix} 2 & 1 & 1 & 2 \\ 0 & 1 & 2 & 1 \\ 3 & 3 & 6 & 1 \\ 2 & 1 & 0 & 4 \end{bmatrix}$$

$$(f) \begin{bmatrix} 2 & 1 & 1 & 1 & 1 \\ 2 & 3 & 2 & 2 & 2 \\ 2 & 3 & 4 & 1 & 3 \\ 2 & 3 & 4 & 2 & 4 \\ 0 & 0 & 0 & 1 & 2 \end{bmatrix}$$

2.2   Prove Theorem 2.2(b) for an upper triangular matrix A.

2.3   Find two square matrices A and B such that

$$\det(A + B) \neq \det(A) + \det(B).$$

2.4   Let $A \in M_{n \times n}$. Show that $\det(xA) = x^n \det(A)$ for any $x \in F$.

2.5   Let V, V_1, ..., V_n be vector spaces over F. A function $f : V_1 \times \cdots \times V_n \mapsto V$ is called a multilinear function if for every $i = 1, \ldots, n$, and for all $\alpha_1 \in V_1, \ldots, \alpha_{i-1} \in V_{i-1}, \alpha_{i+1} \in V_{i+1}, \ldots, \alpha_n \in V_n$,

$$f(\alpha_1, \ldots, \alpha_{i-1}, x\alpha + y\beta, \alpha_{i+1}, \ldots, \alpha_n)$$
$$= xf(\alpha_1, \ldots, \alpha_{i-1}, \alpha, \alpha_{i+1}, \ldots, \alpha_n) + yf(\alpha_1, \ldots, \alpha_{i-1}, \beta, \alpha_{i+1}, \ldots, \alpha_n)$$

for any $\alpha, \beta \in V_i$ and for any $x, y \in F$.

(a) Show that a function $f : V_1 \times \cdots \times V_n \mapsto V$ is multilinear if and only if for each $i = 1, \ldots, n$ and for each $\alpha_1 \in V_1, \ldots, \alpha_{i-1} \in V_{i-1}, \alpha_{i+1} \in V_{i+1}, \ldots, \alpha_n \in V_n$

$$f(\alpha_1, \ldots, \alpha_{i-1}, *, \alpha_{i+1}, \ldots, \alpha_n) \in \text{Hom}(V_i, V)$$

(b) Let $V_i = M_{1 \times n}$ for $i = 1, \ldots, n$. Show that the function $f : M_{1 \times n} \times \cdots \times M_{1 \times n} \mapsto F$ given by $f(R_1, \ldots, R_n) = \det((R_1; \ldots; R_n))$ is multilinear.

(c) Let $V_i = F^n$ for $i = 1, \ldots, n$. Show that the function $f : F^n \times \cdots \times F^n \mapsto F$ given by $f(A_1, \ldots, A_n) = \det([A_1 | \cdots | A_n])$ is multilinear.

2.6   Show that the function $f : F^n \times F^n \mapsto F$ given by $f(\alpha, \beta) = \alpha^t \beta$ is a multilinear function.

2.7   Let $V_i = \mathbb{R}[t]$ for $i = 1, \ldots, n$. Show that the function

$$f: \mathbb{R}[t] \times \cdots \times \mathbb{R}[t] \mapsto \mathbb{R}[t]$$

given by

$$f(p_1, \ldots, p_n) = p_1 p_2 \cdots p_n$$

is a multilinear function.

2.8   A multilinear function $f: V_1 \times \cdots \times V_n \mapsto V$ is said to be alternating if $f(\alpha_1, \ldots, \alpha_n) = 0$ whenever $\alpha_i = \alpha_j$ for some $i \neq j$. Show that the determinant is alternating. Are the two functions given in Exercises 2.6 and 2.7 alternating?

2.9   Let A and B be square matrices (not necessarily of the same size). Show that

$$\det \left[\begin{array}{c|c} A & O \\ \hline C & B \end{array} \right] = \det(A)\det(B)$$

2.10 Generalize the result in Exercise 2.9 as follows:

$$\det \left[\begin{array}{c|c|c|c|c} A_{11} & O & O & \cdots & O \\ \hline A_{21} & A_{22} & O & \cdots & O \\ \hline \vdots & \vdots & \vdots & & \vdots \\ \hline A_{n1} & A_{n2} & A_{n3} & \cdots & A_{nn} \end{array} \right] = \det(A_{11}) \cdots \det(A_{nn})$$

2.11 Suppose

$$\left[\begin{array}{c|c} A & B \\ \hline C & D \end{array} \right]$$

is a partitioned matrix in which A and D are square, and $\det(A) \neq 0$. Show that

$$\left[\begin{array}{c|c} A & B \\ \hline C & D \end{array} \right] = \left[\begin{array}{c|c} I & O \\ \hline CA^{-1} & I \end{array} \right] \left[\begin{array}{c|c} A & O \\ \hline O & D - CA^{-1}B \end{array} \right] \left[\begin{array}{c|c} I & A^{-1}B \\ \hline O & I \end{array} \right]$$

2.12 Use Exercise 2.11 to show that

$$\det \left[\begin{array}{c|c} A & B \\ \hline C & D \end{array} \right] = \det(A)\det(D - CA^{-1}B)$$

What does this equation reduce to if A and C or A and B commute?

2.13 Suppose $f: M_{n \times n} \mapsto F$ is a function with the following properties:
      (a) f is an alternating, multilinear function on the rows of $M_{n \times n}$.

Thus,

(i) $f((R_1; \ldots; xR_i; \ldots; R_n)) = xf((R_1; \ldots; R_n))$ for all $i = 1, \ldots, n$, and for all $x \in F$.

(ii) $f((R_1; \ldots; R_{i-1}; B + C; R_{i+1}; \ldots; R_n))$

$= f((R_1; \ldots; R_{i-1}; B; R_{i+1}, \ldots; R_n))$
$+ f((R_1; \ldots; R_{i-1}; C; R_{i+1}; \ldots; R_n))$

for all $i = 1, \ldots, n$.

(iii) $f((R_1; \ldots; R_n)) = 0$ whenever $R_i = R_j$ for some $i \neq j$.

(b) $f(I_n) = 1$.

Show that $f(A) = \det(A)$ for all $A \in M_{n \times n}$.

2.14 Finish the details in the proof of Theorem 2.8, i.e., show

$$\det((R_{\sigma(1)}; \ldots; R_{\sigma(n)})) = \text{sgn}(\sigma)\det((R_1; \ldots; R_n))$$

for any $\sigma \in S_n$.

2.15 Prove Corollary 2.23.

2.16 Prove Corollary 2.24.

2.17 Let $y = mx + k$ be the equation of the straight line passing through the two points (a, b) and (c, d) in the plane. Show that this equation is also given by

$$\det \begin{bmatrix} x & y & 1 \\ a & b & 1 \\ c & d & 1 \end{bmatrix} = 0$$

2.18 Let (a, b), (c, d), and (e, f) be points in the plane. Show that the area of the triangle determined by these three points is $(1/2)$ the absolute value of

$$\det \begin{bmatrix} a & b & 1 \\ c & d & 1 \\ e & f & 1 \end{bmatrix}$$

3. THE LAPLACE EXPANSION

In this section, we will present another method for computing the determinant. We will use this method to find a formula for the inverse of an invertible matrix. We begin with the following definitions.

Definition 3.1 Let $A = (a_{ij}) \in M_{n \times n}$. For any $i, j = 1, \ldots, n$. $M_{ij}(A)$ will denote the $(n - 1) \times (n - 1)$ submatrix of A obtained by deleting row i and column j of A (here $n > 1$).

The $(n - 1) \times (n - 1)$ matrix $M_{ij}(A)$ is sometimes called the (i, j)th minor of A. More often, the number $\det(M_{ij}(A))$ is called the (i, j)th minor of A.

Definition 3.2 The number $(-1)^{i+j} \det(M_{ij}(A))$ is called the (i, j)th cofactor of A.

We will let $\text{cof}_{ij}(A)$ denote the (i, j)th cofactor of A. Thus, $\text{cof}_{ij}(A) = (-1)^{i+j} \det(M_{ij}(A))$ for all $i, j = 1, \ldots, n$. Notice that there are n^2 submatrices $M_{ij}(A)$ and n^2 cofactors $\text{cof}_{ij}(A)$ of any $n \times n$ matrix A. Consider the following example.

Example 3.3 Let

$$A = \begin{bmatrix} a & b \\ c & d \end{bmatrix}$$

Then

$$M_{11}(A) = [d] \quad \text{and} \quad \text{cof}_{11}(A) = d$$
$$M_{12}(A) = [c] \quad \text{and} \quad \text{cof}_{12}(A) = -c$$
$$M_{21}(A) = [b] \quad \text{and} \quad \text{cof}_{21}(A) = -b$$
$$M_{22}(A) = [a] \quad \text{and} \quad \text{cof}_{22}(A) = a$$

Example 3.4 Let

$$A = \begin{bmatrix} 1 & 2 & 0 \\ 1 & -1 & 1 \\ 1 & 2 & 1 \end{bmatrix}$$

Then A has the following cofactors:

$$\text{cof}_{11}(A) = (-1)^2 \det \begin{bmatrix} -1 & 1 \\ 2 & 1 \end{bmatrix} = -3$$

$$\text{cof}_{12}(A) = (-1)^3 \det \begin{bmatrix} 1 & 1 \\ 1 & 1 \end{bmatrix} = 0$$

$$\text{cof}_{13}(A) = (-1)^4 \det \begin{bmatrix} 1 & -1 \\ 1 & 2 \end{bmatrix} = 3$$

$$\text{cof}_{21}(A) = (-1)^3 \det \begin{bmatrix} 2 & 0 \\ 2 & 1 \end{bmatrix} = -2$$

$$\text{cof}_{22}(A) = (-1)^4 \det \begin{bmatrix} 1 & 0 \\ 1 & 1 \end{bmatrix} = 1$$

$$\text{cof}_{23}(A) = (-1)^5 \det \begin{bmatrix} 1 & 2 \\ 1 & 2 \end{bmatrix} = 0$$

$$\text{cof}_{31}(A) = (-1)^4 \det \begin{bmatrix} 2 & 0 \\ -1 & 1 \end{bmatrix} = 2$$

$$\text{cof}_{32}(A) = (-1)^5 \det \begin{bmatrix} 1 & 0 \\ 1 & 1 \end{bmatrix} = -1$$

$$\text{cof}_{33}(A) = (-1)^6 \det \begin{bmatrix} 1 & 2 \\ 1 & -1 \end{bmatrix} = -3$$

In our first theorem in this section, we show the determinant of an $n \times n$ matrix A is a linear combination of certain cofactors of A. This result is called the Laplace expansion of the determinant.

Theorem 3.5 (Laplace Expansion) Let $A = (a_{ij}) \in M_{n \times n}$.

(a) For any $i = 1, \ldots, n$

$$\det(A) = \sum_{j=1}^{n} (-1)^{i+j} a_{ij} \det(M_{ij}(A))$$

(b) For any $j = 1, \ldots, n$

$$\det(A) = \sum_{i=1}^{n} (-1)^{i+j} a_{ij} \det(M_{ij}(A))$$

Proof. We will prove (a), the row expansion of the determinant, and leave (b), the column expansion of the determinant, as an exercise.

In order to prove (a), we need the following observation:

3.6 If

$$A = \begin{bmatrix} a_{11} & 0 & 0 & \cdots & 0 \\ a_{21} & & & & \\ \vdots & & B & & \\ a_{n1} & & & & \end{bmatrix}$$

then $\det(A) = a_{11} \det(B)$.

The proof of this assertion is similar to the proof of Theorem 2.2(b). If A has the form given in (3.6), then

$$\det(A) = \sum_{\sigma \in S_n} \text{sgn}(\sigma) a_{1\sigma(1)} a_{2\sigma(2)} \cdots a_{n\sigma(n)}$$

with $a_{1\sigma(1)} = 0$ for any $\sigma \in S_n$ with $\sigma(1) \neq 1$. Thus, we need only consider those summands, $\text{sgn}(\sigma) a_{1\sigma(1)} \cdots a_{n\sigma(n)}$ for which

$$\sigma = \begin{bmatrix} 1 & 2 & 3 & \cdots & n \\ 1 & i_2 & i_3 & \cdots & i_n \end{bmatrix}$$

The set $\{\sigma \in S_n \mid \sigma(1) = 1\}$ is precisely the set of all permutations of $\{2, 3, \ldots, n\}$. Therefore,

$$\det(A) = \sum_{\sigma \in S_n} \text{sgn}(\sigma) a_{1\sigma(1)} \cdots a_{n\sigma(n)} = \sum_{\tau} \text{sgn}(\tau) a_{11} a_{2\tau(2)} \cdots a_{n\tau(n)}$$

$$= a_{11} \sum_{\tau} \text{sgn}(\tau) a_{2\tau(2)} \cdots a_{n\tau(n)}$$

Here τ ranges over all permutations of the set $\{2, 3, \ldots, n\}$. Clearly, $\sum_{\tau} \text{sgn}(\tau) a_{2\tau(2)} a_{3\tau(3)} \cdots a_{n\tau(n)} = \det(B)$. Hence, $\det(A) = a_{11} \det(B)$.

We can now prove (a). Fix $i \in \Delta(n)$ and let $A = (R_1; \ldots; R_n)$ be the row partition of A. Interchange R_i with $R_{i-1}, R_{i-2}, \ldots, R_1$ via $i - 1$ transpositions. Corollary 2.16(a) then implies

$$\det(A) = (-1)^{i-1} \det((R_i; R_1; \ldots; R_{i-1}; R_{i+1}; \ldots; R_n))$$

Next, write $R_i = \sum_{j=1}^n a_{ij} \varepsilon_j^t$ where as usual $\{\varepsilon_1, \ldots, \varepsilon_n\}$ is the canonical basis of F^n. Theorem 2.6 implies the following equation.

3.7

$$\det(A) = (-1)^{i-1} \sum_{j=1}^n \det((a_{ij}\varepsilon_j^t; R_1; \ldots; R_{i-1}; R_{i+1}; \ldots; R_n))$$

To finish the proof, we analyze each summand in equation (3.7). If $j = 1$

$$\det((a_{i1}\varepsilon_1^t; R_1; \ldots; R_{i-1}; R_{i+1}; \ldots; R_n)) = \det \begin{bmatrix} a_{i1} & 0 & \cdots & 0 \\ a_{11} & & & \\ \vdots & & & \\ a_{i-11} & & M_{i1}(A) & \\ a_{i+11} & & & \\ \vdots & & & \\ a_{n1} & & & \end{bmatrix}$$

$$= a_{i1} \det(M_{i1}(A)) \quad \text{by equation (3.6)}$$

For the rest of the summands in equation (3.7), we use Corollary 2.24(a). If $j = 2$,

$$\det((a_{i2}\varepsilon_2^t; R_1; \ldots; R_{i-1}; R_{i+1}; \ldots; R_n))$$

$$= \det \begin{bmatrix} 0 & a_{i2} & 0 & \cdots & 0 \\ \hline & & R_1 & & \\ \hline & & \vdots & & \\ \hline & & R_{i-1} & & \\ \hline & & R_{i+1} & & \\ \hline & & \vdots & & \\ \hline & & R_n & & \end{bmatrix} = (-1)\det \begin{bmatrix} a_{i2} & 0 & \cdots & 0 \\ a_{12} & & & \\ \vdots & & & \\ a_{i-12} & & M_{i2}(A) & \\ a_{i+12} & & & \\ \vdots & & & \\ a_{n2} & & & \end{bmatrix}$$

$$= (-1)a_{i2}\det(M_{i2}(A)) \quad \text{by equation (3.6)}$$

In general, we need to move a_{ij} $(j-1)$ slots to the left in order to use equation (3.6). Thus, for any j

$$\det((a_{ij}\varepsilon_j^t; R_1; \ldots; R_{i-1}; R_{i+1}; \ldots; R_n))$$

$$= \det \begin{bmatrix} 0 & \cdots & a_{ij} & \cdots & 0 \\ \hline & & R_1 & & \\ \hline & & \vdots & & \\ \hline & & R_{i-1} & & \\ \hline & & R_{i+1} & & \\ \hline & & \vdots & & \\ \hline & & R_n & & \end{bmatrix} = (-1)^{j-1}\det \begin{bmatrix} a_{ij} & 0 & \cdots & 0 \\ a_{1j} & & & \\ \vdots & & & \\ a_{i-1j} & & M_{ij}(A) & \\ a_{i+1j} & & & \\ \vdots & & & \\ a_{nj} & & & \end{bmatrix}$$

$$= (-1)^{j-1}a_{ij}\det(M_{ij}(A))$$

Substituting these values in equation (3.7), gives the desired result.
3.8

$$\det(A) = (-1)^{i-1}\sum_{j=1}^{n}\det((a_{ij}\varepsilon_j^t; R_1; \ldots; R_{i-1}; R_{i+1}; \ldots; R_n))$$

$$= (-1)^{i-1}\sum_{j=1}^{n}(-1)^{j-1}a_{ij}\det(M_{ij}(A))$$

$$= \sum_{j=1}^{n}(-1)^{i+j}a_{ij}\det(M_{ij}(A)) \qquad \square$$

Using the cofactor notation in Definition 3.2, we can rewrite Theorem 3.5(a) as $\det(A) = \sum_{j=1}^{n}a_{ij}\text{cof}_{ij}(A)$. Theorem 3.5(b) becomes $\det(A) = \sum_{i=1}^{n}a_{ij}\text{cof}_{ij}(A)$. Thus, the determinant of A is a linear combination of cofactors from any fixed row (column) of A with corresponding coefficients from that row (column).

Example 3.9 Suppose

$$A = \begin{bmatrix} a_{11} & a_{12} & a_{13} \\ a_{21} & a_{22} & a_{23} \\ a_{31} & a_{32} & a_{33} \end{bmatrix} \in M_{3 \times 3}$$

The Laplace expansion in Theorem 3.5(a) with $i = 1$, gives the following familiar formula

$$\det(A) = a_{11} \operatorname{cof}_{11}(A) + a_{12} \operatorname{cof}_{12}(A) + a_{13} \operatorname{cof}_{13}(A)$$

$$= a_{11} \det \begin{bmatrix} a_{22} & a_{23} \\ a_{32} & a_{33} \end{bmatrix} - a_{12} \det \begin{bmatrix} a_{21} & a_{23} \\ a_{31} & a_{33} \end{bmatrix}$$

$$+ a_{13} \det \begin{bmatrix} a_{21} & a_{22} \\ a_{31} & a_{32} \end{bmatrix}$$

$$= a_{11}(a_{22}a_{33} - a_{23}a_{32}) - a_{12}(a_{21}a_{33} - a_{23}a_{31})$$

$$+ a_{13}(a_{21}a_{32} - a_{22}a_{31})$$

There are five other expansions of $\det(A)$. We could, for instance, use Theorem 3.5(b) with $j = 2$. We get the same answer, written in a different way

$$\det(A) = a_{12} \operatorname{cof}_{12}(A) + a_{22} \operatorname{cof}_{22}(A) + a_{32} \operatorname{cof}_{32}(A)$$

$$= -a_{12} \det \begin{bmatrix} a_{21} & a_{23} \\ a_{31} & a_{33} \end{bmatrix} + a_{22} \det \begin{bmatrix} a_{11} & a_{13} \\ a_{31} & a_{33} \end{bmatrix}$$

$$- a_{32} \det \begin{bmatrix} a_{11} & a_{13} \\ a_{21} & a_{23} \end{bmatrix}$$

$$= -a_{12}(a_{21}a_{33} - a_{23}a_{31}) + a_{22}(a_{11}a_{33} - a_{13}a_{31})$$

$$- a_{32}(a_{11}a_{23} - a_{13}a_{21})$$

We should point out here that the Laplace expansion is more of a theoretical tool than an actual computing device. If $n \leqslant 3$, then the expansion works well enough. But, for larger values of n, e.g., $n \geqslant 5$, it is better to find the LU-factorization of A and compute the determinant by Corollary 2.17. Most software programs for computers find the determinant of A via the LU-factorization of A.

The next order of business is to introduce the adjoint of a square matrix A.

Definition 3.10   Let $A \in M_{n \times n}$. The adjoint of A, written adj(A), is the $n \times n$ matrix whose (i,j)th entry is defined as follows:

$$[\text{adj}(A)]_{ij} = \text{cof}_{ji}(A) \quad \text{for all } i, j = 1, \ldots, n$$

Thus, for each $i, j = 1, \ldots, n$, $[\text{adj}(A)]_{ij} = (-1)^{i+j} \det(M_{ji}(A))$. Consequently, the adjoint of A is the transpose of the matrix formed from A by replacing each entry of A with the corresponding cofactor. Let us consider two small examples.

Example 3.11   Let

$$A = \begin{bmatrix} a & b \\ c & d \end{bmatrix}$$

Then

$$\text{adj}(A) = \begin{bmatrix} d & -b \\ -c & a \end{bmatrix}$$

This follows directly from the computations in Example 3.3. Notice that $A(\text{adj}(A)) = (\text{adj}(A))A = \det(A)I_2$.

Example 3.12   Let

$$A = \begin{bmatrix} 1 & 2 & 0 \\ 1 & -1 & 1 \\ 1 & 2 & 1 \end{bmatrix}$$

Then

$$\text{adj}(A) = \begin{bmatrix} -3 & -2 & 2 \\ 0 & 1 & -1 \\ 3 & 0 & -3 \end{bmatrix}$$

This last equality follows from the computations in Example 3.4. It is easy to check that $\det(A) = -3$. Note again that $(\text{adj}(A))A = A(\text{adj}(A)) = \det(A)I_3$.

Although the adjoint of A is not easy to compute, it has important uses in linear algebra and commutative ring theory as well. The most important result about the adjoint is the following theorem.

Theorem 3.13 Let $A \in M_{n \times n}$. Then $A \operatorname{adj}(A) = \operatorname{adj}(A)A = \det(A)I_n$.

Proof. There are two theorems here. We will argue that $A \operatorname{adj}(A) = \det(A)I_n$. The other equality, $\operatorname{adj}(A)A = \det(A)I_n$, will be left as an exercise.

In order to show $A \operatorname{adj}(A) = \det(A)I_n$, compare the (i, j)th entries of each side of this equation. Let $A = (a_{ij})$. Then for every $i, j = 1, \ldots, n$

3.14
$$[A \operatorname{adj}(A)]_{ij} = \sum_{k=1}^{n} a_{ik}[\operatorname{adj}(A)]_{kj}$$
$$= \sum_{k=1}^{n} a_{ik}(-1)^{k+j}\det(M_{jk}(A))$$

If $i = j$, then the sum in equation (3.14) becomes

$$\sum_{k=1}^{n} a_{ik}(-1)^{i+k}\det(M_{ik}(A)).$$

This sum is the Laplace expansion of $\det(A)$ along the ith row of A (Theorem 3.5(a)). Therefore,

$$[A \operatorname{adj}(A)]_{ii} = \det(A) = [\det(A)I_n]_{ii}.$$

Fix i and j, and suppose $i \neq j$. Let $A = (R_1; \ldots; R_n)$ be the row partition of A. Let B be the $n \times n$ matrix obtained from A by replacing the jth row of A with the ith row of A. Thus, $B = (R_1; \ldots; R_{j-1}; R_i; R_{j+1}; \ldots; R_n)$. Since $\operatorname{Row}_i(B) = R_i = \operatorname{Row}_j(B)$ and $i \neq j$, Lemma 2.9 implies $\det(B) = 0$. Expanding $\det(B)$ along row j using Theorem 3.5(a), we have

$$0 = \det(B) = \sum_{k=1}^{n} (-1)^{j+k}[B]_{jk}\det(M_{jk}(B))$$
$$= \sum_{k=1}^{n} a_{ik}(-1)^{k+j}\det(M_{jk}(A))$$
$$= \sum_{k=1}^{n} a_{ik}[\operatorname{adj}(A)]_{kj} = [A \operatorname{adj}(A)]_{ij}$$

Since $[\det(A)I_n]_{ij} = 0$, we conclude that $[A \operatorname{adj}(A)]_{ij} = [\det(A)I_n]_{ij}$ when $i \neq j$.

Thus, $[A \operatorname{adj}(A)]_{ij} = [\det(A)I_n]_{ij}$ for all $i, j = 1, \ldots, n$. This completes the proof of Theorem 3.13. $\qquad \square$

When $\det(A) \neq 0$, Theorem 3.13 implies the following corollary.

Corollary 3.15 Let $A \in M_{n \times n}$. If A is nonsingular, then $A^{-1} = (\det(A))^{-1} \operatorname{adj}(A)$.

Example 3.16 Let

$$A = \begin{bmatrix} a & b \\ c & d \end{bmatrix}$$

and assume $\Delta = ad - bc \neq 0$. In this case, Corollary 3.15 implies

$$A^{-1} = \Delta^{-1} \operatorname{adj}(A) = \begin{bmatrix} d/\Delta & -b/\Delta \\ -c/\Delta & a/\Delta \end{bmatrix}$$

Example 3.17 Let

$$A = \begin{bmatrix} 1 & 2 & 0 \\ 1 & -1 & 1 \\ 1 & 2 & 1 \end{bmatrix}$$

We have seen in Example 3.12 that $\det(A) = -3$. Therefore

$$A^{-1} = (-1/3) \operatorname{adj}(A) = \begin{bmatrix} 1 & 2/3 & -2/3 \\ 0 & -1/3 & 1/3 \\ -1 & 0 & 1 \end{bmatrix}$$

We should caution the reader that using Corollary 3.15 to compute A^{-1} even when n is small, e.g., $n = 3$, is not very efficient. The algorithm given in (5.14) of Chapter I is both faster and easier to calculate than computing the adjoint.

Another important application of the adjoint is Cramer's rule for solving square systems of equations.

Theorem 3.18 (Cramer's rule) Consider the following system of n equations in n unknowns $x_1, \ldots, x_n$

$$a_{11}x_1 + a_{12}x_2 + \cdots + a_{1n}x_n = b_1$$
$$\vdots \qquad \vdots \qquad \qquad \vdots \qquad \vdots$$
$$a_{n1}x_1 + a_{n2}x_2 + \cdots + a_{nn}x_n = b_n$$

Let $A = (a_{ij})$ be the coefficient matrix of this system, and partition A into columns, $A = [A_1 | A_2 | \cdots | A_n]$. Set $B = [b_1 \ \cdots \ b_n]^t$. If A is nonsingular, then the (unique) solution $\xi = [x_1 \cdots x_n]^t$ to this system is given by

$$x_i = \frac{\det([A_1 | \cdots | A_{i-1} | B | A_{i+1} | \cdots | A_n])}{\det(A)} \quad \text{for all } i = 1, \ldots, n$$

Proof. The system of equations can be written as $AX = B$ where $X = [x_1 \cdots x_n]^t$. Since A is nonsingular, $X = A^{-1}B$. Corollary 3.15 implies $X = (\det(A))^{-1} \operatorname{adj}(A)B$. Therefore, to prove the theorem, we examine the entries in the column vector $(\det(A))^{-1} \operatorname{adj}(A)B$.

Fix $i = 1, \ldots, n$. Then

$$x_i = [(\det(A))^{-1} \operatorname{adj}(A)B]_i$$

$$= (\det(A))^{-1} \sum_{j=1}^{n} [\operatorname{adj}(A)]_{ij}[B]_j$$

$$= (\det(A))^{-1} \sum_{j=1}^{n} b_j(-1)^{i+j} \det(M_{ji}(A))$$

The sum

$$\sum_{j=1}^{n} b_j(-1)^{i+j} \det(M_{ji}(A))$$

is the Laplace expansion (down column i) for

$$\det([A_1 | \cdots | A_{i-1} | B | A_{i+1} | \cdots | A_n]).$$

Thus,

$$x_i = (\det(A))^{-1} \det([A_1 | \cdots | A_{i-1} | B | A_{i+1} | \cdots | A_n]) \qquad \square$$

We again caution the reader that Cramer's rule works well when n is small. For large values of n, it is quicker and less work to use Gaussian elimination and back substitution.

Example 3.19 Solve

$$ax + by = e$$
$$cx + dy = f$$

when $\Delta = ad - bc \neq 0$. By Cramer's rule, we have

$$x = \Delta^{-1} \det \begin{bmatrix} e & b \\ f & d \end{bmatrix} \quad \text{and} \quad y = \Delta^{-1} \det \begin{bmatrix} a & e \\ c & f \end{bmatrix}$$

Example 3.20 Solve

$$\begin{aligned} x + y &= 1 \\ x - y + z &= 0 \\ x + 2y + z &= 1 \end{aligned}$$

Here

$$A = \begin{bmatrix} 1 & 1 & 0 \\ 1 & -1 & 1 \\ 1 & 2 & 1 \end{bmatrix} \quad \text{and} \quad \det(A) = -3$$

Therefore, Cramer's rule implies

$$x = (-1/3)\det \begin{bmatrix} 1 & 1 & 0 \\ 0 & -1 & 1 \\ 1 & 2 & 1 \end{bmatrix} = 2/3,$$

$$y = (-1/3)\det \begin{bmatrix} 1 & 1 & 0 \\ 1 & 0 & 1 \\ 1 & 1 & 1 \end{bmatrix} = 1/3$$

$$z = (-1/3)\det \begin{bmatrix} 1 & 1 & 1 \\ 1 & -1 & 0 \\ 1 & 2 & 1 \end{bmatrix} = -1/3$$

We finish this section with a method for computing the rank of an $m \times n$ matrix A using determinants. Suppose $A \in M_{m \times n}$. Recall that the rank of A, rk(A), is the maximum number of linearly independent rows (or columns) of A. Given the matrix A, we can examine all square submatrices B of A and compute $\det(B)$. Let $r(A)$ denote the size of the largest square submatrix B of A for which $\det(B) \neq 0$. Therefore, $r(A) = \max\{t \mid A$ contains a $t \times t$ submatrix with nonzero determinant\}. If $A = O$, then we set $r(A) = 0$. If $A \neq O$, then clearly $1 \leqslant r(A) \leqslant \min\{m, n\}$. Let us consider a few examples before proceeding further.

Example 3.21 Let

$$A_1 = [1 \quad 2 \quad 3], \quad A_2 = \begin{bmatrix} 1 & 2 & 3 \\ 2 & 4 & 6 \end{bmatrix}, \quad A_3 = \begin{bmatrix} 1 & 1 & 0 \\ 1 & -1 & 1 \\ 1 & 2 & 1 \end{bmatrix}$$

Then $r(A_1) = 1$, $r(A_2) = 1$, and $r(A_3) = 3$.

In each of these examples, $r(A) = \text{rk}(A)$. This is true in general.

Theorem 3.22 Let $A \in M_{m \times n}$.

$$\text{rk}(A) = \max\left\{ t \left| \begin{array}{l} A \text{ contains a } t \times t \text{ submatrix} \\ \text{with nonzero determinant} \end{array} \right. \right\}$$

Proof. Set $r(A) = \max\{t \mid A \text{ contains a } t \times t \text{ submatrix with nonzero}$ determinant$\}$. If $A = O$, then $r(A) = \text{rk}(A) = O$, and the result is trivial. Hence, we can assume $A \neq O$. Then $r(A)$ and $\text{rk}(A)$ are both positive integers.

Suppose $\text{rk}(A) = r$. There are r columns of A, say $A_{i_1}, \ldots, A_{i_r}$, such that $A_{i_1}, \ldots, A_{i_r}$ are linearly independent in F^m. We can assume these columns have been labeled so that $1 \leqslant i_1 < i_2 < \cdots < i_r \leqslant n$. In particular, $A_1 = [A_{i_1} | A_{i_2} | \cdots | A_{i_r}]$ is an $m \times r$ submatrix of A with $\text{rk}(A_1) = r$. Since $\text{rk}(A_1) = r$, A_1 contains r linearly independent rows. Hence, there exists integers $j_1, \ldots, j_r$ such that $1 \leqslant j_1 < j_2 < \cdots < j_r \leqslant m$, and such that $\text{Row}_{j_1}(A_1), \ldots, \text{Row}_{j_r}(A_1)$ are linearly independent. The matrix $A_2 = (\text{Row}_{j_1}(A_1); \ldots; \text{Row}_{j_r}(A_1))$ is an $r \times r$ submatrix of A. Since the rows of A_2 are linearly independent, A_2 is nonsingular. Theorem 2.19 implies $\det(A_2) \neq 0$. In particular, $r(A) \geqslant r = \text{rk}(A)$.

Suppose $s = r(A)$. Then A contains an $s \times s$ submatrix B such that $\det(B) \neq 0$. Let $B = (B_1; \ldots; B_s)$ be the row partition of B. Then each B_j is part of a row of A. Suppose B_j is part of row A_{i_j} of A for $j = 1, \ldots, s$. Then $1 \leqslant i_1 < i_2 < \cdots < i_s \leqslant m$. We claim the row vectors $A_{i_1}, \ldots, A_{i_s}$ are linearly independent in $M_{1 \times n}$. Suppose $x_1 A_{i_1} + x_2 A_{i_2} + \cdots + x_s A_{i_s} = O$. Then $x_1 B_1 + x_2 B_2 + \cdots + x_s B_s = O$ in $M_{1 \times s}$. Since $\det(B) \neq 0$, B is nonsingular. In particular, $B_1, \ldots, B_s$ are linearly independent. Thus, $x_1 = x_2 = \cdots = x_s = 0$. Therefore, $A_{i_1}, A_{i_2}, \ldots, A_{i_s}$ are linearly independent. In particular, $\text{rk}(A) \geqslant s = r(A)$. Thus, $\text{rk}(A) = r(A)$, and the proof of Theorem 3.22 is complete. $\qquad\qquad\qquad\qquad\qquad\qquad\qquad\square$

Theorem 3.22 is sometimes used as an alternate definition of the rank of A. Thus, the rank of A is sometimes defined to be the size of the largest square submatrix of A with nonzero determinant.

Exercises for Section 3

3.1   Compute the determinant of the matrices listed in (a) through (e) in Exercise 2.1 using the Laplace expansion.

3.2   Prove Theorem 3.5(b).

3.3   Compute the adjoint of the following matrices:

(a) $\begin{bmatrix} 1 & 0 & 0 \\ 2 & 1 & 0 \\ 1 & -1 & 3 \end{bmatrix}$

(b) $\begin{bmatrix} -2 & 1 & 3 \\ 0 & -1 & 1 \\ 1 & 2 & 0 \end{bmatrix}$

(c) $\begin{bmatrix} 1 & 2 & 0 \\ 0 & 1 & -1 \\ 0 & 0 & 2 \end{bmatrix}$

(d) $\begin{bmatrix} 1 & 0 & -1 & 1 \\ 0 & 1 & 2 & 1 \\ -1 & 1 & 1 & 0 \\ 1 & 2 & 1 & -1 \end{bmatrix}$

3.4   Compute the inverses of the matrices listed in Exercise 3.3 using Corollary 3.15.

3.5   How many computations are needed to compute the determinant of A using the Laplace expansion.

3.6   Estimate how many operations it takes to compute A^{-1} using the algorithm in (5.14) of Chapter I. Compare this estimate with the number of operations it takes to compute A^{-1} using Corollary 3.15.

3.7   Show that

$$\det \begin{bmatrix} 1 & x & x^2 \\ 1 & y & y^2 \\ 1 & z & z^2 \end{bmatrix} = (z - x)(z - y)(y - x)$$

3.8   Generalize Exercise 3.7 to the following formula for the determinant of the Vandermonde matrix:

$$\det \begin{bmatrix} 1 & a_0 & a_0^2 & \cdots & a_0^n \\ 1 & a_1 & a_1^2 & \cdots & a_1^n \\ \vdots & \vdots & \vdots & & \vdots \\ 1 & a_n & a_n^2 & \cdots & a_n^n \end{bmatrix} = \prod_{i<j}(a_j - a_i)$$

3.9   Let $A = (a_{ij}) \in M_{n \times n}$. Show that

$$\det(xI_n - A) = \det \begin{bmatrix} x-a_{11} & -a_{12} & \cdots & -a_{1n} \\ -a_{21} & x-a_{22} & \cdots & -a_{2n} \\ \vdots & \vdots & & \vdots \\ -a_{n1} & -a_{n2} & \cdots & x-a_{nn} \end{bmatrix}$$

$$= x^n + b_1 x^{n-1} + \cdots + b_{n-1} x + b_n$$

where $b_1 = -\sum_{i=1}^{n} a_{ii}$ and $b_n = (-1)^n \det(A)$.

3.10   Let $A \in M_{n \times n}$. Prove the following identities:
(a) $\mathrm{adj}(A^t) = (\mathrm{adj}(A))^t$.
(b) $\mathrm{adj}(A)A = \det(A)I_n$.
(c) $\det(\mathrm{adj}(A)) = \det(A)^{n-1}$.

3.11   Solve the following systems of equations using Cramer's rule:
(a) $x - 2y = 6$
       $2x + 3y = 5$

(b) $-2x + y + 3z = 0$
       $-y + z = 2$
       $x + 2y = -1$

(c) $x_1 \qquad -x_3 + x_4 = 3$
       $x_2 + 2x_3 + x_4 = 1$
       $-x_1 + x_2 + x_3 = -2$
       $x_1 + 2x_2 + x_3 - 2x_4 = 0$

3.12   A matrix $A \in M_{n \times n}(\mathbb{R})$ is said to be skew symmetric if $A^t = -A$. Show that $\det(A) = 0$ for any skew symmetric matrix when n is odd. What happens when n is even?

3.13   A matrix $A \in M_{n \times n}(\mathbb{R})$ is orthogonal if $AA^t = I_n$. Show that $\det(A) = \pm 1$ for any orthogonal matrix. Give an example of an orthogonal matrix A such that $n > 1$, and $\det(A) = -1$.

3.14   A matrix $A \in M_{n \times n}(\mathbb{C})$ is said to be unitary if $AA^* = I_n$. Show that the determinant of any unitary matrix is a complex number whose length is 1.

3.15   Compute the rank of the following matrices by using Theorem 3.22:

(a) $\begin{bmatrix} 1 & 2 & 0 & 1 & 3 \\ -2 & -4 & 0 & -2 & -6 \end{bmatrix}$

(b) $\begin{bmatrix} 1 & 0 & 0 & 1 & 2 \\ 2 & 1 & 3 & 1 & 4 \\ 5 & 3 & 9 & 2 & 10 \end{bmatrix}$

3.16   Let V denote a finite dimensional vector space over F. Suppose $T \in \mathrm{Hom}(V, V)$. Define the determinant of T, written $\det(T)$, as follows: $\det(T) = \det(\Gamma(B, B)(T))$ where B is any basis of V. Show that this definition makes sense, i.e., $\det(T)$ does not depend on B.

3.17   Let $A \in M_{n \times n}$. A induces a linear transformation $T : F^n \mapsto F^n$ given by $T(\xi) = A\xi$. Compute $\det(T)$.

3.18   Let $A \in M_{n \times n}$. Define $T : M_{n \times n} \mapsto M_{n \times n}$ by $T(B) = AB$. Show T is a linear transformation, and $\det(T) = (\det(A))^n$.

3.19   In Exercise 3.18, let $S : M_{n \times n} \mapsto M_{n \times n}$ be given by $S(B) = BA - AB$. Show that S is a linear transformation for which $\det(S) = 0$.

3.20   Let $A = (a_{ij}) \in M_{n \times n}$ $(n > 1)$. Suppose $\sum_{j=1}^n a_{ij} = 0$ for every $i = 1, \ldots, n$, and $\sum_{i=1}^n a_{ij} = 0$ for every $j = 1, \ldots, n$. Show there exists a constant K such that $|\det(M_{ij}(A))| = K$ for all $i, j = 1, \ldots, n$.

4. EIGENVALUES AND THE SPECTRUM OF A MATRIX

In this last section of Chapter III, we will discuss one of the principal applications of the determinant, namely computing the spectrum of a matrix. We begin with the definition of an eigenvalue.

Definition 4.1   Let $A \in M_{n \times n}$. A scalar $d \in F$ is called an eigenvalue (or characteristic value) of A if there exists a nonzero vector $\xi \in F^n$ such that $A\xi = d\xi$.

We have seen in Chapter II that left multiplication by A induces a linear transformation from F^n to F^n. If a nonzero vector is mapped by this transformation into a multiple of itself, then this multiple is an eigenvalue of A.

Definition 4.2   The set of eigenvalues of A will be called the spectrum of A and written $\mathscr{S}_F(A)$.

Thus, $\mathscr{S}_F(A) = \{d \in F \mid d \text{ is an eigenvalue of } A\}$. As we will soon see, the spectrum of A is always a finite (possibly empty) subset of F.

In order to compute the spectrum of A, we need to introduce the characteristic polynomial of A. Let t denote an indeterminate (i.e., a variable) over F, and consider the vector space $F[t]$. As a set, $F[t]$ consists of all polynomials in t with coefficients in F $(= \mathbb{R}$ or $\mathbb{C})$. We will let

$M_{n \times n}(F[t])$ denote the set of all $n \times n$ matrices whose entries are polynomials from $F[t]$. Thus, a typical element $C = (f_{ij}(t))$ in $M_{n \times n}(F[t])$ is an $n \times n$ matrix whose (i,j)th entry is a polynomial $f_{ij}(t) \in F[t]$. Since polynomials can be added and multiplied by scalars, $M_{n \times n}(F[t])$ forms a vector space over F with the same definitions of addition and scalar multiplication as those given in Definitions I.2.1 and I.2.4. Thus, if $C = (f_{ij}(t))$ and $D = (g_{ij}(t))$ are two matrices in $M_{n \times n}(F[t])$, then $C + D$ is the $n \times n$ matrix in $M_{n \times n}(F[t])$ whose (i,j)th entry is given by $[C + D]_{ij} = f_{ij}(t) + g_{ij}(t)$. Similarly, if $c \in F$, then $[cD]_{ij} = cg_{ij}(t)$. It is easy to check that $M_{n \times n}(F[t])$ with these definitions of vector addition and scalar multiplication is a vector space over F. See Exercise 4.17 at the end of this section.

Two polynomials in $F[t]$ can be multiplied together to obtain a third polynomial in $F[t]$. In particular, the definition of matrix multiplication can be extended to $M_{n \times n}(F[t])$ in the obvious way: If $C = (f_{ij}(t))$ and $D = (g_{ij}(t))$, then $CD = (\sum_{k=1}^{n} f_{ik}g_{kj})$. The basic properties of a matrix multiplication (i.e., Theorem I.2.24) remain valid for this more general definition. We can also extend the definition of the determinant to $M_{n \times n}(F[t])$ in the obvious way: If $C = (f_{ij}(t)) \in M_{n \times n}(F[t])$, define

$$\det(C) = \sum_{\sigma \in S_n} \text{sgn}(\sigma) f_{1\sigma(1)}(t) f_{2\sigma(2)}(t) \cdots f_{n\sigma(n)}(t).$$

Notice the determinant of $C \in M_{n \times n}(F[t])$ is a polynomial in $F[t]$. Many of the theorems in Sections 2 and 3 of this chapter are valid for determinants of matrices in $M_{n \times n}(F[t])$. In particular, Theorems 2.2, 2.6, 2.8, 2.12, 2.20 3.5, and 3.13 as well as Corollaries 2.16, 2.23, and 2.24 are all true for matrices in $M_{n \times n}(F[t])$. The proofs of these results for matrices in $M_{n \times n}(F[t])$ are virtually identical to the proofs already given for matrices in $M_{n \times n}(F)$.

Scalars in F are polynomials in t of degree zero, i.e., constants. Thus, $F \subseteq F[t]$. In particular, a matrix with entries from F can be viewed as a matrix with entries in $F[t]$. Therefore, $M_{n \times n}(F) \subseteq M_{n \times n}(F[t])$. Now suppose $A = (a_{ij}) \in M_{n \times n}(F)$. Consider the $n \times n$ matrix $tI_n - A \in M_{n \times n}(F[t])$:

4.3

$$tI_n - A = \begin{bmatrix} t - a_{11} & -a_{12} & \cdots & -a_{1n} \\ -a_{21} & t - a_{22} & \cdots & -a_{2n} \\ \vdots & \vdots & & \vdots \\ -a_{n1} & -a_{n2} & \cdots & t - a_{nn} \end{bmatrix}$$

The matrix $tI_n - A$ given in (4.3) is usually called the characteristic matrix of A. The determinant of the characteristic matrix of A will be very important in what follows.

Definition 4.4 Let $A \in M_{n \times n}(F)$. $C_A(t) = \det(tI_n - A)$ is called the characteristic polynomial of the matrix A.

If $A \in M_{n \times n}(\mathbb{R})$, then $C_A(t) \in \mathbb{R}[t]$. If $A \in M_{n \times n}(\mathbb{C})$, then $C_A(t) \in \mathbb{C}[t]$. Consider the following examples.

Example 4.5 If

$$A = \begin{bmatrix} a & b \\ c & d \end{bmatrix}$$

then

$$C_A(t) = \det \begin{bmatrix} t - a & -b \\ -c & t - d \end{bmatrix}$$

$$= (t - a)(t - d) - cb = t^2 - (a + d)t + (ad - bc).$$

For instance, if

$$A = \begin{bmatrix} 1 & 2 \\ 1 & 3 \end{bmatrix}$$

then $C_A(t) = t^2 - 4t + 1$. If

$$A = \begin{bmatrix} 1 + i & 3 \\ i & 1 + i \end{bmatrix}$$

then $C_A(t) = t^2 - (2 + 2i)t - i$.

Example 4.6 Let

$$A = \begin{bmatrix} 1 & 2 & 0 \\ 1 & 1 & 1 \\ 1 & 2 & 1 \end{bmatrix}$$

Then

$$tI_3 - A = \begin{bmatrix} t - 1 & -2 & 0 \\ -1 & t - 1 & -1 \\ -1 & -2 & t - 1 \end{bmatrix}$$

Using Laplace's expansion,

$$C_A(t) = (t - 1)[(t - 1)^2 - 2] + 2[(1 - t) - 1] = t^3 - 3t^2 - t + 1.$$

Example 4.7   Suppose $A = \text{Diag}(d_1, \ldots, d_n)$. Then $tI_n - A = \text{Diag}(t - d_1, \ldots, t - d_n)$. Therefore, $C_A(t) = \prod_{i=1}^{n} (t - d_i)$. In particular, if $A = O$, then $C_A(t) = t^n$ and if $A = I_n$, then $C_A(t) = (t - 1)^n$.

We have seen in Exercise 3.9 that $C_A(t) = \det(tI_n - A)$ is always a monic polynomial of degree n in t. This means the characteristic polynomial of A always has the following form.

4.8 $C_A(t) = t^n + b_1 t^{n-1} + b_2 t^{n-2} + \cdots + b_{n-1} t + b_n$

The coefficients $b_1, \ldots, b_n$ in equation (4.8) are constants in F. In particular, the degree of $C_A(t)$ is n for any $n \times n$ matrix A. The connection between the spectrum of A and the characteristic polynomial of A is very simple. The spectrum of A is precisely the set of roots of $C_A(t)$ which lie in F.

Theorem 4.9   Let $A \in M_{n \times n}(F)$. Then $\mathscr{S}_F(A) = \{d \in F \mid C_A(d) = 0\}$.

Proof.   Suppose $d \in \mathscr{S}_F(A)$. Then there exists a nonzero vector $\xi \in F^n$ such that $A\xi = d\xi$. We can rewrite this equation as $(dI_n - A)\xi = O$. Since $\xi \neq 0$, $NS(dI_n - A) \neq (O)$. Thus, $v(dI_n - A) = \dim NS(dI_n - A) \neq 0$. Now Theorem II.3.23 implies $dI_n - A$ is singular. But, if $dI_n - A$ is singular, then Theorem 2.19 implies $\det(dI_n - A) = 0$. Thus, $C_A(d) = \det(dI_n - A) = 0$. Therefore, any eigenvalue of A is a root of $C_A(t)$.

Conversely, suppose $d \in F$ such that $C_A(d) = 0$. Then $\det(dI_n - A) = 0$ and $dI_n - A$ is singular. Consequently, $NS(dI_n - A) \neq (O)$. If ξ is any nonzero vector in $NS(dI_n - A)$, then $A\xi = d\xi$. Thus, d is an eigenvalue of A.   □

Since the degree of $C_A(t)$ is n, $C_A(t)$ can have at most n roots in F. Hence, we have the following corollary to Theorem 4.9.

Corollary 4.10   Let $A \in M_{n \times n}(F)$. There are at most n distinct eigenvalues of A in F.

Thus, the spectrum of an $n \times n$ matrix A can contain at most n distinct numbers. There could very well be no eigenvalues of A in F. Consider the following example.

Example 4.11   Let

$$A = \begin{bmatrix} 0 & -1 \\ 1 & 0 \end{bmatrix} \in M_{2 \times 2}(\mathbb{R})$$

Then $C_A(t) = t^2 + 1$ has no roots in $\mathbb{R}$. Therefore, Theorem 4.9 implies $\mathscr{S}_\mathbb{R}(A) = \varnothing$.

This is one of the few places in the text where the base field F is important. The spectrum of A depends on the field F. Return to Example 4.11, and consider A as a complex matrix in $M_{2 \times 2}(\mathbb{C})$. The characteristic polynomial of A is still $C_A(t) = t^2 + 1$. But, now $C_A(t)$ has two roots i and $-i$ in $\mathbb{C}$. Therefore, $\mathscr{S}_\mathbb{C}(A) = \{i, -i\}$, while $\mathscr{S}_\mathbb{R}(A) = \varnothing$.

The Fundamental Theorem of Algebra says that every nonconstant polynomial $f(t) \in \mathbb{C}[t]$ has a root in $\mathbb{C}$, i.e., there exists a complex number z such that $f(z) = 0$. It follows easily from this statement that any polynomial with real or complex coefficients has all of its roots in $\mathbb{C}$. In particular, if $A \in M_{n \times n}(F)$, then $C_A(t)$ has all of its roots in $\mathbb{C}$. Thus, whether $F = \mathbb{R}$ or $\mathbb{C}$, if A is regarded as a matrix with complex entries, then $\mathscr{S}_\mathbb{C}(A) \neq \varnothing$.

It follows from elementary calculus that any polynomial $f(t) \in \mathbb{R}[t]$ of odd degree has at least one real root. Thus, if $A \in M_{n \times n}(\mathbb{R})$, and n is odd, then A has at least one real eigenvalue (i.e., $\mathscr{S}_\mathbb{R}(A) \neq \varnothing$). If n is even, then we have seen in Example 4.11 that all eigenvalues of A could very well be complex and not real.

There is one more corollary to Theorem 4.9 which is important when discussing linear transformations.

Corollary 4.12 Let $A, B \in M_{n \times n}(F)$. If A and B are similar, then $\mathscr{S}_F(A) = \mathscr{S}_F(B)$.

Proof. If A and B are similar, then there exists an invertible matrix $P \in M_{n \times n}(F)$ such that $P^{-1}AP = B$. Then

$$P^{-1}(tI_n - A)P = P^{-1}(tI_n)P - P^{-1}AP = tP^{-1}P - B = tI_n - B$$

Theorem 2.12 implies

$$C_B(t) = \det(tI_n - B) = \det(P^{-1}(tI_n - A)P)$$

$$= \det(P^{-1})\det(tI_n - A)\det(P) = \det(P)^{-1}\det(P)\det(tI_n - A)$$

$$= \det(tI_n - A) = C_A(t)$$

Since $C_A(t) = C_B(t)$, $\mathscr{S}_F(A) = \mathscr{S}_F(B)$ by Theorem 4.9. $\qquad\square$

We can use Corollary 4.12 to compute the spectrum of an arbitrary linear transformation T.

Definition 4.13   Let V denote a finite dimensional vector space over F. Suppose $T \in \text{Hom}(V, V)$. A scalar $d \in F$ is called an eigenvalue of T if $T(\xi) = d\xi$ for some nonzero vector $\xi \in V$. The set of eigenvalues of T in F is called the spectrum of T (in F).

We will let $\mathscr{S}_F(T)$ denote the spectrum of T. Thus, $\mathscr{S}_F(T) = \{d \in F \mid d$ is an eigenvalue of $T\}$. We have seen in equation (5.27) of Chapter II that any two matrix representations of T are similar. Thus, by Corollary 4.12, T determines a unique set of numbers $\mathscr{S}_F(A)$ where A is any matrix representation of T. It is easy to see (using diagram (5.9) in Chapter II) that $\mathscr{S}_F(T) = \mathscr{S}_F(A)$ for any matrix representation A of T. Thus, the spectrum of a linear transformation T is computed from any matrix representation of T.

The entire theory of eigenvalues can be discussed on the linear transformation level or on the matrix level. In this book, we will treat the theory of eigenvalues from the matrix point of view. In $[2, 3, 5]$, the discussion of eigenvalues is primarily on the linear transformation level. No matter which approach is taken, the way one computes the eigenvalues of a linear transformation is to compute the roots of the characteristic polynomial of a matrix representation of the linear transformation. So we eventually return to matrices no matter what the setting of the original problem.

Suppose $d \in \mathscr{S}_F(A)$ and $A\xi = d\xi$ for some nonzero vector $\xi \in F^n$. The vector ξ is called an eigenvector of A. More precisely, we have the following definition.

Definition 4.14   Let $A \in M_{n \times n}(F)$. A nonzero vector $\xi \in F^n$ is called an eigenvector (or characteristic vector) of A belonging to d if $A\xi = d\xi$.

Thus, if $d \in \mathscr{S}_F(A)$, any nonzero vector in $NS(dI_n - A)$ is an eigenvector of A belonging to d. Notice that eigenvectors are always nonzero. If $\mathscr{S}_F(A) = \varnothing$, then of course A has no eigenvectors. Consider the following examples.

Example 4.15

(a)   Suppose $A = O$. Then $C_A(t) = t^n$ and $\mathscr{S}_F(A) = \{0\}$. Clearly, any nonzero vector in F^n is an eigenvector of A (belonging to 0) since $A\xi = O = 0\xi$.

(b)   Suppose $A = I_n$. Then $C_A(t) = (t - 1)^n$ and $\mathscr{S}_F(A) = \{1\}$. Any nonzero vector in F^n is an eigenvector of I_n (belonging to 1) since $I_n\xi = \xi = 1\xi$.

(c) Suppose $A = \text{Diag}(d_1, \ldots, d_n)$. Then $C_A(t) = \prod_{i=1}^{n} (t - d_i)$ and $\mathscr{S}_F(A) = \{d_1, \ldots, d_n\}$ (The d_i need not be distinct here). Let $\underline{\varepsilon} = \{\varepsilon_1, \ldots, \varepsilon_n\}$ be the canonical basis of F^n. Then $A\varepsilon_i = \text{Col}_i(A) = d_i\varepsilon_i$ for all $i = 1, \ldots, n$. Thus, ε_i is an eigenvector of A belonging to d_i for each $i = 1, \ldots, n$.

Example 4.16   Let

$$A = \begin{bmatrix} 0 & -1 \\ 1 & 0 \end{bmatrix}$$

If $F = \mathbb{R}$, then $A \in M_{2 \times 2}(\mathbb{R})$. We have seen in Example 4.11 that $\mathscr{S}_\mathbb{R}(A) = \varnothing$, and, consequently, A has no eigenvectors in $\mathbb{R}^2$.

If $F = \mathbb{C}$, then $A \in M_{2 \times 2}(\mathbb{C})$. In this case, $\mathscr{S}_\mathbb{C}(A) = \{i, -i\}$. Since

$$\begin{bmatrix} 0 & -1 \\ 1 & 0 \end{bmatrix} \begin{bmatrix} 1 \\ -i \end{bmatrix} = i \begin{bmatrix} 1 \\ -i \end{bmatrix}$$

$\begin{bmatrix} 1 \\ -i \end{bmatrix}$ is an eigenvector of A belonging to i

Since

$$\begin{bmatrix} 0 & -1 \\ 1 & 0 \end{bmatrix} \begin{bmatrix} -1 \\ -i \end{bmatrix} = -i \begin{bmatrix} -1 \\ -i \end{bmatrix}$$

$\begin{bmatrix} -1 \\ -i \end{bmatrix}$ is an eigenvector of A belonging to $-i$

Example 4.17   Let

$$A = \begin{bmatrix} 0 & 0 & 6 \\ 1 & 0 & -11 \\ 0 & 1 & 6 \end{bmatrix} \in M_{3 \times 3}(\mathbb{R})$$

It is easy to check that

$$C_A(t) = t^3 - 6t^2 + 11t - 6 = (t - 1)(t - 2)(t - 3).$$

Therefore, $\mathscr{S}_\mathbb{R}(A) = \{1, 2, 3\}$.

If $d = 1$, then

$$dI_3 - A = \begin{bmatrix} 1 & 0 & -6 \\ -1 & 1 & 11 \\ 0 & -1 & -5 \end{bmatrix} \underset{r}{\sim} \begin{bmatrix} 1 & 0 & -6 \\ 0 & 1 & 5 \\ 0 & 0 & 0 \end{bmatrix}$$

Therefore, $NS(I_3 - A) = L([6, -5, 1]^t)$. In particular

$$\xi_1 = \begin{bmatrix} 6 \\ -5 \\ 1 \end{bmatrix}$$

is an eigenvector of A belonging to 1.

If $d = 2$, then

$$dI_3 - A = \begin{bmatrix} 2 & 0 & -6 \\ -1 & 2 & 11 \\ 0 & -1 & -4 \end{bmatrix} \underset{r}{\sim} \begin{bmatrix} 1 & 0 & -3 \\ 0 & 1 & 4 \\ 0 & 0 & 0 \end{bmatrix}$$

Therefore, $NS(2I_3 - A) = L([3, -4, 1]^t)$. In particular

$$\xi_2 = \begin{bmatrix} 3 \\ -4 \\ 1 \end{bmatrix}$$

is an eigenvector of A belonging to 2.

If $d = 3$, then

$$dI_3 - A = \begin{bmatrix} 3 & 0 & -6 \\ -1 & 3 & 11 \\ 0 & -1 & -3 \end{bmatrix} \underset{r}{\sim} \begin{bmatrix} 1 & 0 & -2 \\ 0 & 1 & 3 \\ 0 & 0 & 0 \end{bmatrix}$$

Therefore, $NS(3I_3 - A) = L([2, -3, 1]^t)$. In particular,

$$\xi_3 = \begin{bmatrix} 2 \\ -3 \\ 1 \end{bmatrix}$$

is an eigenvector of A belonging to 3.

The three vectors ξ_1, ξ_2, and ξ_3 in Example 4.17 are linearly independent, and hence form a basis for F^3. Our next theorem implies the matrix A in Example 4.17 is similar to the diagonal matrix $D = \text{Diag}(1, 2, 3)$.

Theorem 4.18   Let $A \in M_{n \times n}(F)$. A is similar to a diagonal matrix if and only if F^n has a basis consisting of eigenvectors of A.

Proof.   Suppose A is similar to a diagonal matrix $D = \text{Diag}(d_1, \ldots, d_n)$.

Then there exists an invertible matrix $P \in M_{n \times n}(F)$ such that $P^{-1}AP = D$. Let $P = [\xi_1 | \xi_2 | \cdots | \xi_n]$ be the column partition of P. Since P is invertible, $\text{rk}(P) = n$. Thus, the vectors $\xi_1, \ldots, \xi_n$ are linearly independent in F^n. In particular, $\{\xi_1, \ldots, \xi_n\}$ is a basis of F^n.

By Theorem I.3.10, $AP = A[\xi_1 | \xi_2 | \cdots | \xi_n] = [A\xi_1 | A\xi_2 | \cdots | A\xi_n]$. On the other hand,

$$AP = PD = [\xi_1 | \cdots | \xi_n] \text{Diag}(d_1, \ldots, d_n) = [d_1\xi_1 | \cdots | d_n\xi_n].$$

Thus, $A\xi_i = d_i\xi_i$ for all $i = 1, \ldots, n$. Hence, F^n has a basis $\{\xi_1, \ldots, \xi_n\}$ consisting of eigenvectors of A.

Conversely, suppose F^n has a basis $\{\xi_1, \ldots, \xi_n\}$ consisting of eigenvectors of A. Suppose $A\xi_i = d_i\xi_i$ for all $i = 1, \ldots, n$. Set $P = [\xi_1 | \xi_2 | \cdots | \xi_n]$. Since $\xi_1, \ldots, \xi_n$ are linearly independent, $\text{rk}(P) = n$. Therefore, P is invertible.

$$AP = A[\xi_1 | \cdots | \xi_n] = [A\xi_1 | \cdots | A\xi_n] = [d_1\xi_1 | \cdots | d_n\xi_n] = [\xi_1 | \cdots | \xi_n]$$
$$\text{Diag}(d_1, \ldots, d_n)$$

Therefore, $P^{-1}AP = \text{Diag}(d_1, \ldots, d_n)$ and A is similar to a diagonal matrix. $\qquad \square$

Let us summarize the important parts of the computation in the proof of Theorem 4.18.

4.19 Let $A \in M_{n \times n}(F)$. If $\xi_1, \ldots, \xi_n$ are eigenvectors of A belonging to $d_1, \ldots, d_n$, respectively, and $P = [\xi_1 | \xi_2 | \cdots | \xi_n]$ is invertible, then $P^{-1}AP = \text{Diag}(d_1, \ldots, d_n)$.

Let us return to Examples 4.16 and 4.17 for an illustration of these ideas.

Example 4.20 Let

$$A = \begin{bmatrix} 0 & -1 \\ 1 & 0 \end{bmatrix} \in M_{2 \times 2}(\mathbb{C})$$

We have seen in Example 4.16 that

$$\xi_1 = \begin{bmatrix} 1 \\ -i \end{bmatrix} \quad \text{and} \quad \xi_2 = \begin{bmatrix} -1 \\ -i \end{bmatrix}$$

are eigenvectors of A belonging to i and $-i$, respectively. The vectors ξ_1 and ξ_2 are obviously linearly independent. Hence, the matrix

$$P = \begin{bmatrix} 1 & -1 \\ -i & -i \end{bmatrix}$$

is invertible. The inverse of P is easily seen to be

$$P^{-1} = \begin{bmatrix} 1/2 & i/2 \\ -1/2 & i/2 \end{bmatrix}$$

Our remarks in (4.19) imply

$$\begin{bmatrix} 1/2 & i/2 \\ -1/2 & i/2 \end{bmatrix}\begin{bmatrix} 0 & -1 \\ 1 & 0 \end{bmatrix}\begin{bmatrix} 1 & -1 \\ -i & -i \end{bmatrix} = \begin{bmatrix} i & 0 \\ 0 & -i \end{bmatrix} = \mathrm{Diag}(i, -i).$$

Example 4.21   Let

$$A = \begin{bmatrix} 0 & 0 & 6 \\ 1 & 0 & -11 \\ 0 & 1 & 6 \end{bmatrix}$$

We have seen in Example 4.17 that $\xi_1 = [6 \; -5 \; 1]^t$, $\xi_2 = [3 \; -4 \; 1]^t$, and $\xi_3 = [2 \; -3 \; 1]^t$ are a basis of F^3 consisting of eigenvectors of A belonging to 1, 2, and 3 respectively. Thus,

$$P = \begin{bmatrix} 6 & 3 & 2 \\ -5 & -4 & -3 \\ 1 & 1 & 1 \end{bmatrix}$$

is invertible. Our remarks in (4.19) imply $P^{-1}AP = \mathrm{Diag}(1, 2, 3)$.

It is often difficult to tell when a matrix A is similar to a diagonal matrix. Our next theorem leads to some sufficient conditions on A which imply A is similar to a diagonal matrix.

Theorem 4.22   Let $A \in M_{n \times n}(F)$. Suppose $d_1, \ldots, d_r$ are distinct eigenvalues of A (in F). Let $\xi_1, \ldots, \xi_r$ be eigenvectors of A belonging to $d_1, \ldots, d_r$, respectively. Then $\xi_1, \ldots, \xi_r$ are linearly independent.

Proof.   If $r = 1$, then the result is trivial. Hence, we can assume $r > 1$. We suppose $\xi_1, \ldots, \xi_r$ are linearly dependent and derive a contradiction. Suppose $y_1\xi_1 + \cdots + y_r\xi_r = O$ is a nontrivial linear dependence relation among the ξ_i. Among all such nontrivial dependence relations, we can select a relation which has the fewest number of ξ_i appearing in it. After relabeling the ξ_i (and d_i) if need be, we can assume this "smallest" relation is $x_1\xi_1 + x_2\xi_2 + \cdots + x_s\xi_s = O$. Thus, the s here is the smallest number of ξ_i which can appear in any nontrivial linear dependence relation among the ξ_i.

Since the relation $x_1\xi_1 + \cdots + x_s\xi_s = O$ is nontrivial and has the fewest number of ξ_i appearing in it, no x_i is zero. Thus, $x_1 \neq 0, \ldots, x_s \neq 0$. Since no ξ_i is zero, $2 \leqslant s \leqslant r$.

Since $A\xi_i = d_i\xi_i$ for $i = 1, \ldots, s$,

$$O = AO = A(x_1\xi_1 + \cdots + x_s\xi_s) = x_1A\xi_1 + \cdots + x_sA\xi_s$$
$$= (x_1d_1)\xi_1 + \cdots + (x_sd_s)\xi_s.$$

Multiplying the equation $x_1\xi_1 + \cdots + x_s\xi_s = O$ by d_1, gives

$$(x_1d_1)\xi_1 + \cdots + (x_sd_1)\xi_s = O.$$

Thus, we have the following two equations.

4.23
$$(x_1d_1)\xi_1 + (x_2d_2)\xi_2 + \cdots + (x_sd_s)\xi_s = O$$
$$(x_1d_1)\xi_1 + (x_2d_1)\xi_2 + \cdots + (x_sd_1)\xi_s = O$$

Subtracting the two equations in (4.23), we get a new dependence relation, $x_2(d_2 - d_1)\xi_2 + \cdots + x_s(d_s - d_1)\xi_s = O$. Since, $d_1, \ldots, d_s$ are all distinct, no coefficient in this new relation is zero. We have constructed a nontrivial dependence relation in which fewer than s of the ξ_i appear. This is contrary to our definition of s.

We conclude that no linear dependence relation is possible among the vectors $\xi_1, \ldots, \xi_r$. Therefore, $\xi_1, \ldots, \xi_r$ are linearly independent. $\quad\square$

There is one application of Theorem 4.22 which gives sufficient conditions on a matrix A to guarantee that A is similar to a diagonal matrix.

Corollary 4.24 Let $A \in M_{n \times n}(F)$. Suppose $C_A(t)$ has n distinct roots in F, then A is similar to a diagonal matrix.

Proof. Suppose $C_A(t)$ has n distinct roots $d_1, \ldots, d_n$ in F. Then Theorem 4.9 implies $d_1, \ldots, d_n$ are eigenvalues of A. If ξ_i is an eigenvector of A belonging to d_i, then Theorem 4.22 implies $\xi_1, \ldots, \xi_n$ are linearly independent. Set $P = [\xi_1 | \xi_2 | \cdots | \xi_n]$. Then $\mathrm{rk}(P) = n$ and P is invertible. Our remarks in (4.19) imply A is similar to the diagonal matrix $D = \mathrm{Diag}(d_1, \ldots, d_n)$. $\square$

For instance, in Example 4.17, the roots of $C_A(t)$ are 1, 2, and 3. Therefore, A is similar to the diagonal matrix $\mathrm{Diag}(1, 2, 3)$.

We finish this section with a couple of applications of the theorems concerning eigenvalues. Our first application is purely an algebraic one. How do we compute large powers of a matrix? Suppose $A \in M_{n \times n}(F)$. We want to compute $\mathbf{A}^k$ for large values of k. Due to the complexity of matrix multiplication, this is not an easy problem. However, if A is similar to a diagonal matrix D, then this problem is easy to solve. Suppose $P \in M_{n \times n}(F)$ is an invertible matrix such that $P^{-1}AP = D = \mathrm{Diag}(d_1, \ldots, d_n)$. Then $A = PDP^{-1}$ and

$$A^k = (PDP^{-1})^k = (PDP^{-1})(PDP^{-1}) \cdots (PDP^{-1}) = PD^k P^{-1}$$

Since D is a diagonal matrix, $D^k = \mathrm{Diag}(d_1^k, d_2^k, \ldots, d_n^k)$. We have established the following result.

4.25   If $P^{-1}AP = \mathrm{Diag}(d_1, \ldots, d_n)$, then for any integer $k \geqslant 0$,

$$A^k = P\,\mathrm{Diag}(d_1^k, d_2^k, \ldots, d_n^k)P^{-1}$$

Of course, we use the material developed in this section to compute P when it exists. Consider the following example.

Example 4.26   Let

$$A = \begin{bmatrix} 0 & 0 & 6 \\ 1 & 0 & -11 \\ 0 & 1 & 6 \end{bmatrix}$$

Compute A^{10}. We have seen in Examples 4.17 and 4.21 that A is similar to the diagonal matrix $D = \mathrm{Diag}(1, 2, 3)$.

$$P = \begin{bmatrix} 6 & 3 & 2 \\ -5 & -4 & -3 \\ 1 & 1 & 1 \end{bmatrix} \quad \text{and} \quad P^{-1} = \begin{bmatrix} 1/2 & 1/2 & 1/2 \\ -1 & -2 & -4 \\ 1/2 & 3/2 & 9/2 \end{bmatrix}$$

Thus, equation (4.25) implies

$$A^{10} = \begin{bmatrix} 6 & 3 & 2 \\ -5 & -4 & -3 \\ 1 & 1 & 1 \end{bmatrix} \begin{bmatrix} 1 & 0 & 0 \\ 0 & 2^{10} & 0 \\ 0 & 0 & 3^{10} \end{bmatrix} \begin{bmatrix} 1/2 & 1/2 & 1/2 \\ -1 & -2 & -4 \\ 1/2 & 3/2 & 9/2 \end{bmatrix}$$

$$= \begin{bmatrix} 55{,}980 & 171{,}006 & 519{,}156 \\ -84.480 & -257{,}531 & -780{,}780 \\ 28{,}501 & 86{,}526 & 261{,}625 \end{bmatrix}$$

There are many types of iteration problems in which A^k must be computed for all $k \geqslant 1$. We will give one example from the theory of linear difference equations. Suppose $S_1, \ldots, S_n$ are n sequences of real (or complex) numbers. The terms of the ith sequence S_i are labeled as follows:

$$S_i = \{x_{i0}, x_{i1}, x_{i2}, \ldots\} = \{x_{ik}\}_{k=0}^{\infty}.$$

Suppose for every $k \geqslant 1$, the kth terms of $S_1, \ldots, S_n$ are linearly dependent on the $(k-1)$th terms of $S_1, \ldots, S_n$ as follows.

4.27
$$\begin{aligned}
x_{1k} &= a_{11}x_{1(k-1)} &+& \quad a_{12}x_{2(k-1)} + \cdots + a_{1n}x_{n(k-1)} \\
x_{2k} &= a_{21}x_{1(k-1)} &+& \quad a_{22}x_{2(k-1)} + \cdots + a_{2n}x_{n(k-1)} \\
&\;\vdots & & \qquad\quad \vdots \\
x_{nk} &= a_{n1}x_{1(k-1)} &+& \quad a_{n2}x_{2(k-1)} + \cdots + a_{nn}x_{n(k-1)}
\end{aligned}$$

In the equations in (4.27), the a_{ij} are fixed constants from F which do not depend on k. The problem is to find n sequences $S_1, \ldots, S_n$ whose terms satisfy the equations in (4.27), and whose initial terms, $x_{10}, \ldots, x_{n0}$, are some given values $b_1, \ldots, b_n$, respectively. These types of problems are called linear difference equations.

As the reader might expect, the matrix $A = (a_{ij})$ of coefficients in (4.27) holds the key to whether the problem is solvable. If the matrix A is similar to a diagonal matrix D, then we can find a solution to the linear difference equations. To see how this is done, set $X_k = [x_{1k} \cdots x_{nk}]^t$ for all $k \geqslant 1$ and $B = [b_1 \cdots b_n]^t$. Then the equations in (4.27) can be rewritten in the following matrix form.

4.28
$$\begin{aligned} X_k &= AX_{k-1} \quad \text{for all } k \geqslant 1 \\ X_0 &= B \end{aligned}$$

The problem is then to find X_k for all $k \geqslant 1$. This is easily done. Equation (4.28) implies $X_0 = B$, $X_1 = AX_0 = AB$, $X_2 = AX_1 = A^2B$, etc.

In general, $X_k = A^k B$ for all $k \geqslant 0$. The real problem is how to compute $A^k B$ for all k? We have seen how to make this computation if A is similar to a diagonal matrix. If $P^{-1}AP = \text{Diag}(d_1, \ldots, d_n)$, then $A^k = P\,\text{Diag}(d_1^k, \ldots, d_n^k)P^{-1}$ by (4.25). In this case, a solution to (4.27) is as follows.

4.29 $$X_k = P\,\text{Diag}(d_1^k, \ldots, d_n^k)P^{-1}B \quad \text{for all } k \geqslant 0$$

The formula for X_k in equation (4.29) is a solution to the finite difference equations in (4.27) when A is similar to the diagonal matrix $\text{Diag}(d_1, \ldots, d_n)$. If A is not similar to a diagonal matrix, then different techniques will have to be employed to compute $A^k B$ for all $k \geqslant 0$.

Example 4.30 Find two sequences $S_1 = \{x_{1k}\}_{k=0}^{\infty}$ and $S_2 = \{x_{2k}\}_{k=0}^{\infty}$ such that

$$\begin{aligned} x_{1k} &= 2x_{1(k-1)} - x_{2(k-1)} \\ x_{2k} &= -x_{1(k-1)} + 2x_{2(k-1)} \end{aligned} \quad \text{for all } k \geqslant 1 \qquad (*)$$

and $x_{10} = 3$, $x_{20} = -1$.

The matrix of $(*)$ is

$$A = \begin{bmatrix} 2 & -1 \\ -1 & 2 \end{bmatrix}.$$

$C_A(t) = t^2 - 4t + 3$. Therefore, $\mathscr{S}_F(A) = \{1, 3\}$. Thus, Corollary 4.24 implies A is similar to $\text{Diag}(1, 3)$. It is easy to see that $\xi_1 = [1 \ \ 1]^t$ is an eigenvector of A belonging to 1. $\xi_2 = [1 \ -1]^t$ is an eigenvector of A belonging to 3. Thus, if

$$P = \begin{bmatrix} 1 & 1 \\ 1 & -1 \end{bmatrix}$$

our remarks in (4.19) imply $P^{-1}AP = \text{Diag}(1, 3)$. Then

$$P^{-1} = \begin{bmatrix} 1/2 & 1/2 \\ 1/2 & -1/2 \end{bmatrix}$$

The solution to $(*)$ is given by equation (4.29): $X_k = P\,\text{Diag}(1, 3^k)$ $P^{-1}[3 \ -1]^t$. Hence

$$\begin{bmatrix} x_{1k} \\ x_{2k} \end{bmatrix} = \begin{bmatrix} 1 & 1 \\ 1 & -1 \end{bmatrix}\begin{bmatrix} 1 & 0 \\ 0 & 3^k \end{bmatrix}\begin{bmatrix} 1/2 & 1/2 \\ 1/2 & -1/2 \end{bmatrix}\begin{bmatrix} 3 \\ -1 \end{bmatrix} = \begin{bmatrix} 1 + 2(3^k) \\ 1 - 2(3^k) \end{bmatrix}$$

In other words, $S_1 = \{1 + 2(3^k)\}_{k=0}^{\infty}$ and $S_2 = \{1 - 2(3^k)\}_{k=0}^{\infty}$.

Our second application of eigenvalues is in differential equations. Suppose $f_1(x), \ldots, f_n(x)$ are continuously differentiable functions on an open interval $I \subseteq \mathbb{R}$. We assume $0 \in I$. Suppose the derivatives $f'_1, \ldots, f'_n$ satisfy the following linear system of equations.

4.31

$$
\begin{aligned}
f'_1 &= a_{11}f_1 + a_{12}f_2 + \cdots + a_{1n}f_n \\
f'_2 &= a_{21}f_1 + a_{22}f_2 + \cdots + a_{2n}f_n \\
&\vdots \qquad \vdots \qquad \vdots \qquad \vdots \\
f'_n &= a_{n1}f_1 + a_{n2}f_2 + \cdots + a_{nn}f_n
\end{aligned} \qquad \text{for all } x \in I
$$

The coefficients a_{ij} in (4.31) are fixed constants in $\mathbb{R}$. The problem is to find a solution $f_1, \ldots, f_n$ to the equations in (4.31) which satisfies some given initial condition $f_1(0) = b_1, \ldots, f_n(0) = b_n$.

If the matrix $A = (a_{ij})$ is similar to a diagonal matrix $D = \mathrm{Diag}(d_1, \ldots, d_n)$, then we can easily construct a solution to (4.31). We need to introduce some notation. Suppose $Y(x) = [g_1(x) \cdots g_n(x)]^t$ is a column vector whose components are differentiable functions $g_1(x), \ldots, g_n(x)$ on I. Define the derivative dY/dx and the value $Y(0)$ of $Y(x)$ at zero as follows.

4.32

$$
\frac{dY}{dx} = \begin{bmatrix} g'_1 \\ g'_2 \\ \vdots \\ g'_n \end{bmatrix} \quad \text{and} \quad Y(0) = \begin{bmatrix} g_1(0) \\ g_2(0) \\ \vdots \\ g_n(0) \end{bmatrix}
$$

Set $B = [b_1 \cdots b_n]^t$ and $Y(x) = [f_1(x) \cdots f_n(x)]^t$. Then the system of equations in (4.31) becomes the following problem.

4.33 Solve $dY/dx = AY$ with $Y(0) = B$.

Suppose $\xi \in F^n$ is an eigenvector of A belonging to d. Then $Y(x) = e^{dx}\xi$ is a solution to the differential equation $dY/dx = AY$. To see this, suppose $\xi = [c_1 \cdots c_n]^t$. Then $Y(x) = [c_1 e^{dx} \cdots c_n e^{dx}]^t$. Therefore,

$$
dY/dx = [c_1 d e^{dx} \cdots c_n d e^{dx}]^t = d e^{dx}\xi.
$$

On the other hand,

$$
AY(x) = A(e^{dx}\xi) = e^{dx}A\xi = e^{dx}d\xi.
$$

In particular, $dY/dx = AY(x)$.

If A is similar to $D = \mathrm{Diag}(d_1, \ldots, d_n)$, there exists an invertible matrix P such that $P^{-1}AP = D$. If $P = [\xi_1 | \cdots | \xi_n]$, then we have seen in the proof of Theorem 4.18 that $A\xi_i = d_i\xi_i$ for all $i = 1, \ldots, n$. Thus, $\{\xi_1, \ldots, \xi_n\}$ is a basis of F^n consisting of eigenvectors of A. We can now present a solution to the problem in (4.33).

4.34
$$Y(x) = \sum_{i=1}^{n} z_i e^{d_i x}\xi_i$$

where $[z_1 \cdots z_n]^t = P^{-1}B$ is a solution to (4.33).

To see this, first observe that $[z_1 \cdots z_n]^t = P^{-1}B$ implies that

$$B = P[z_1 \cdots z_n]^t = z_1\xi_1 + \cdots + z_n\xi_n.$$

Therefore, $Y(0) = \sum_{i=1}^{n} z_i\xi_i = B$ as required. We also have

$$dY/dx = \sum_{i=1}^{n} z_i\, d/dx(e^{d_i x}\xi_i).$$

Since ξ_i is an eigenvalue of A belonging to d_i, we have seen that

$$d/dx(e^{d_i x}\xi_i) = A(e^{d_i x}\xi_i).$$

Therefore,

$$dY/dx = \sum_{i=1}^{n} z_i A(e^{d_i x}\xi_i) = A\left(\sum_{i=1}^{n} z_i e^{d_i x}\xi_i\right) = AY(x)$$

Consider the following example.

Example 4.35 Solve the following system of differential equations:

$$\begin{array}{lll} f'_1 = 2f_1 + f_2 + f_3 & & f_1(0) = 1 \\ f'_2 = 2f_1 + 3f_2 + 2f_3 & \text{with} & f_2(0) = 1 \\ f'_3 = f_1 + f_2 + 2f_3 & & f_3(0) = -1 \end{array} \qquad (*)$$

The matrix of coefficients in $(*)$ is

$$A = \begin{bmatrix} 2 & 1 & 1 \\ 2 & 3 & 2 \\ 1 & 1 & 2 \end{bmatrix}$$

Then

$$C_A(t) = t^3 - 7t^2 + 11t - 5$$
$$\mathscr{S}_\mathbb{R}(A) = \{1, 5\}$$

Since A has only two distinct eigenvalues, we cannot conclude so quickly that A is similar to a diagonal matrix. We must compute all eigenvectors of A.

If $d = 1$, then

$$I_3 - A = \begin{bmatrix} -1 & -1 & -1 \\ -2 & -2 & -2 \\ -1 & -1 & -1 \end{bmatrix} \underset{r}{\sim} \begin{bmatrix} 1 & 1 & 1 \\ 0 & 0 & 0 \\ 0 & 0 & 0 \end{bmatrix}$$

Therefore, $NS(I_3 - A) = L(\xi_1 = [1 \ -1 \ 0]^t, \ \xi_2 = [1 \ 0 \ -1]^t)$. In particular, ξ_1 and ξ_2 are two linearly independent eigenvectors of A belonging to 1.

If $d = 5$, then

$$5I_3 - A = \begin{bmatrix} 3 & -1 & -1 \\ -2 & 2 & -2 \\ -1 & -1 & 3 \end{bmatrix} \underset{r}{\sim} \begin{bmatrix} 1 & 0 & -1 \\ 0 & 1 & -2 \\ 0 & 0 & 0 \end{bmatrix}$$

Therefore, $NS(5I_3 - A) = L(\xi_3 = [1 \ 2 \ 1]^t)$ and ξ_3 is an eigenvector of A belonging to 5.

It is easy to check that ξ_1, ξ_2, ξ_3 are linearly independent, and consequently, a basis of F^3. Thus, our remarks in (4.19) imply $P^{-1}AP = D = \text{Diag}(1, 1, 3)$ where

$$P = [\xi_1 \,|\, \xi_2 \,|\, \xi_3] = \begin{bmatrix} 1 & 1 & 1 \\ -1 & 0 & 2 \\ 0 & -1 & 1 \end{bmatrix}$$

and

$$P^{-1} = \begin{bmatrix} 1/2 & -1/2 & 1/2 \\ 1/4 & 1/4 & -3/4 \\ 1/4 & 1/4 & 1/4 \end{bmatrix}$$

We can now compute the solution to $(*)$ given in equation (4.34). The column vector $Z = [z_1 \ z_2 \ z_3]^t$ is given by

$$Z = P^{-1}B = \begin{bmatrix} 1/2 & -1/2 & 1/2 \\ 1/4 & 1/4 & -3/4 \\ 1/4 & 1/4 & 1/4 \end{bmatrix} \begin{bmatrix} 1 \\ 1 \\ -1 \end{bmatrix} = \begin{bmatrix} -1/2 \\ 5/4 \\ 1/4 \end{bmatrix}$$

Thus

$$Y(x) = (-1/2)e^x \begin{bmatrix} 1 \\ -1 \\ 0 \end{bmatrix} + (5/4)e^x \begin{bmatrix} 1 \\ 0 \\ -1 \end{bmatrix} + (1/4)e^{5x} \begin{bmatrix} 1 \\ 2 \\ 1 \end{bmatrix}$$

Therefore

$$\begin{aligned} f_1(x) &= (3/4)e^x + (1/4)e^{5x} \\ f_2(x) &= (1/2)e^x + (1/2)e^{5x} \\ f_3(x) &= (-5/4)e^x + (1/4)e^{5x} \end{aligned}$$

One final word about these applications is in order here. Our solutions to the problems in (4.28) and (4.33) are based upon the assumption that the coefficient matrix A is similar to a diagonal matrix. The problems make sense for any square matrix A. If A is not similar to a diagonal matrix, then different methods must be employed to solve the problem. For example, the reader can consult [2] or [4] for a complete solution to (4.33) in all cases.

Exercises for Section 4

4.1   Find all the eigenvalues and eigenvectors for the following matrices:

(a) $\begin{bmatrix} -38 & 20 \\ -63 & 33 \end{bmatrix}$

(b) $\begin{bmatrix} 8 & -3 \\ 15 & -6 \end{bmatrix}$

(c) $\begin{bmatrix} 11 & -2 & -6 \\ 8 & 0 & -5 \\ 17 & -4 & -9 \end{bmatrix}$

(d) $\begin{bmatrix} -2 & -1 & 0 \\ 0 & 1 & 1 \\ -2 & -2 & -1 \end{bmatrix}$

(e) $\begin{bmatrix} 1 & 0 & 0 & 0 \\ -1 & 2 & 0 & 0 \\ 0 & 0 & 0 & -1 \\ 0 & 0 & 1 & 1 \end{bmatrix}$

4.2   Let $f(t) \in F[t]$ and suppose $f(t)$ has degree $n \geqslant 1$. Show f can have at most n distinct roots in $\mathbb{C}$. Can f have less than n roots?

4.3   Let

$$
A = \begin{bmatrix}
0 & 0 & 0 & \cdots & 0 & 0 \\
1 & 0 & 0 & \cdots & 0 & 0 \\
0 & 1 & 0 & \cdots & 0 & 0 \\
\vdots & \vdots & \vdots & & \vdots & \vdots \\
0 & 0 & 0 & \cdots & 1 & 0
\end{bmatrix} \in M_{n \times n}(F)
$$

Compute $C_A(t)$ and $\mathscr{S}_F(A)$.

4.4   Let $A \in M_{n \times n}(F)$. Show that $0 \in \mathscr{S}_F(A)$ if and only if A is singular.

4.5   Let V denote a finite dimensional vector space over F, and let $T \in \mathrm{Hom}(V, V)$. Show that $\mathscr{S}_F(T) = \mathscr{S}_F(A)$ for any matrix representation A of T.

4.6   Let $D: \mathscr{P}_n \mapsto \mathscr{P}_n$ be ordinary differentiation. Thus, $D(p) = dp/dt$. Compute $\mathscr{S}_\mathbb{R}(D)$ and determine all eigenvectors of D (if any).

4.7   Let $T: \mathbb{R}^2 \mapsto \mathbb{R}^2 \mapsto$ be the linear transformation given in Exercise 5.17 of Chapter II. Compute $\mathscr{S}_\mathbb{R}(T)$ and all eigenvectors of T.

4.8   Let $T: M_{n \times n} \mapsto M_{n \times n}$ be given by $T(A) = A^t$. Compute $\mathscr{S}_F(T)$ and all eigenvectors of T.

4.9   Show that the converse of Corollary 4.24 is false.

4.10 Give an example of an $n \times n$ matrix A such that $C_A(t)$ has all of its roots in F, but A is not similar to any diagonal matrix.

4.11 Show that any solution to (4.28) is necessarily unique.

4.12 Is the solution given in (4.34) unique?

4.13 Check the answers given in Examples 4.30 and 4.35 by direct substitution into the original equations.

4.14 Compute

$$
\begin{bmatrix} 0 & -1 \\ 1 & 0 \end{bmatrix}^{25}
$$

using the information in Example 4.16. Can you compute

$$
\begin{bmatrix} 0 & -1 \\ 1 & 0 \end{bmatrix}^{k}
$$

for any k?

4.15 Solve the following linear difference equations:

(a) $x_{1k} = 8x_{1(k-1)} - 3x_{2(k-1)}$ with $x_{10} = 1 = x_{20}$

$x_{2k} = 15x_{1(k-1)} - 6x_{2(k-1)}$

(b) $x_{1k} = 11x_{1(k-1)} - 2x_{2(k-1)} - 6x_{3(k-1)}$ with $x_{10} = 1$

$x_{2k} = 8x_{1(k-1)} \phantom{- 2x_{2(k-1)}} - 5x_{3(k-1)}$ $x_{20} = -1$

$x_{3k} = 17x_{1(k-1)} - 4x_{2(k-1)} - 9x_{3(k-1)}$ $x_{30} = 2$

(c) $x_{1k} = -2x_{1(k-1)} - x_{2(k-1)}$ $x_{10} = -1$

$x_{2k} = \phantom{-2x_{1(k-1)}} x_{2(k-1)} + x_{3(k-1)}$ with $x_{20} = -2$

$x_{3k} = -2x_{1(k-1)} - 2x_{2(k-1)} - x_{3(k-1)}$ $x_{30} = 1$

4.16 Solve the following systems of differential equations:

(a) $f_1' = -38f_1 + 20f_2$ with $f_1(0) = f_2(0) = 2$

$f_2' = -63f_1 + 33f_2$

(b) $x_1' = 4x_1 - x_2 + x_3$ $x_1(0) = 1$

$x_2' = -x_1 + 4x_2 - x_3$ with $x_2(0) = 1$

$x_3' = x_1 - x_2 + 4x_3$ $x_3(0) = 1$

(c) $u' = u + 2v - w$ $u(0) = 2$

$v' = 2u + v - w$ with $v(0) = 1$

$w' = -u - v + 2w$ $w(0) = -3$

4.17 Let $M_{n \times n}(F[t])$ denote the set of all $n \times n$ matrices whose entries are polynomials in $F[t]$.

(a) Show that $M_{n \times n}(F[t])$ is a vector space over F with addition and scalar multiplication defined as follows:

$$(f_{ij}(t)) + (g_{ij}(t)) = (f_{ij}(t) + g_{ij}(t))$$
$$c(f_{ij}(t)) = (cf_{ij}(t))$$

(b) Is $M_{n \times n}(F[t])$ a finite dimensional vector space over F?

(c) We can define scalar multiplication between vectors in $M_{n \times n}(F[t])$ and polynomials in $F[t]$ in the obvious way: $f(t)(g_{ij}(t)) = (f(t)g_{ij}(t))$. Show that axioms V5 through V8 (Chapter II, Section 1) are satisfied for this scalar multiplication.

(d) Which of the theorems and corollaries in Sections 2 and 3 of this chapter remain true for the scalar multiplication introduced above in (c)?

(e) If $A = (g_{ij}(t)) \in M_{n \times n}(F[t])$ and $\det(A) \neq 0$, can we conclude that A is invertible, i.e., there exists a matrix $B \in M_{n \times n}(F[t])$ such that $AB = BA = I_n$?

Determinants   **215**

4.18   Let $f(t) = t^n - b_{n-1}t^{n-1} - \cdots - b_1 t - b_0 \in F[t]$. Let

$$A = \begin{bmatrix} 0 & 0 & \cdots & 0 & b_0 \\ 1 & 0 & \cdots & 0 & b_1 \\ 0 & 1 & \cdots & 0 & b_2 \\ \vdots & \vdots & & \vdots & \vdots \\ 0 & 0 & \cdots & 1 & b_{n-1} \end{bmatrix}$$

Show that $C_A(t) = f(t)$. Thus, every monic polynomial is the characteristic polynomial of some matrix A. The matrix A here is called the companion matrix of $f(t)$.

4.19   Let

$$A = \begin{bmatrix} A_{11} & \cdots & 0 \\ \vdots & & \vdots \\ 0 & \cdots & A_{nn} \end{bmatrix}$$

Here $A_{11}, A_{22}, \ldots, A_{nn}$ are square matrices of various sizes. Show that $C_A(t) = C_{A_{11}}(t) C_{A_{22}}(t) \cdots C_{A_{nn}}(t)$.

IV
INNER PRODUCT SPACES

1. REAL INNER PRODUCT SPACES

In this chapter, we will explore some of the extra structures which are present on a vector space V. An inner product on V is one of the most useful tools in linear algebra. Since the definitions for an inner product are somewhat different for $\mathbb{R}$ and $\mathbb{C}$, we will divide our treatment of this subject into two parts. In this section, we will discuss real inner product spaces. In the next section, we will treat complex inner product spaces.

Throughout this section, V will denote a real vector space. Thus, V is a vector space over $\mathbb{R}$. An inner product on V is the natural generalization of the familiar "dot product" on $\mathbb{R}^n$.

Definition 1.1 An inner product on V is a function $f : V \times V \mapsto \mathbb{R}$ which satisfies the following properties:

(a) $f(x\alpha + y\beta, \gamma) = xf(\alpha, \gamma) + yf(\beta, \gamma)$.
(b) $f(\gamma, x\alpha + y\beta) = xf(\gamma, \alpha) + yf(\gamma, \beta)$.
(c) $f(\alpha, \beta) = f(\beta, \alpha)$.
(d) $f(\alpha, \alpha) > 0$ for all nonzero vectors $\alpha \in V$.

The conditions (a), (b), and (c) are to hold for all vectors α, β, $\gamma \in V$ and all scalars x, $y \in \mathbb{R}$.

If the reader has worked Exercise 2.5 of Chapter III, then you will recognize an inner product on V is a special type of multilinear function from $V \times V$ to $\mathbb{R}$. Any multilinear function from $V \times V$ to $\mathbb{R}$ is called a bilinear form. If the bilinear form satisfies axiom (c) in Definition 1.1, then the form is said to be symmetric. Thus, an inner product on V is a symmetric, bilinear form which satisfies condition (d).

It is customary when dealing with an inner product f on V to adopt some special notation for the image $f(\alpha, \beta)$ of $(\alpha, \beta) \in V \times V$. We will set $f(\alpha, \beta) = \langle \alpha, \beta \rangle$ for all $\alpha, \beta \in V$. When $f(\alpha, \beta)$ is replaced with the notation $\langle \alpha, \beta \rangle$, then $\langle *, * \rangle$ is a function from $V \times V$ to $\mathbb{R}$ whose value on the ordered pair $(\alpha, \beta) \in V \times V$ is denoted by $\langle \alpha, \beta \rangle$. The function $\langle *, * \rangle$ satisfies the following properties.

1.2   (a) $\langle x\alpha + y\beta, \gamma \rangle = x\langle \alpha, \gamma \rangle + y\langle \beta, \gamma \rangle$.
        (b) $\langle \gamma, x\alpha + y\beta \rangle = x\langle \gamma, \alpha \rangle + y\langle \gamma, \beta \rangle$.
        (c) $\langle \alpha, \beta \rangle = \langle \beta, \alpha \rangle$.
        (d) $\langle \alpha, \alpha \rangle$ is positive for all nonzero vectors $\alpha \in V$.

These equations are to hold for all vectors $\alpha, \beta, \gamma \in V$ and all scalars $x, y \in \mathbb{R}$.

Notice that the axioms listed in (1.2) are not independent of each other. For example, (b) follows from (a) and (c).

A vector space V over $\mathbb{R}$ together with some inner product $\langle *, * \rangle : V \times V \mapsto \mathbb{R}$ will be called a real inner product space. The most important single example of a real inner product space is the familiar one from the calculus.

Example 1.3   Let $V = \mathbb{R}^n$. Define a function $\langle *, * \rangle : \mathbb{R}^n \times \mathbb{R}^n \mapsto \mathbb{R}$ by the following formula: $\langle \alpha, \beta \rangle = \alpha^t \beta$. Here we have identified the 1×1 matrix $\alpha^t \beta$ with its single entry. Thus, $\alpha^t \beta \in \mathbb{R}$. If $\alpha = [a_1 \cdots a_n]^t$, and $\beta = [b_1 \cdots b_n]^t$, then clearly $\langle \alpha, \beta \rangle = \sum_{i=1}^n a_i b_i$. $\langle \alpha, \beta \rangle$ is just the familiar dot product of α and β in $\mathbb{R}^n$.

The fact that $\langle *, * \rangle$ satisfies the axioms listed in (1.2) is a well-known exercise in the calculus. For instance,

$$\langle \alpha, \beta \rangle = \alpha^t \beta = \sum_{i=1}^n a_i b_i = \sum_{i=1}^n b_i a_i = \beta^t \alpha = \langle \beta, \alpha \rangle.$$

Thus, (c) is satisfied. If $\alpha = [a_1 \cdots a_n]^t$, then $\langle \alpha, \alpha \rangle = \sum_{i=1}^n a_i^2 > 0$ if and only if $\alpha \neq 0$. Thus, (d) is satisfied. We leave (a) and (b) as exercises.

We will always refer to the inner product on $\mathbb{R}^n$ given by $\langle \alpha, \beta \rangle = \alpha^t \beta$ as the standard inner product on $\mathbb{R}^n$. The vector space $\mathbb{R}^n$ also has many nonstandard inner products on it. For instance, if $a_1, \ldots, a_n$ are positive numbers, then it is easy to check that

$$\langle [x_1 \cdots x_n]^t, [y_1 \cdots y_n]^t \rangle = \sum_{i=1}^{n} a_i x_i y_i$$

is an inner product on $\mathbb{R}^n$. We get the standard inner product on $\mathbb{R}^n$ when $a_1 = \cdots = a_n = 1$.

Example 1.4   Let $V = \mathbb{R}^2$. Define $\langle *, * \rangle_1 : \mathbb{R}^2 \times \mathbb{R}^2 \mapsto \mathbb{R}$ as follows:

$$\left\langle \begin{bmatrix} a \\ b \end{bmatrix}, \begin{bmatrix} c \\ d \end{bmatrix} \right\rangle_1 = 2ac + ad + bc + bd$$

It is easy to see that $\langle *, * \rangle_1$ satisfies (a), (b), and (c) in (1.2). To see that (d) is satisfied, notice that

$$\left\langle \begin{bmatrix} a \\ b \end{bmatrix}, \begin{bmatrix} a \\ b \end{bmatrix} \right\rangle_1 = 2a^2 + 2ab + b^2 = a^2 + (a + b)^2$$

This is positive if and only if $[a \ b]^t \neq 0$. Thus, $\langle *, * \rangle_1$ is an inner product on $\mathbb{R}^2$.

Examples 1.3 and 1.4 as well as our remarks after Example 1.3 indicate that a given vector space V can have infinitely many different inner products on it. These examples show that it is important to emphasize that a real inner product space is really an ordered pair $(V, \langle *, * \rangle)$ consisting of a real vector space V and some specified function $\langle *, * \rangle : V \times V \mapsto \mathbb{R}$ which satisfies the conditions listed in (1.2). In this book, we will always write an inner product space as an ordered pair $(V, \langle *, * \rangle)$ to remind the reader that $\langle *, * \rangle$ as well as V must be specified.

Our first two examples of real inner product spaces were finite dimensional vector spaces. Most of the interesting examples in analysis are infinite dimensional vector spaces.

Example 1.5   Let $V = C([a, b])$, the vector space of all real valued, continuous functions on the interval $[a, b]$. Define an inner product on V by setting $\langle f, g \rangle = \int_a^b f(t)g(t)\, dt$. It is easy to verify that $\langle f, g \rangle$ satisfies the conditions listed in (1.2). Hence, $(C([a, b]), \langle *, * \rangle)$ is an infinite dimensional, real inner product space.

New examples of real inner product spaces can be obtained from known examples by using injective linear transformations. Suppose V and W are real vector spaces. Let $\langle *, * \rangle$ be an inner product on V and suppose $T: W \mapsto V$ is an injective linear transformation. Then we can define an inner product $\langle *, * \rangle_1$ on W as follows.

1.6 $\langle \alpha, \beta \rangle_1 = \langle T(\alpha), T(\beta) \rangle$ for all $\alpha, \beta \in W$.

Since T is a linear transformation, $\langle *, * \rangle_1$ clearly satisfies (a), (b), and (c) in (1.2). Since T is injective, (d) is also satisfied. Thus, W inherits the structure of an inner product space via the injective, linear transformation T.

The formula given in (1.6) implies that any finite dimensional, real vector space V has an inner product on it. For if B is a basis of V, and $\dim V = n$, then $[*]_B: V \mapsto \mathbb{R}^n$ is an isomorphism. If $\langle *, * \rangle$ denotes the standard inner product on $\mathbb{R}^n$, then $\langle \alpha, \beta \rangle_1 = \langle [\alpha]_B, [\beta]_B \rangle$ defines an inner product on V by (1.6). Consider the following example.

Example 1.7 Let $V = \mathscr{P}_n(\mathbb{R}) = \{a_0 + a_1 t + \cdots + a_n t^n \mid a_i \in \mathbb{R}\}$. We have seen in Chapter II that $B = \{1, t, t^2, \ldots, t^n\}$ is a basis for the vector space V. Thus, $[*]_B: \mathscr{P}_n(\mathbb{R}) \mapsto \mathbb{R}^{n+1}$ is an isomorphism. Let $\langle *, * \rangle$ denote the standard inner product on $\mathbb{R}^{n+1}$. Then our comments after (1.6) imply

$$\langle p(t), q(t) \rangle_1 = \langle [p(t)]_B, [q(t)]_B \rangle$$

is an inner product on $\mathscr{P}_n$.

For instance, if $n = 3$, $p(t) = 1 + t + t^2 - t^3$ and $q(t) = 2 - t - 3t^3$, then

$$\langle p(t), q(t) \rangle_1 = \langle [1 \quad 1 \quad 1 \quad -1]^t, [2 \quad -1 \quad 0 \quad -3]^t \rangle = 4.$$

Example 1.8 Again let $V = \mathscr{P}_n(\mathbb{R})$. Every polynomial in $\mathscr{P}_n(\mathbb{R})$ defines a continuous, real valued function on $\mathbb{R}$. In particular, $\mathscr{P}_n(\mathbb{R}) \subseteq C([0, 1])$. The inclusion map of $\mathscr{P}_n$ into $C([0, 1])$ is clearly an injective, linear transformation. Let $\langle *, * \rangle$ denote the inner product on $C([0, 1])$ given in Example 1.5. Then equation (1.6) implies

$$\langle p(t), q(t) \rangle_2 = \int_0^1 p(t) q(t) \, dt$$

defines an inner product on $\mathscr{P}_n(\mathbb{R})$.

For instance, if $n = 3$, $p(t) = 1 + t + t^2 - t^3$ and $q(t) = 2 - t - 3t^3$, then

$$\langle p(t), q(t) \rangle_2 = \int_0^1 (1 + t + t^2 - t^3)(2 - t - 3t^3)\,dt = 181/210.$$

Example 1.8 is a special case of (1.6) which is worth emphasizing. Suppose $(V, \langle *, * \rangle)$ is a real inner product space. If W is a subspace of V, then we can let $T: W \mapsto V$ be the inclusion map of W into V. We can define an inner product $\langle *, * \rangle_1$ on W by setting $\langle \alpha, \beta \rangle_1 = \langle T(\alpha), T(\beta) \rangle = \langle \alpha, \beta \rangle$ for all $\alpha, \beta \in W$. We will call the function $\langle *, * \rangle_1$ the restriction of $\langle *, * \rangle$ to W. We will also drop the subscript 1 from our notation when dealing with the restriction of $\langle *, * \rangle$ to W. Thus, if $(V, \langle *, * \rangle)$ is a real inner product space and W is a subspace of V, the restriction of $\langle *, * \rangle$ to W makes $(W, \langle *, * \rangle)$ a real inner product space. Example 1.8 is a typical example of restricting an inner product to a smaller space.

Suppose $(V, \langle *, * \rangle)$ is an arbitrary real inner product space. Notice that (a) of (1.2) implies that $\langle o, \gamma \rangle = 0$ for any $\gamma \in V$. Thus, (d) of (1.2) implies $\langle \alpha, \alpha \rangle \geqslant 0$ for all $\alpha \in V$. Furthermore, $\langle \alpha, \alpha \rangle = 0$ if and only if $\alpha = o$. The nonnegative square root of $\langle \alpha, \alpha \rangle$, i.e., $\sqrt{\langle \alpha, \alpha \rangle}$, is called the length of the vector α. We will let $\|\alpha\|$ denote the length of α. Thus, $\|\alpha\| = \sqrt{\langle \alpha, \alpha \rangle}$ for all $\alpha \in V$. The length of α of course depends on the inner product being used.

Example 1.9 Let $V = \mathbb{R}^n$ and let $\langle \alpha, \beta \rangle = \alpha^t \beta$ be the standard inner product on $\mathbb{R}^n$. If $\alpha = [x_1 \ \cdots \ x_n]^t \in \mathbb{R}^n$, then $\|\alpha\| = \sqrt{\alpha^t \alpha} = \{\sum_{i=1}^n x_i^2\}^{1/2}$. This is just the ordinary length of a vector familiar from the calculus.

Example 1.10 Let $V = \mathbb{R}^2$. Let $\langle *, * \rangle_1$ denote the inner product given in Example 1.4. If $\alpha = [x_1 \ x_2]^t \in \mathbb{R}^2$, then

$$\|\alpha\| = \sqrt{\langle \alpha, \alpha \rangle_1} = \{x_1^2 + (x_1 + x_2)^2\}^{1/2}.$$

Thus, the vector $[1 \ 1]^t \in \mathbb{R}^2$ has length $\sqrt{5}$ relative to the inner product $\langle *, * \rangle_1$ and length $\sqrt{2}$ relative to the standard inner product on $\mathbb{R}^2$.

The following rules concerning the lengths of vectors are valid in any real inner product space $(V, \langle *, * \rangle)$.

1.11 (a) $\|\alpha\| \geqslant 0$ for all $\alpha \in V$.
      (b) $\|\alpha\| = 0$ if and only if $\alpha = o$.
      (c) $\|x\alpha\| = |x| \, \|\alpha\|$ for all $x \in \mathbb{R}$ and $\alpha \in V$.

We will have much more to say about lengths in Section 3 of this chapter.

Finite dimensional, inner product spaces have certain types of bases, called orthogonal bases, which are very convenient to use when making computations. In order to discuss these bases, we need the following definitions which make sense in any real inner product space.

Definition 1.12 Let $(V, \langle *, * \rangle)$ be a real inner product space.

(a) Two vectors $\alpha, \beta \in V$ are said to be orthogonal if $\langle \alpha, \beta \rangle = 0$.

(b) Vectors $\alpha_1, \ldots, \alpha_n \in V$ are said to be pairwise orthogonal if $\langle \alpha_i, \alpha_j \rangle = 0$ whenever $i \neq j$.

Sometimes we say α is orthogonal to β, meaning $\langle \alpha, \beta \rangle = 0$. Since $\langle \alpha, \beta \rangle = \langle \beta, \alpha \rangle$, clearly "$\alpha$ is orthogonal to β", "β is orthogonal to α" and "α and β are orthogonal" are all the same statement. A nonzero vector α cannot be orthogonal to itself since $\langle \alpha, \alpha \rangle = 0$ implies $\alpha = o$. In particular, only the zero vector o is orthogonal to every vector in V. The importance of pairwise orthogonal vectors will become apparent in our first theorem in this section.

Theorem 1.13 Let $(V, \langle *, * \rangle)$ be a real inner preduct space. Suppose $\alpha_1, \ldots, \alpha_n$ are pairwise orthogonal, nonzero vectors in V. Then

(a) $\alpha_1, \ldots, \alpha_n$ are linearly independent over $\mathbb{R}$.

(b) If $\gamma \in L(\alpha_1, \ldots, \alpha_n)$, then

$$\gamma = \sum_{i=1}^{n} (\langle \gamma, \alpha_i \rangle / \langle \alpha_i, \alpha_i \rangle) \alpha_i.$$

Proof. (a) Suppose $x_1 \alpha_1 + \cdots + x_n \alpha_n = o$ for some scalars $x_1, \ldots, x_n \in \mathbb{R}$. Using (a) of (1.2) and the fact that the α_k are pairwise orthogonal, we have

$$0 = \langle o, \alpha_j \rangle = \langle x_1 \alpha_1 + \cdots + x_n \alpha_n, \alpha_j \rangle$$

$$= \sum_{i=1}^{n} x_i \langle \alpha_i, \alpha_j \rangle = x_j \langle \alpha_j, \alpha_j \rangle = x_j \| \alpha_j \|^2.$$

Since $\alpha_j \neq o$, $\| \alpha_j \|^2 \neq 0$. Thus, $x_j = 0$. Since j is arbitrary here, $x_1 = \cdots = x_n = 0$. Hence, $\alpha_1, \ldots, \alpha_n$ are linearly independent.

(b) Suppose $\gamma \in L(\alpha_1, \ldots, \alpha_n)$. Since the α_k are linearly independent, γ is a unique linear combination of $\alpha_1, \ldots, \alpha_n$. Suppose $\gamma = x_1 \alpha_1 + \cdots + x_n \alpha_n$. Fix $i \in \{1, 2, \ldots, n\}$. Then

$$\langle \gamma, \alpha_i \rangle = \left\langle \sum_{k=1}^{n} x_k \alpha_k, \alpha_i \right\rangle = \sum_{k=1}^{n} x_k \langle \alpha_k, \alpha_i \rangle = x_i \langle \alpha_i, \alpha_i \rangle.$$

Since $\alpha_i \neq o$, $\langle \alpha_i, \alpha_i \rangle \neq 0$. Thus, $x_i = \langle \gamma, \alpha_i \rangle / \langle \alpha_i, \alpha_i \rangle$ as required.   □

The coefficients $\langle \gamma, \alpha_i \rangle / \langle \alpha_i, \alpha_i \rangle$ appearing in Theorem 1.13(b) are called the Fourier coefficients of γ with respect to $\alpha_1, \ldots, \alpha_n$. If $B = \{\alpha_1, \ldots, \alpha_n\}$, then the Fourier coefficients of γ with respect to $\alpha_1, \ldots, \alpha_n$ are the entries of the column vector $[\gamma]_B$ in $\mathbb{R}^n$. In particular, the B-skeleton of any vector in $L(\alpha_1, \ldots, \alpha_n)$ is just a simple computation of Fourier coefficients. Consider the following example.

Example 1.14 Let $V = \mathbb{R}^3$, and let $\langle \alpha, \beta \rangle = \alpha^t \beta$ denote the standard inner product on $\mathbb{R}^3$. Suppose

$$\alpha_1 = \begin{bmatrix} 1 \\ 2 \\ 1 \end{bmatrix}, \quad \alpha_2 = \begin{bmatrix} 1 \\ 1 \\ -3 \end{bmatrix}, \quad \alpha_3 = \begin{bmatrix} -7 \\ 4 \\ -1 \end{bmatrix}$$

It is easy to see that these vectors are pairwise orthogonal. Thus, Theorem 1.13(a) implies $B = \{\alpha_1, \alpha_2, \alpha_3\}$ is a basis for $\mathbb{R}^3$.

Consider the coordinate map $[*]_B : \mathbb{R}^3 \mapsto \mathbb{R}^3$. Let $\gamma = [x \ y \ z]^t$. The Fourier coefficients of γ with respect to $\alpha_1, \alpha_2, \alpha_3$ are as follows:

$$\langle \gamma, \alpha_1 \rangle / \langle \alpha_1, \alpha_1 \rangle = (x + 2y + z)/6$$
$$\langle \gamma, \alpha_2 \rangle / \langle \alpha_2, \alpha_2 \rangle = (x + y - 3z)/11$$
$$\langle \gamma, \alpha_3 \rangle / \langle \alpha_3, \alpha_3 \rangle = (-7x + 4y - z)/66$$

Therefore, the B-skeleton of any vector $\gamma \in \mathbb{R}^3$ is given by

$$\left[\begin{bmatrix} x \\ y \\ z \end{bmatrix} \right]_B = \begin{bmatrix} (x + 2y + z)/6 \\ (x + y - 3z)/11 \\ (-7x + 4y - z)/66 \end{bmatrix}$$

Example 1.15 Let $V = \mathscr{P}_2(\mathbb{R})$. Let the inner product on $\mathscr{P}_2$ be given by

$$\langle p, q \rangle = \int_0^1 p(t)q(t)\, dt.$$

Set

$$p_1(t) = 1, \ p_2(t) = -1 + 2t, \quad \text{and} \quad p_3(t) = 1 - 6t + 6t^2$$

Then

$$\langle p_1, p_2 \rangle = \int_0^1 (-1 + 2t)\, dt = 0, \qquad \langle p_1, p_3 \rangle = \int_0^1 (1 - 6t + 6t^2)\, dt = 0,$$

and

$$\langle p_2, p_3 \rangle = \int_0^1 (-1 + 8t - 18t^2 + 12t^3)\, dt = 0$$

Thus, the vectors $p_1(t)$, $p_2(t)$, $p_3(t)$ are pairwise orthogonal in the real inner product space $(\mathscr{P}_2, \langle *, * \rangle)$. In particular, Theorem 1.13(a) implies that

$$B = \{p_1(t), p_2(t), p_3(t)\}$$

is a basis of $\mathscr{P}_2$.

For the coordinate map $[*]_B : \mathscr{P}_2 \mapsto \mathbb{R}^3$, we compute the Fourier coefficients of $q(t) = a + bt + ct^2$ with respect to p_1, p_2, p_3. The reader can easily check that

$$\frac{\langle q, p_1 \rangle}{\langle p_1, p_1 \rangle} = \frac{\displaystyle\int_0^1 (a + bt + ct^2)\, dt}{\displaystyle\int_0^1 1\, dt} = \frac{6a + 3b + 2c}{6}$$

$$\frac{\langle q, p_2 \rangle}{\langle p_2, p_2 \rangle} = \frac{\displaystyle\int_0^1 (-a + (2a - b)t + (2b - c)t^2 + 2ct^3)\, dt}{\displaystyle\int_0^1 (1 - 4t + 4t^2)\, dt} = \frac{b + c}{2}$$

$$\frac{\langle q, p_3 \rangle}{\langle p_3, p_3 \rangle} = \frac{\displaystyle\int_0^1 (a + (b - 6a)t + (6a - 6b + c)t^2 + (6b - 6c)t^3 + 6ct^4)\, dt}{\displaystyle\int_0^1 (1 - 12t + 48t^2 - 72t^3 + 36t^4)\, dt}$$

$$= \frac{c}{6}$$

Therefore

$$[a + bt + ct^2]_B = \begin{bmatrix} (6a + 3b + 2c)/6 \\ (b + c)/2 \\ c/6 \end{bmatrix}$$

If W is a finite dimensional subspace of some real inner product space $(V, \langle *, * \rangle)$, then we can always construct a basis of W consisting of pairwise orthogonal vectors. Such a basis is called an orthogonal basis of W. The construction of an orthogonal basis of W is called the Gram–Schmidt process.

Theorem 1.16 (Gram–Schmidt) Let $(V, \langle *, * \rangle)$ be a real inner product space. Suppose W is a finite dimensional subspace of V. Then W has an orthogonal basis.

Proof. If $W = (o)$, then the empty set $\varnothing$ is an orthogonal basis of W. Hence, we may assume $W \neq (o)$. Let $B = \{\alpha_1, \ldots, \alpha_n\}$ be a basis of W. Consider the following list of vectors.

1.17
$$\lambda_1 = \alpha_1$$
$$\lambda_2 = \alpha_2 - (\langle \alpha_2, \lambda_1 \rangle / \langle \lambda_1, \lambda_1 \rangle)\lambda_1$$
$$\lambda_3 = \alpha_3 - (\langle \alpha_3, \lambda_1 \rangle / \langle \lambda_1, \lambda_1 \rangle)\lambda_1 - (\langle \alpha_3, \lambda_2 \rangle / \langle \lambda_2, \lambda_2 \rangle)\lambda_2$$
$$\vdots$$
$$\lambda_i = \alpha_i - \sum_{k=1}^{i-1} (\langle \alpha_i, \lambda_k \rangle / \langle \lambda_k, \lambda_k \rangle)\lambda_k \quad \text{for } i = 2, \ldots, n$$

We claim the vectors $\lambda_1, \ldots, \lambda_n$ are all well-defined, nonzero vectors in W.

Since B is a basis of W, $\alpha_1 \neq o$. In particular, $\langle \lambda_1, \lambda_1 \rangle = \langle \alpha_1, \alpha_1 \rangle \neq 0$. Therefore, the equation defining λ_2 in (1.17) makes sense. Notice that $\lambda_2 = \alpha_2 - c_{21}\alpha_1$ where $c_{21} = \langle \alpha_2, \lambda_1 \rangle / \langle \lambda_1, \lambda_1 \rangle$. If $\lambda_2 = o$, then $\alpha_2 = c_{21}\alpha_1$ which is impossible since the vectors $\alpha_1, \ldots, \alpha_n$ are linearly independent. Thus, $\lambda_2 \neq o$. Clearly $\lambda_1, \lambda_2 \in W$.

We can now proceed by induction. Suppose we have shown that $\lambda_1, \ldots, \lambda_i$ (as defined in (1.17)) are all well-defined, nonzero vectors in W with each

$$\lambda_k = \alpha_k - c_{k1}\alpha_1 - \cdots - c_{k(k-1)}\alpha_{(k-1)}, \quad k = 1, \ldots, i$$

Here the c_{kj} are scalars in $\mathbb{R}$. Since the vectors $\lambda_1, \ldots, \lambda_i$ are all nonzero, $\langle \lambda_k, \lambda_k \rangle \neq 0$ for $k = 1, \ldots, i$. In particular, the formula for λ_{i+1} in (1.17), namely

$$\lambda_{i+1} = \alpha_{i+1} - \sum_{k=1}^{i} \left(\frac{\langle \alpha_{i+1}, \lambda_k \rangle}{\langle \lambda_k, \lambda_k \rangle} \right) \lambda_k$$

is a well-defined vector in W. Since each

$$\lambda_k = \alpha_k - c_{k1}\alpha_1 - \cdots - c_{k(k-1)}\alpha_{(k-1)}$$

we have

$$\lambda_{i+1} = \alpha_{i+1} - c_{(i+1)1}\alpha_1 - \cdots - c_{(i+1)i}\alpha_i$$

for some scalars $c_{(i+1)1}, \ldots, c_{(i+1)i} \in \mathbb{R}$. If $\lambda_{i+1} = o$, then

$$\alpha_{i+1} = c_{(i+1)1}\alpha_1 + \cdots + c_{(i+1)i}\alpha_i$$

This is impossible since B is a basis of W. Therefore, $\lambda_{i+1} \neq o$. It now follows from induction that $\lambda_1, \ldots, \lambda_n$ are all well-defined, nonzero vectors in W.

We now claim that $\{\lambda_1, \ldots, \lambda_n\}$ is an orthogonal basis of W. Since $\dim W = n$, it suffices to show that $\lambda_1, \ldots, \lambda_n$ are pairwise orthogonal. Then Theorem 1.13(a) implies $\{\lambda_1, \ldots, \lambda_n\}$ is a basis of W.

We again proceed by induction.

$$\begin{aligned}
\langle \lambda_1, \lambda_2 \rangle &= \langle \lambda_1, \alpha_2 - (\langle \alpha_2, \lambda_1 \rangle / \langle \lambda_1, \lambda_1 \rangle)\lambda_1 \rangle \\
&= \langle \lambda_1, \alpha_2 \rangle - (\langle \alpha_2, \lambda_1 \rangle / \langle \lambda_1, \lambda_1 \rangle)\langle \lambda_1, \lambda_1 \rangle \\
&= \langle \lambda_1, \alpha_2 \rangle - \langle \alpha_2, \lambda_1 \rangle = 0
\end{aligned}$$

Thus, λ_1 and λ_2 are orthogonal. Suppose we have shown that $\lambda_1, \ldots, \lambda_i$ are pairwise orthogonal. Then $\langle \lambda_p, \lambda_q \rangle = 0$ whenever $p \neq q$ and $1 \leqslant p, q \leqslant i$. Fix $j \in \{1, 2, \ldots, i\}$. Consider $\langle \lambda_j, \lambda_{i+1} \rangle$. We have

$$\begin{aligned}
\langle \lambda_j, \lambda_{i+1} \rangle &= \left\langle \lambda_j, \alpha_{i+1} - \sum_{k=1}^{i} \frac{\langle \alpha_{i+1}, \lambda_k \rangle}{\langle \lambda_k, \lambda_k \rangle} \lambda_k \right\rangle \\
&= \langle \lambda_j, \alpha_{i+1} \rangle - \sum_{k=1}^{i} \frac{\langle \alpha_{i+1}, \lambda_k \rangle}{\langle \lambda_k, \lambda_k \rangle} \langle \lambda_j, \lambda_k \rangle \\
&= \langle \lambda_j, \alpha_{i+1} \rangle - \frac{\langle \alpha_{i+1}, \lambda_j \rangle}{\langle \lambda_j, \lambda_j \rangle} \langle \lambda_j, \lambda_j \rangle \\
&= \langle \lambda_j, \alpha_{i+1} \rangle - \langle \alpha_{i+1}, \lambda_j \rangle = 0
\end{aligned}$$

Thus, the vectors $\lambda_1, \ldots, \lambda_{i+1}$ are pairwise orthogonal. It now follows from induction that $\lambda_1, \ldots, \lambda_n$ are pairwise orthogonal. $\qquad\square$

Example 1.18 Let $V = \mathbb{R}^4$ and let $\langle *, * \rangle$ denote the standard inner product on $\mathbb{R}^4$. Suppose $\alpha_1, \alpha_2, \alpha_3$ are the following vectors:

$$\alpha_1 = \begin{bmatrix} 1 \\ 0 \\ 1 \\ 0 \end{bmatrix}, \quad \alpha_2 = \begin{bmatrix} 1 \\ 1 \\ 2 \\ 1 \end{bmatrix}, \quad \alpha_3 = \begin{bmatrix} 0 \\ 1 \\ 1 \\ 2 \end{bmatrix}$$

It is easy to check that α_1, α_2, α_3 are linearly independent. Hence, we can apply the Gram–Schmidt process to the vectors α_1, α_2, α_3. From equation (1.17), we have

$$\lambda_1 = \begin{bmatrix} 1 \\ 0 \\ 1 \\ 0 \end{bmatrix}$$

$$\lambda_2 = \alpha_2 - \frac{\langle \alpha_2, \lambda_1 \rangle}{\langle \lambda_1, \lambda_1 \rangle} \lambda_1 = \begin{bmatrix} 1 \\ 1 \\ 2 \\ 1 \end{bmatrix} - \frac{3}{2} \begin{bmatrix} 1 \\ 0 \\ 1 \\ 0 \end{bmatrix} = \begin{bmatrix} -1/2 \\ 1 \\ 1/2 \\ 1 \end{bmatrix}$$

$$\lambda_3 = \alpha_3 - \frac{\langle \alpha_3, \lambda_1 \rangle}{\langle \lambda_1, \lambda_1 \rangle} \lambda_1 - \frac{\langle \alpha_3, \lambda_2 \rangle}{\langle \lambda_2, \lambda_2 \rangle} \lambda_2$$

$$= \begin{bmatrix} 0 \\ 1 \\ 1 \\ 2 \end{bmatrix} - \frac{1}{2} \begin{bmatrix} 1 \\ 0 \\ 1 \\ 0 \end{bmatrix} - \frac{7/2}{5/2} \begin{bmatrix} -1/2 \\ 1 \\ 1/2 \\ 1 \end{bmatrix} = \begin{bmatrix} 1/5 \\ -2/5 \\ -1/5 \\ 3/5 \end{bmatrix}$$

In particular, $\{\lambda_1, \lambda_2, \lambda_3\}$ is an orthogonal basis of $W = L(\alpha_1, \alpha_2, \alpha_3)$.

Suppose $B = \{\lambda_1, \ldots, \lambda_n\}$ is an orthogonal basis of a subspace W of $(V, \langle *, * \rangle)$. If we multiply each vector λ_i in B with a nonzero real number c_i, we obtain a new orthogonal basis $B' = \{c_1\lambda_1, \ldots, c_n\lambda_n\}$ of W. In this way, we can remove all denominators from rational expressions generated by the Gram–Schmidt process. Thus, in Example 1.18,

$$B' = \{\lambda_1' = [1 \quad 0 \quad 1 \quad 0]^t, \quad \lambda_2' = [-1 \quad 2 \quad 1 \quad 2]^t,$$
$$\lambda_3 = [1 \quad -2 \quad -1 \quad 3]^t\}$$

is also an orthogonal basis of W.

In some applications of this theory, it is important to construct an orthogonal basis of W consisting of vectors all of length 1, i.e., unit vectors.

Definition 1.19 Let $(V, \langle *, * \rangle)$ be a real inner product space. Let W be a finite dimensional subspace of V. A set of vectors $\{\lambda_1, \ldots, \lambda_n\}$ in W is called an orthonormal basis of W, if $\{\lambda_1, \ldots, \lambda_n\}$ is an orthogonal basis of W and $\|\lambda_i\| = 1$ for all $i = 1, \ldots, n$.

For example, the canonical basis $\underline{\varepsilon} = \{\varepsilon_1, \ldots, \varepsilon_n\}$ is an orthonormal basis of $(\mathbb{R}^n, \langle *, * \rangle)$ when $\langle *, * \rangle$ is the standard inner product. The Gram–Schmidt process guarantees that every finite dimensional subspace of a real inner product space has an orthonormal basis.

Corollary 1.20 Let $(V, \langle *, * \rangle)$ denote a real inner product space. Suppose W is a finite dimensional subspace of V. Then W has an orthonormal basis.

Proof. If $W = (o)$, then the empty set $\varnothing$ is an orthonormal basis of W. Suppose $W \neq (o)$. By the Gram–Schmidt process, W has an orthogonal basis $\{\lambda_1, \ldots, \lambda_n\}$. Each λ_i is nonzero, and, therefore, $\|\lambda_i\| \neq 0$. Equation (1.11)(c) implies $\|\lambda_i\|^{-1}\lambda_i$ is a unit vector in V. Thus, $\{\|\lambda_1\|^{-1}\lambda_1, \ldots, \|\lambda_n\|^{-1}\lambda_n\}$ is an orthonormal basis of W. $\square$

Example 1.21 Find an orthonormal basis of the subspace W in Example 1.18. We have seen that $\lambda_1' = [1\ 0\ 1\ 0]^t$, $\lambda_2' = [-1\ 2\ 1\ 2]^t$, and $\lambda_3' = [1\ -2\ -1\ 3]^t$ form an orthogonal basis of W. Thus, an orthonormal basis of W is as follows

$$\gamma_1 = \begin{bmatrix} 1/\sqrt{2} \\ 0 \\ 1/\sqrt{2} \\ 0 \end{bmatrix}, \quad \gamma_2 = \begin{bmatrix} -1/\sqrt{10} \\ 2/\sqrt{10} \\ 1/\sqrt{10} \\ 2/\sqrt{10} \end{bmatrix}, \quad \gamma_3 = \begin{bmatrix} 1/\sqrt{15} \\ -2/\sqrt{15} \\ -1/\sqrt{15} \\ 3/\sqrt{15} \end{bmatrix}$$

Before introducing the last topic in this section, we need the following definition.

Definition 1.22 Let $(V, \langle *, * \rangle)$ be a real inner product space. Two subsets S_1 and S_2 of V are said to be orthogonal if $\langle \alpha, \beta \rangle = 0$ for all $\alpha \in S_1$ and all $\beta \in S_2$.

Thus, the sets S_1 and S_2 are orthogonal if every vector in S_1 is orthogonal to every vector in S_2. Recall that

$$S_1 + S_2 = \{\alpha + \beta \mid \alpha \in S_1, \text{ and } \beta \in S_2\}$$

We can use the Gram–Schmidt Theorem to prove the following result.

Theorem 1.23 Let $(V, \langle *, * \rangle)$ be a finite dimensional, real inner product space. Let W be a subspace of V. Then there exists a unique subspace W'

with the following properties:

(a) W' is orthogonal to W.
(b) $W' \cap W = (o)$.
(c) $W' + W = V$.

Proof. If $W = (o)$, then $W' = V$, and the result is trivial. If $W = V$, then $W' = (o)$, and again the result is trivial. Hence, we may assume W is a proper subspace of V. Let $\{\alpha_1, \ldots, \alpha_r\}$ be a basis of W. By Theorem II.2.9, there exists a basis of V of the form $\{\alpha_1, \ldots, \alpha_r, \alpha_{r+1}, \ldots, \alpha_n\}$. Since W is a proper subspace of V, $1 \leqslant r < n$.

Now apply the Gram–Schmidt process to the basis $\{\alpha_1, \ldots, \alpha_n\}$. We construct an orthogonal basis $B = \{\lambda_1, \ldots, \lambda_n\}$ of V. We have seen in the proof of Theorem 1.16 that (1.17) implies $L(\alpha_1, \ldots, \alpha_i) = L(\lambda_1, \ldots, \lambda_i)$ for all $i = 1, \ldots, n$. In particular, $W = L(\lambda_1, \ldots, \lambda_r)$. Set $W' = L(\lambda_{r+1}, \ldots, \lambda_n)$. Since the vectors in B are pairwise orthogonal, it is easy to see that W and W' are orthogonal. Clearly, $B \subseteq W + W'$. Therefore, $W + W' = V$. If $\alpha \in W \cap W'$, then $\langle \alpha, \alpha \rangle = 0$. Thus, $\alpha = o$. Hence, W' satisfies (a), (b), and (c).

Suppose S is subspace of V such that S is orthogonal to W, $S + W = V$ and $S \cap W = (o)$. We claim $S = W'$. Let $\gamma \in S$. Since $W + W' = V$, $\gamma = \alpha + \beta$ for some $\alpha \in W$ and $\beta \in W'$. Then $\langle \gamma, \alpha \rangle = \langle \alpha + \beta, \alpha \rangle = \langle \alpha, \alpha \rangle + \langle \beta, \alpha \rangle$. Since W is orthogonal to both S and W', $\langle \gamma, \alpha \rangle = \langle \beta, \alpha \rangle = 0$. Therefore, $\langle \alpha, \alpha \rangle = 0$. Thus, $\alpha = o$ and $\gamma = \beta \in W'$. We have now shown that $S \subseteq W'$. Reversing the roles of S and W' in this argument gives $W' \subseteq S$. Thus, W' is the unique subspace of V satisfying (a), (b), and (c). $\square$

The unique subspace W' constructed in Theorem 1.23 is called the orthogonal complement of W. We will let $W^{\perp}$ denote the orthogonal complement of W. Thus, $W^{\perp}$ is the unique subspace of V which is orthogonal to W and has the property that $W + W^{\perp} = V$.

Example 1.24   Let $V = \mathbb{R}^3$, and suppose $\langle *, * \rangle$ is the standard inner product. Suppose W is a plane passing through the origin $O = [0\ 0\ 0]^t$ in $\mathbb{R}^3$. Then W is just a two-dimensional subspace of $\mathbb{R}^3$. Let $\{\alpha_1, \alpha_2\}$ be a basis of W. Expand this basis to a basis $\{\alpha_1, \alpha_2, \alpha_3\}$ of $\mathbb{R}^3$. Applying the Gram–Schmidt process to $\alpha_1, \alpha_2, \alpha_3$, we produce an orthogonal basis $\{\lambda_1, \lambda_2, \lambda_3\}$ of $\mathbb{R}^3$ in which $W = L(\lambda_1, \lambda_2)$. Then $W^{\perp} = L(\lambda_3)$. Thus, $W^{\perp}$ is the unique line in $\mathbb{R}^3$ which passes through the origin and is perpendicular to W.

If W is a subspace of a finite dimensional, real inner product space $(V, \langle *, * \rangle)$, then every vector $\gamma \in V$ can be written uniquely in the form $\gamma = \alpha + \beta$ where $\alpha \in W$ and $\beta \in W^\perp$. To see this, let $\gamma \in V$. Since $W + W^\perp = V, \gamma = \alpha + \beta$ for some $\alpha \in W$ and $\beta \in W^\perp$. Suppose $\gamma = \alpha_1 + \beta_1$ for some $\alpha_1 \in W$ and $\beta_1 \in W^\perp$. Then $\alpha - \alpha_1 = \beta_1 - \beta \in W \cap W^\perp = (o)$. So, $\alpha = \alpha_1$ and $\beta = \beta_1$. Thus, the decomposition $\gamma = \alpha + \beta$ ($\alpha \in W$, $\beta \in W^\perp$) is unique. This orthogonal decomposition of vectors is used often in problems in physics.

There are many interesting applications of Theorem 1.23 in linear algebra. We finish this section with one example.

Theorem 1.25 Let $A, B \in M_{m \times n}(\mathbb{R})$. If $NS(A) = NS(B)$, then $A \underset{r}{\sim} B$.

Proof. Let $V = M_{1 \times n}(\mathbb{R})$. Define an inner product on V by the following formula.

1.26
$$\langle [x_1, \ldots, x_n], [y_1, \ldots, y_n] \rangle = \sum_{i=1}^{n} x_i y_i$$

Let $W = \{\xi^t \mid \xi \in NS(A)\}$. Clearly, W is a subspace of V. Let $\xi \in NS(A)$. Then $A\xi = O$. In particular, $\langle \text{Row}_i(A), \xi^t \rangle = 0$ for all $i = 1, \ldots, m$. Thus, the row space of A, i.e., $RS(A)$, is orthogonal to W in V. Since $RS(A)$ is orthogonal to W, $RS(A) \cap W = (O)$.

The map $\xi \mapsto \xi^t$ is an isomorphism of $\mathbb{R}^n$ onto V. In particular, $NS(A) \cong W$ implies $\dim W = \dim NS(A) = \nu(A)$. On the other hand, $\dim RS(A) = \text{rk}(A)$ by Theorem II.3.3. If $\{\xi_1^t, \ldots, \xi_r^t\}$ is a basis of W ($r = \nu(A)$) and $\{\eta_1, \ldots, \eta_s\}$ is a basis of $RS(A)$ ($s = \text{rk}(A)$), then $\{\xi_1^t, \ldots, \xi_r^t, \eta_1, \ldots, \eta_s\}$ is a set of linearly independent vectors. This follows easily from the fact that $RS(A) \cap W = (O)$. But, $r + s = \text{rk}(A) + \nu(A) = n$ by Theorem II.3.22. Since $n = \dim V, \{\xi_1^t, \ldots, \xi_r^t, \eta_1, \ldots, \eta_s\}$ is a basis of V. In particular, $RS(A) + W = V$. It now follows from Theorem 1.23 that $RS(A)$ is the orthogonal complement of W in V.

We can now prove the theorem. Since $NS(A) = NS(B)$, we have

$$RS(A) = \{\xi^t \mid \xi \in NS(A)\}^\perp = \{\xi^t \mid \xi \in NS(B)\}^\perp = RS(B)$$

The result now follows from Corollary II.3.20. □

Exercises for Section 1

1.1 Verify that the standard inner product $\langle \alpha, \beta \rangle = \alpha^t \beta$ on $\mathbb{R}^n$ satisfies (a) and (b) of (1.2).

1.2 Show $\langle f, g \rangle = \int_a^b f(t)g(t)\, dt$ is an inner product on $C([a, b])$.

1.3   Show the formula given in equation (1.6) defines an inner product on W.

1.4   Show every basis B of $\mathbb{R}^n$ determines an inner product on $\mathbb{R}^n$ via equation (1.6).

1.5   Let $a_1, \ldots, a_n$ be positive real numbers. Show

$$\langle [x_1 \quad \cdots \quad x_n]^t, [y_1 \quad \cdots \quad y_n]^t \rangle = \sum_{i=1}^{n} a_i x_i y_i$$

is an inner product on $\mathbb{R}^n$.

1.6   Show $\langle *, * \rangle_1$ given in Example 1.4 satisfies (a), (b), and (c) of (1.2).

1.7   Let $V = M_{1 \times 2}(\mathbb{R})$. Show

$$\langle (x_1, x_2), (y_1, y_2) \rangle = x_1 y_1 - x_2 y_1 - x_1 y_2 + 4 x_2 y_2$$

is an inner product on V.

1.8   Let $V = M_{n \times n}(\mathbb{R})$. Define the trace of a matrix $A \in M_{n \times n}(\mathbb{R})$ (written $\mathrm{Tr}(A)$) as follows: $\mathrm{Tr}(A) = \sum_{i=1}^{n} [A]_{ii}$.
       (a)   Show that $\mathrm{Tr}(*): M_{n \times n}(\mathbb{R}) \mapsto \mathbb{R}$ is a linear transformation.
       (b) Show $\langle A, B \rangle = \mathrm{Tr}(AB^t)$ is an inner product on $M_{n \times n}(\mathbb{R})$.

1.9   Describe explicitly all inner products on $\mathbb{R}$.

1.10   Let $(V, \langle *, * \rangle)$ be a real inner product space. Prove the following polarization identity:

$$\langle \alpha, \beta \rangle = (1/4) \|\alpha + \beta\|^2 - (1/4) \|\alpha - \beta\|^2$$

1.11   Let $(V, \langle *, * \rangle)$ be a real inner product space. Prove the parallelogram law

$$\|\alpha + \beta\|^2 + \|\alpha - \beta\|^2 = 2\|\alpha\|^2 + 2\|\beta\|^2$$

1.12   Let $\langle *, * \rangle$ denote the standard inner product on $\mathbb{R}^4$. Use the Gram–Schmidt process to find an orthogonal basis for the following subspaces of $\mathbb{R}^4$:

(a)
$$W = CS \begin{bmatrix} 1 & 1 & 1 \\ 1 & 0 & 2 \\ 1 & -1 & 1 \\ 1 & 1 & 3 \end{bmatrix}$$

(b)
$$W = CS \begin{bmatrix} 1 & 0 & -1 & 0 \\ 0 & 1 & 1 & 2 \\ 1 & 1 & -1 & 1 \\ 1 & 1 & 1 & 3 \end{bmatrix}$$

Find an orthonormal basis of these two subspaces.

1.13 Let $V = \mathscr{P}_2(\mathbb{R})$. Set $\langle p, q \rangle = \int_0^1 p(t)q(t)\, dt$. Apply the Gram–Schmidt process to the basis $1, t, t^2$.

1.14 In Exercise 1.13, find an orthogonal basis of $\mathscr{P}_2$ which contains $p(t) = 1 + t$.

1.15 If W_1 and W_2 are orthogonal complements in $(V, \langle *, * \rangle)$, then every vector $\gamma \in V$ can be written uniquely in the form $\gamma = \alpha + \beta$ where $\alpha \in W_1$ and $\beta \in W_2$. Is this true for any two subspaces of V whose sum is all of V?

1.16 Let W_1 and W_2 be subspaces of a vector space V. Suppose $\alpha_1, \ldots, \alpha_r$ are linearly independent vectors in W_1 and $\beta_1, \ldots, \beta_s$ are linearly independent vectors in W_2. Show if $W_1 \cap W_2 = (o)$, then $\alpha_1, \ldots, \alpha_r, \beta_1, \ldots, \beta_s$ are linearly independent in V.

1.17 Give an example which shows Exercise 1.16 is not true if $W_1 \cap W_2 \neq (o)$.

1.18 Let $(V, \langle *, * \rangle)$ be a finite dimensional, real inner product space. Suppose W is a subspace of V. Show $W^\perp = \{\xi \in V \mid \langle \xi, \alpha \rangle = 0 \text{ for all } \alpha \in W\}$.

1.19 Let $(V, \langle *, * \rangle)$ be a finite dimensional, real inner product space. Suppose W_1 and W_2 are subspaces of V. Show the following relations are true:
(a) If $W_1 \subseteq W_2$, then $W_2^\perp \subseteq W_1^\perp$.
(b) $W_1^{\perp\perp} = W_1$.
(c) $(W_1 + W_2)^\perp = W_1^\perp \cap W_2^\perp$.

1.20 Let $V = \{y(t) \in C^2([0, \pi]) \mid y''(t) + 4y(t) = 0\}$.
(a) Show that V is a two-dimensional vector space over $\mathbb{R}$.
(b) Show that $\langle y_1(t), y_2(t) \rangle = \int_0^\pi y_1(t)y_2(t)\, dt$ defines an inner product on V.
(c) Find an orthonormal basis of V.

2. COMPLEX INNER PRODUCT SPACES

Before discussing the definition of a complex inner product space, it would be wise to review some basic arithmetic facts about the complex numbers $\mathbb{C}$. As a set, $\mathbb{C} = \{a + bi \mid a, b \in \mathbb{R}\}$. Here $i^2 = -1$. Complex numbers are added and multiplied as follows:

$$(a + bi) + (a' + b'i) = (a + a') + (b + b')i$$

and

$$(a + bi)(a' + b'i) = (aa' - bb') + (ab' + a'b)i$$

The real number x is identified with the complex number $x + 0i$. In this way, $\mathbb{R}$ is a subset of $\mathbb{C}$. Notice that $\mathbb{C}$ is a two-dimensional vector space over $\mathbb{R}$ with basis $\{1, i\}$.

Let $z = a + bi \in \mathbb{C}$. The real number a is called the real part of z and written $\mathrm{Re}(z)$. The real number b is called the imaginary part of z and written $\mathrm{Im}(z)$. The definition of complex addition implies that both functions $\mathrm{Re}(*)$, $\mathrm{Im}(*) : \mathbb{C} \mapsto \mathbb{R}$ are linear transformations when $\mathbb{C}$ is viewed as a vector space over $\mathbb{R}$. Thus, $\mathrm{Re}(z_1 + z_2) = \mathrm{Re}(z_1) + \mathrm{Re}(z_2)$ and $\mathrm{Re}(xz) = x\,\mathrm{Re}(z)$ for all $z, z_1, z_2 \in \mathbb{C}$ and all $x \in \mathbb{R}$. Similar equations hold for $\mathrm{Im}(z)$.

If $z = a + bi \in \mathbb{C}$, then the nonnegative real number $\sqrt{a^2 + b^2}$ is called the modulus (or length) of z. We will let $|z|$ denote the modulus of z. Thus, $|a + bi| = (a^2 + b^2)^{1/2}$. Clearly $|z| \geq 0$ for all $z \in \mathbb{C}$. Furthermore, $|z| = 0$ if and only if $z = 0$. There are two important inequalities which relate $|z|$, $\mathrm{Re}(z)$, and $\mathrm{Im}(z)$.

2.1
$$0 \leq |\mathrm{Re}(z)| \leq |z| \quad \text{for all } z \in \mathbb{C}$$
$$0 \leq |\mathrm{Im}(z)| \leq |z|$$

Since $0 \leq |a|$, $|b| \leq (a^2 + b^2)^{1/2}$ for any two real numbers a, b, the inequalities in (2.1) are obvious.

If $z = a + bi \in \mathbb{C}$, then $\bar{z} = a - bi$ is called the conjugate of z. The following rules about forming conjugates are all easy to prove.

2.2 (a) $\overline{(z_1 + z_2)} = \bar{z}_1 + \bar{z}_2$.

(b) $\overline{z_1 z_2} = \bar{z}_1 \bar{z}_2$.

(c) $z = \bar{z}$ if and only if $z \in \mathbb{R}$.

(d) $\bar{\bar{z}} = z$.

(e) $z\bar{z} = |z|^2$.

For the proofs of these assertions, see Exercise 2.13 of Chapter I. Notice that (a) through (d) imply complex conjugation $z \mapsto \bar{z}$ is an isomorphism of the $\mathbb{R}$-vector space $\mathbb{C}$ which preserves complex multiplication.

The definition of a real inner product $\langle *, * \rangle$ contains, and in fact is motivated by, the properties of the standard inner product on $\mathbb{R}^n$. The standard inner product, $\langle \alpha, \beta \rangle = \alpha^t \beta$, on $\mathbb{R}^n$ is a symmetric bilinear form satisfying condition (d) of (1.2). Switching to $\mathbb{C}^n$, $\langle \alpha, \beta \rangle = \alpha^t \beta$ still defines a symmetric, bilinear form on $\mathbb{C}^n$. However, $\langle \alpha, \beta \rangle = \alpha^t \beta$ fails to satisfy condition (d) of (1.2). For example, if $\alpha = [1 \; i \; 0 \; \cdots \; 0]^t \in \mathbb{C}^n$, then $\alpha^t \alpha = 0$, but $\alpha \neq 0$.

Axiom (d) of (1.2) is very important for real inner product spaces. It leads to the notions of length, $\|\alpha\| = \sqrt{\langle \alpha, \alpha \rangle}$ and distance,

$d(\alpha, \beta) = \|\alpha - \beta\|$. In the calculus, limits, derivatives, etc. all depend on our ability to measure the distance between two vectors in some way. Thus, we would like to keep axiom (d) of (1.2) in our definition of a complex inner product on $\mathbb{C}^n$. We can do this if we weaken axioms (b) and (c) of 1.2 slightly.

Suppose we define a function $\langle *, * \rangle : \mathbb{C}^n \times \mathbb{C}^n \mapsto \mathbb{C}$ as follows:

$$\langle [z_1 \quad \cdots \quad z_n]^t, [w_1 \quad \cdots \quad w_n]^t \rangle = \sum_{k=1}^{n} z_k \bar{w}_k.$$

We can shorten our notation here by adopting the following convention: if $\beta = [w_1 \quad \cdots \quad w_n]^t \in \mathbb{C}^n$, then let $\bar{\beta} = [\bar{w}_1 \quad \cdots \quad \bar{w}_n]^t$. We will call $\bar{\beta}$ the conjugate of β. If $\alpha = [z_1 \quad \cdots \quad z_n]^t$ and $\beta = [w_1 \quad \cdots \quad w_n]^t$ are vectors in $\mathbb{C}^n$, then

$$\langle [z_1 \quad \cdots \quad z_n]^t, [w_1 \quad \cdots \quad w_n]^t \rangle = \sum_{k=1}^{n} z_k \bar{w}_k = \alpha^t \bar{\beta}.$$

Lemma 2.3 The function $\langle *, * \rangle : \mathbb{C}^n \times \mathbb{C}^n \mapsto \mathbb{C}$ given by $\langle \alpha, \beta \rangle = \alpha^t \bar{\beta}$ satisfies the following conditions:

(a) $\langle z_1 \alpha_1 + z_2 \alpha_2, \beta \rangle = z_1 \langle \alpha_1, \beta \rangle + z_2 \langle \alpha_2, \beta \rangle$.
(b) $\langle \beta, z_1 \alpha_1 + z_2 \alpha_2 \rangle = \bar{z}_1 \langle \beta, \alpha_1 \rangle + \bar{z}_2 \langle \beta, \alpha_2 \rangle$.
(c) $\langle \alpha, \beta \rangle = \overline{\langle \beta, \alpha \rangle}$.
(d) $\langle \alpha, \alpha \rangle$ is a positive real number for all nonzero $\alpha \in \mathbb{C}^n$.

These equations hold for all $\alpha, \alpha_1, \alpha_2, \beta \in \mathbb{C}^n$ and all $z_1, z_2 \in \mathbb{C}$.

Proof. (a) and (b) follow directly from the definition $\langle \alpha, \beta \rangle = \alpha^t \bar{\beta}$. For (c), suppose $\alpha = [z_1 \quad \cdots \quad z_n]^t$ and $\beta = [w_1 \quad \cdots \quad w_n]^t$. Then

$$\langle \alpha, \beta \rangle = \sum_{k=1}^{n} z_k \bar{w}_k = \sum_{k=1}^{n} \overline{(\bar{z}_k w_k)} = \overline{\sum_{k=1}^{n} \bar{z}_k w_k} = \overline{\langle \beta, \alpha \rangle}$$

If $\alpha = [z_1 \quad \cdots \quad z_n]^t \in \mathbb{C}^n$, then

$$\langle \alpha, \alpha \rangle = \sum_{k=1}^{n} z_k \bar{z}_k = \sum_{k=1}^{n} |z_k|^2.$$

Thus, $\langle \alpha, \alpha \rangle$ is a nonnegative real number which is zero if and only if $\alpha = 0$. $\square$

Complex analysts call the function $\langle *, * \rangle : \mathbb{C}^n \times \mathbb{C}^n \mapsto \mathbb{C}$ defined in Lemma 2.3 the standard inner product on $\mathbb{C}^n$. Notice that the restriction of the standard inner product on $\mathbb{C}^n$ to vectors in $\mathbb{R}^n$ is just the standard inner product on $\mathbb{R}^n$. As in Section 1, any function on the cross product of a complex vector space with itself which satisfies the same axioms as the standard inner product on $\mathbb{C}^n$ will be called a complex inner product.

Definition 2.4 Let V be a complex vector space, i.e., a vector space over $\mathbb{C}$. A complex inner product on V is a function $\langle *, * \rangle : V \times V \mapsto \mathbb{C}$ which satisfies the following conditions:

(a) $\langle z_1\alpha_1 + z_2\alpha_2, \beta \rangle = z_1\langle \alpha_1, \beta \rangle + z_2\langle \alpha_2, \beta \rangle$.
(b) $\langle \beta, z_1\alpha_1 + z_2\alpha_2 \rangle = \bar{z}_1\langle \beta, \alpha_1 \rangle + \bar{z}_2\langle \beta, \alpha_2 \rangle$.
(c) $\langle \alpha, \beta \rangle = \overline{\langle \beta, \alpha \rangle}$.
(d) $\langle \alpha, \alpha \rangle$ is a positive real number for any nonzero $\alpha \in V$.

The conditions listed in (a) through (c) are to hold for all vectors α, α_1, α_2, $\beta \in V$ and all scalars $z_1, z_2 \in \mathbb{C}$. A complex vector space V together with some complex inner product $\langle *, * \rangle$ on V will be called a complex inner product space and written $(V, \langle *, * \rangle)$.

Of course, our principal example of a complex inner product space is $(\mathbb{C}^n, \langle *, * \rangle)$ where $\langle * * \rangle$ is the standard inner product $(\langle \alpha, \beta \rangle = \alpha^t\bar{\beta})$ on $\mathbb{C}^n$. The real examples given in Section 1 all have complex analogs.

Example 2.5 Let $V = \mathbb{C}^2$. Define

$$\left\langle \begin{bmatrix} z_1 \\ z_2 \end{bmatrix}, \begin{bmatrix} w_1 \\ w_2 \end{bmatrix} \right\rangle_1 = 2z_1\bar{w}_1 + z_1\bar{w}_2 + z_2\bar{w}_1 + z_2\bar{w}_2$$

It is easy to verify that $\langle *, * \rangle_1$ is a complex inner product on $\mathbb{C}^2$ different from the standard inner product on $\mathbb{C}^2$.

Example 2.6 Let V denote the set of all continuous, complex valued functions on the real interval $[a, b] \subseteq \mathbb{R}$. V is a complex vector space with the usual definitions of pointwise addition and scalar multiplication. Define a complex inner product on V by setting

$$\langle f, g \rangle = \int_a^b f(t)\overline{g(t)}\, dt.$$

It is easy to see that $(V, \langle *, * \rangle)$ is a complex inner product space.

Our remarks concerning equation (1.6) are equally valid for complex spaces. Suppose $(V, \langle *, * \rangle)$ is a complex inner product space. Let W be a complex vector space, and suppose $T: W \mapsto V$ is an injective, linear transformation. Then $\langle \alpha, \beta \rangle_1 = \langle T(\alpha), T(\beta) \rangle$ defines a complex inner product on W.

Example 2.7 Let $V = \mathscr{P}_n(\mathbb{C}) = \{z_0 + z_1 t + \cdots + z_n t^n \mid z_i \in \mathbb{C}\}$. Thus, V is the set of all polynomials in t (of degree at most n) with complex coefficients. $B = \{1, t, t^2, \ldots, t^n\}$ is a basis of V over $\mathbb{C}$. Every complex polynomial $p(t) \in V$ defines a continuous, complex valued function on $[0, 1]$. Therefore,

2.8
$$\langle p(t), q(t) \rangle_1 = [p(t)]_B^t \overline{[q(t)]_B}$$

$$\langle p(t), q(t) \rangle_2 = \int_0^1 p(t)\overline{q(t)}\, dt$$

are both well-defined, complex inner products on $\mathscr{P}_n(\mathbb{C})$.

For instance, if $n = 2$, $p(t) = 2i + t$, and $q(t) = 1 + (1 + i)t + (1 - i)t^2$, then

$$\langle p(t), q(t) \rangle_1 = [2i,\ 1,\ 0] \begin{bmatrix} 1 \\ 1 - i \\ 1 + i \end{bmatrix} = 1 + i$$

$$\langle p(t), q(t) \rangle_2 = \int_0^1 (2i + t)\overline{(1 + (1 + i)t + (1 - i)t^2)}\, dt$$

$$= \int_0^1 (2i + t)(1 + (1 - i)t + (1 + i)t^2)\, dt$$

$$= \int_0^1 (2i + (3 + 2i)t + (-1 + i)t^2 + (1 + i)t^3)\, dt$$

$$= 2i + (3 + 2i)/2 + (-1 + i)/3 + (1 + i)/4$$
$$= (17 + 43i)/12$$

Notice that a complex inner product $\langle *, * \rangle$ on V is not a bilinear function from $V \times V$ to $\mathbb{C}$. The function $\langle *, * \rangle$ is linear in its first variable, but only conjugate linear in its second variable. We still have $\langle o, \alpha \rangle = \langle \alpha, o \rangle = 0$ for all $\alpha \in V$. This follows from (a) and (c) of 2.4. Consequently, $\langle \alpha, \alpha \rangle \geqslant 0$ for all $\alpha \in V$. Furthermore, $\langle \alpha, \alpha \rangle = 0$ if and only if $\alpha = o$. The length of a vector in $(V, \langle *, * \rangle)$ is defined as before: $\|\alpha\| = \sqrt{\langle \alpha, \alpha \rangle}$. The rules listed in (1.11) are valid in any complex inner product space. The statement for (c) of (1.11) in a complex inner product space is $\|z\alpha\| = |z| \|\alpha\|$ where $|z|$ is the modulus of the complex number z. To see this, we have

$$\|z\alpha\| = \sqrt{\langle z\alpha, z\alpha \rangle} = \sqrt{z\bar{z}\langle \alpha, \alpha \rangle} = (|z|^2 \langle \alpha, \alpha \rangle)^{1/2} = |z| \|\alpha\|.$$

The definitions for orthogonal vectors, sets and bases are the same as in Section 1.

Definition 2.9   Let $(V, \langle *, * \rangle)$ be a complex inner product space.

(a)   Two vectors α, $\beta \in V$ are said to be orthogonal if $\langle \alpha, \beta \rangle = 0$.

(b)   Two sets S_1 and S_2 in V are orthogonal if $\langle \alpha, \beta \rangle = 0$ for all $\alpha \in S_1$ and $\beta \in S_2$.

(c)   Vectors α_1, ..., $\alpha_n \in V$ are said to be pairwise orthogonal if $\langle \alpha_j, \alpha_k \rangle = 0$ whenever $j \neq k$.

(d) $B = \{\alpha_1, \ldots, \alpha_n\}$ is an orthogonal basis of $W \subseteq V$ if B is a basis of W and the vectors in B are pairwise orthogonal.

(e)   An orthogonal basis $B = \{\alpha_1, \ldots, \alpha_n\}$ of W is an orthonormal basis if $\|\alpha_j\| = 1$ for all $j = 1, \ldots, n$.

In a real inner product space, the function $\langle *, * \rangle$ is symmetric. In particular, α is orthogonal to β if and only if β is orthogonal to α. In a complex inner product space, we also have α is orthogonal to β if and only if β is orthogonal to α. To see this, suppose α is orthogonal to β. Then $\langle \alpha, \beta \rangle = 0$. Thus, $\langle \beta, \alpha \rangle = \overline{\langle \alpha, \beta \rangle} = \bar{0} = 0$. Hence, β is orthogonal to α. In particular, the relationship of two vectors being orthogonal does not depend on the order of the vectors.

We have the analog of Theorem 1.13 for complex inner product spaces.

Theorem 2.10   Let $(V, \langle *, * \rangle)$ be a complex inner product space. Suppose $\alpha_1, \ldots, \alpha_n$ are nonzero, pairwise orthogonal vectors in V. Then

(a) $\alpha_1, \ldots, \alpha_n$ are linearly independent over $\mathbb{C}$.

(b)   If $\gamma \in L(\alpha_1, \ldots, \alpha_n)$, then $\gamma = \sum_{k=1}^{n} (\langle \gamma, \alpha_k \rangle / \langle \alpha_k, \alpha_k \rangle) \alpha_k$.

The proof of Theorem 2.10 is exactly the same as the proof of Theorem 1.13. The complex numbers

$$\{(\langle \gamma, \alpha_k \rangle / \langle \alpha_k, \alpha_k \rangle) \mid k = 1, \ldots, n\}$$

are called the Fourier coefficients of γ with respect to $\alpha_1, \ldots, \alpha_n$.

The Gram–Schmidt process is also valid in any complex inner product space.

Theorem 2.11   Let $(V, \langle *, * \rangle)$ be a complex inner product space. Suppose W is a finite dimensional subspace of V. Then W has an orthogonal basis.

Proof.   The process is the same as in the real case. Let $\{\alpha_1, \ldots, \alpha_n\}$ be any basis of W. Define $\lambda_1, \ldots, \lambda_n \in W$ as follows.

2.12
$$\lambda_1 = \alpha_1$$

$$\lambda_j = \alpha_j - \sum_{k=1}^{j-1} \frac{\langle \alpha_j, \lambda_k \rangle}{\langle \lambda_k, \lambda_k \rangle} \lambda_k \quad \text{for } j = 2, \ldots, n$$

The vectors $\lambda_1, \ldots, \lambda_n$ are all well-defined, nonzero vectors in W. The argument for this is the same as in the real case (replace $\mathbb{R}$ with $\mathbb{C}$ in the proof of Theorem 1.16).

The argument that $\lambda_1, \ldots, \lambda_n$ are pairwise orthogonal is only slightly different from that given in Theorem 1.16. We have

$$\langle \lambda_1, \lambda_2 \rangle = \overline{\langle \lambda_2, \lambda_1 \rangle} = \overline{\left\langle \alpha_2 - \frac{\langle \alpha_2, \lambda_1 \rangle}{\langle \lambda_1, \lambda_1 \rangle} \lambda_1, \lambda_1 \right\rangle}$$

$$= \overline{\langle \alpha_2, \lambda_1 \rangle - \langle \alpha_2, \lambda_1 \rangle} = \bar{0} = 0.$$

The general induction step is as follows:

$$\langle \lambda_j, \lambda_{i+1} \rangle = \overline{\langle \lambda_{i+1}, \lambda_j \rangle}$$

$$= \overline{\left\langle \alpha_{i+1} - \sum_{k=1}^{i} \frac{\langle \alpha_{i+1}, \lambda_k \rangle}{\langle \lambda_k, \lambda_k \rangle} \lambda_k, \lambda_j \right\rangle} = \overline{\langle \alpha_{i+1}, \lambda_j \rangle - \langle \alpha_{i+1}, \lambda_j \rangle}$$

$$= \bar{0} = 0. \qquad \square$$

An orthonormal basis of W is constructed by replacing $\{\lambda_1, \ldots, \lambda_n\}$ with

$$\{\|\lambda_1\|^{-1}\lambda_1, \ldots, \|\lambda_n\|^{-1}\lambda_n\}$$

in (2.12).

We can also construct the orthogonal complement $W^\perp$ of a subspace W of a finite dimensional, complex inner product space $(V, \langle *, * \rangle)$. We leave this construction to the exercises.

Exercises for Section 2

2.1 Show that $\text{Re}(*), \text{Im}(*) \in \text{Hom}_\mathbb{R}(\mathbb{C}, \mathbb{R})$.

2.2 Is the modulus $|*| : \mathbb{C} \mapsto \mathbb{R}$ a linear transformation?

2.3 Show the modulus has the following properties:
(a) $|z| > 0$ for any $z \in \mathbb{C} - (0)$

(b) $|0| = 0$

(c) $|z_1 z_2| = |z_1| |z_2|$

(d) $|1/z| = 1/|z|$

2.4 Show the function $f(\alpha, \beta) = \alpha^t \beta$ is a symmetric, bilinear form from $\mathbb{C}^n \times \mathbb{C}^n$ to $\mathbb{C}$.

2.5 Let $a_1, \ldots, a_n$ be positive real numbers. Show

$$\langle [z_1 \quad \cdots \quad z_n]^t, [w_1 \quad \cdots \quad w_n]^t \rangle = \sum_{k=1}^{n} a_k z_k \bar{w}_k$$

is a complex inner product on $\mathbb{C}^n$.

2.6 Is Exercise 2.5 true if $a_1, \ldots, a_n$ are nonzero, complex numbers?

2.7 Show the function defined in Example 2.5 is a complex inner product on $\mathbb{C}^2$.

2.8 Show $\langle A, B \rangle = \text{Tr}(AB^*)$ defines a complex inner product on $M_{n \times n}(\mathbb{C})$.

2.9 Determine all complex inner products on $\mathbb{C}$.

2.10 Let $\underline{\varepsilon} = \{\varepsilon_1, \ldots, \varepsilon_n\}$ be the canonical basis of $\mathbb{C}^n$. Show $\underline{\varepsilon}$ is an orthonormal basis of $\mathbb{C}^n$ relative to the standard inner product on $\mathbb{C}^n$.

2.11 Is $\underline{\varepsilon}$ an orthonormal basis of $\mathbb{C}^n$ with respect to any complex inner product on $\mathbb{C}^n$?

2.12 Use the Gram–Schmidt process to find an orthogonal basis of each of the following complex inner product spaces:

(a)

$$W = CS \begin{bmatrix} i & 1-i & 1 \\ 1+i & i & 2 \\ 1 & 2+i & 1+i \\ 0 & i & i \end{bmatrix} \quad \text{in } (\mathbb{C}^4, \langle \alpha, \beta \rangle = \alpha^t \bar{\beta})$$

(b) $\mathscr{P}_2(\mathbb{C})$ with $(p, q) = \int_0^1 p(t) \overline{q(t)} \, dt$.

(c) $(\mathbb{C}^2, \langle *, * \rangle_1)$ as in Example 2.5.

(d) $(M_{2 \times 2}(\mathbb{C}), \langle *, * \rangle)$ as in Exercise 2.8.

2.13 Show the conjugate map $\alpha \mapsto \bar{\alpha}$ from $\mathbb{C}^n$ to $\mathbb{C}^n$ given by

$$\alpha = [z_1 \quad \cdots \quad z_n]^t \mapsto \bar{\alpha} = [\bar{z}_1 \quad \cdots \quad \bar{z}_n]^t$$

has the following properties:

(a) $\overline{\alpha + \beta} = \bar{\alpha} + \bar{\beta}$.

(b) $\bar{\bar{\alpha}} = \alpha$.

(c) $\overline{z\alpha} = \bar{z}\bar{\alpha}$.

(d) $\alpha \mapsto \bar{\alpha}$ is a 1-1 and onto map.

2.14 Let $(V, \langle *, * \rangle)$ be a complex inner product space. Prove the parallelogram law:

$$\|\alpha + \beta\|^2 + \|\alpha - \beta\|^2 = 2\|\alpha\|^2 + 2\|\beta\|^2$$

2.15 Let $(V, \langle *, * \rangle)$ be a complex inner product space. Prove the following polarization identity:

$$\langle \alpha, \beta \rangle = (1/4)\|\alpha + \beta\|^2 - (1/4)\|\alpha - \beta\|^2 + (i/4)\|\alpha + i\beta\|^2$$
$$- (i/4)\|\alpha - i\beta\|^2$$

2.16 Let $(V, \langle *, * \rangle)$ be a finite dimensional, complex inner product space. Let $B = \{\alpha_1, \ldots, \alpha_n\}$ be a basis of V. Suppose $c_1, \ldots, c_n \in \mathbb{C}$. Show there exists a unique vector $\alpha \in V$ such that $\langle \alpha, \alpha_j \rangle = c_j$ for all $j = 1, \ldots, n$.

2.17 Let $(V, \langle *, * \rangle)$ be a finite dimensional, complex inner product space. Let W be a subspace of V. Show there exists a unique subspace $W^\perp$ in V such that
(a) W and $W^\perp$ are orthogonal.
(b) $W \cap W^\perp = (0)$.
(c) $W + W^\perp = V$.
($W^\perp$ is called the orthogonal complement of W.)

2.18 Let W be the subspace of $M_{n \times n}(\mathbb{C})$ consisting of all diagonal matrices. Let $\langle A, B \rangle = \mathrm{Tr}(AB^*)$. Find $W^\perp$.

2.19 Let $n = 2$ in Example 2.7. Let $W = L(2i + t)$. Compute $W^\perp$ in both inner products $\langle *, * \rangle_1$ and $\langle *, * \rangle_2$.

3. LEAST SQUARES PROBLEMS

In this section, we will develop a series of theorems about distances which are valid in any inner product space, complex or real. Consequently, we return to our notation in previous chapters. We will let V denote a vector space over F. The field F can be either $\mathbb{R}$ or $\mathbb{C}$. We assume $(V, \langle *, * \rangle)$ is an inner product space via some fixed inner product $\langle *, * \rangle$ on V. If $F = \mathbb{R}$, i.e., $(V, \langle *, * \rangle)$ is a real inner product space, then $\langle *, * \rangle$ satisfies the conditions given in (1.2). If $F = \mathbb{C}$, i.e., $(V, \langle *, * \rangle)$ is a complex inner product space, then $\langle *, * \rangle$ satisfies the conditions given in Definition 2.4. We will prove all theorems in this section by assuming $\langle *, * \rangle$ is a complex inner product. The corresponding proofs for real spaces are obtained by erasing the conjugate symbol.

If $(V, \langle *, * \rangle)$ is an inner product space over F, then we have seen in Sections 1 and 2 that there is a well-defined function $\| * \| : V \mapsto \mathbb{R}$ given by

$\|\alpha\| = \sqrt{\langle \alpha, \alpha \rangle}$ for all $\alpha \in V$. The real number $\|\alpha\|$ is called the length of the vector α. We should point out again that $\|\alpha\|$ depends on the particular inner product $\langle *, * \rangle$ being used. Whether $(V, \langle *, * \rangle)$ is real or complex, the length function has the following properties.

3.1 (a) $\|\alpha\| > 0$ for all $\alpha \in V - (o)$.
     (b) $\|o\| = 0$.
     (c) $\|z\| = |z| \, \|\alpha\|$ for all $z \in F$ and all $\alpha \in V$.

Our first two theorems in this section add two more results to the list in (3.1). We need a preliminary result from algebra before stating our first theorem.

Lemma 3.2 Suppose $p(t) = at^2 + 2bt + c \in \mathbb{R}[t]$ with $a > 0$. If $p(x) \geqslant 0$ for all $x \in \mathbb{R}$, then $b^2 \leqslant ac$.

Proof. Complete the square on $p(t)$. Thus,

$$at^2 + 2bt + c = (1/a)(at + b)^2 + [c - (b^2/a)].$$

Since $p(x) \geqslant 0$ for all $x \in \mathbb{R}$, $p(-b/a) \geqslant 0$. Thus, $c - (b^2/a) \geqslant 0$. Thus, $b^2 \leqslant ac$. $\square$

We can now prove an inequality which is called the Cauchy–Schwarz inequality.

Theorem 3.3 Let $(V, \langle *, * \rangle)$ be an inner product space over F. Then $|\langle \alpha, \beta \rangle| \leqslant \|\alpha\| \, \|\beta\|$ for all $\alpha, \beta \in V$.

Proof. Let $\alpha, \beta \in V$. If either α or β is zero, then $\langle \alpha, \beta \rangle = 0$, and $\|\alpha\| \, \|\beta\| = 0$. Hence, the inequality is true in these cases. Thus, we can assume that both α and β are nonzero vectors in V. We divide the proof into two cases.

Let us first suppose $\langle \alpha, \beta \rangle$ is a real number. For each $x \in \mathbb{R}$ consider the function $p(x) = \langle x\alpha + \beta, x\alpha + \beta \rangle$. By Definition 2.4(d), $p(x) \geqslant 0$ for all $x \in \mathbb{R}$. Since $\langle \alpha, \beta \rangle$ is real, $\langle \beta, \alpha \rangle = \overline{\langle \alpha, \beta \rangle} = \langle \alpha, \beta \rangle$. Hence, for any $x \in \mathbb{R}$, $p(x) = \langle \alpha, \alpha \rangle x^2 + 2\langle \alpha, \beta \rangle x + \langle \beta, \beta \rangle$. Set $a = \langle \alpha, \alpha \rangle$, $b = \langle \alpha, \beta \rangle$, and $c = \langle \beta, \beta \rangle$. Then $a, b, c \in \mathbb{R}$ and $a > 0$ since $\alpha \neq o$. The polynomial $p(t) = at^2 + 2bt + c$ is nonnegative at every real number x. Thus, Lemma 3.2 implies $b^2 \leqslant ac$. Therefore, $\langle \alpha, \beta \rangle^2 \leqslant \langle \alpha, \alpha \rangle \langle \beta, \beta \rangle$. Taking square roots, $|\langle \alpha, \beta \rangle| \leqslant \|\alpha\| \, \|\beta\|$.

Suppose $\langle \alpha, \beta \rangle$ is not a real number. Since 0 is real, $z = \langle \alpha, \beta \rangle$ is a nonzero, complex number which is not real. In particular, $z^{-1} \in \mathbb{C}$, and

$\langle z^{-1}\alpha, \beta \rangle = z^{-1}\langle \alpha, \beta \rangle = z^{-1}z = 1$. Since $\langle z^{-1}\alpha, \beta \rangle$ is a real number, the reasoning in the second paragraph of this proof implies the Cauchy–Schwarz inequality is valid for the two vectors $z^{-1}\alpha$ and β. Thus

$$1 = |\langle z^{-1}\alpha, \beta \rangle| \leqslant \|z^{-1}\alpha\| \, \|\beta\| = |z^{-1}| \, \|\alpha\| \, \|\beta\|.$$

Since $|z^{-1}| = |z|^{-1}$, we have $|z| \leqslant \|\alpha\| \, \|\beta\|$, i.e., $|\langle \alpha, \beta \rangle| \leqslant \|\alpha\| \, \|\beta\|$. □

Although the Cauchy–Schwarz inequality is easy to prove, its conclusions in specific examples are not at all obvious. For instance, suppose $V = \mathbb{R}^n$ and $\langle \alpha, \beta \rangle = \alpha^t\beta$ is the standard inner product on $\mathbb{R}^n$. Then the Cauchy–Schwarz inequality implies the following algebraic result.

3.4 For any real numbers $x_1, \ldots, x_n$ and $y_1, \ldots, y_n$

$$(x_1 y_1 + \cdots + x_n y_n)^2 \leqslant (x_1^2 + \cdots + x_n^2)(y_1^2 + \cdots + y_n^2)$$

If $(V, \langle *, * \rangle)$ is the inner product space given in Example 2.6, then the Cauchy–Schwarz inequality implies

3.5

$$\left| \int_a^b f(t)\overline{g(t)} \, dt \right|^2 \leqslant \left(\int_a^b |f(t)|^2 \, dt \right)\left(\int_a^b |g(t)|^2 \, dt \right)$$

The inequality in (3.5) holds for any complex valued, continuous functions $f(t)$ and $g(t)$ on $[a, b]$.

Our next theorem is a simple consequence of the Cauchy–Schwarz inequality.

Theorem 3.6 Let $(V, \langle *, * \rangle)$ be an inner product space over F. Then

$$\|\alpha + \beta\| \leqslant \|\alpha\| + \|\beta\|$$

for all $\alpha, \beta \in V$.

Proof. Fix $\alpha, \beta \in V$. Then

$$\|\alpha + \beta\|^2 = \langle \alpha + \beta, \alpha + \beta \rangle = \langle \alpha, \alpha \rangle + \langle \alpha, \beta \rangle + \langle \beta, \alpha \rangle + \langle \beta, \beta \rangle$$
$$= \|\alpha\|^2 + (\langle \alpha, \beta \rangle + \overline{\langle \alpha, \beta \rangle}) + \|\beta\|^2$$
$$= \|\alpha\|^2 + 2\,\mathrm{Re}(\langle \alpha, \beta \rangle) + \|\beta\|^2$$
$$\leqslant \|\alpha\|^2 + 2|\mathrm{Re}(\langle \alpha, \beta \rangle)| + \|\beta\|^2$$
$$\leqslant \|\alpha\|^2 + 2|\langle \alpha, \beta \rangle| + \|\beta\|^2 \quad \text{(by 2.1)}$$
$$\leqslant \|\alpha\|^2 + 2\|\alpha\| \, \|\beta\| + \|\beta\|^2 = (\|\alpha\| + \|\beta\|)^2 \quad \text{(by Theorem 3.3)}$$

Thus, $\|\alpha + \beta\|^2 \leqslant (\|\alpha\| + \|\beta\|)^2$. Taking square roots completes the proof. $\square$

The inequality in Theorem 3.6 is called the triangle inequality. If $V = \mathbb{R}^n$, and $\langle *, * \rangle$ is the standard inner product on $\mathbb{R}^n$, then $\|\alpha + \beta\| \leqslant \|\alpha\| + \|\beta\|$ is the familiar statement that the length of one side of a triangle is less than or equal to the sum of the lengths of the other two sides.

We have now shown that the length function $\|*\| : V \mapsto \mathbb{R}$ satisfies the following four properties.

3.7   (a) $\|\alpha\| > 0$ for any $\alpha \in V - (o)$.
      (b) $\|o\| = 0$.
      (c) $\|z\alpha\| = |z| \|\alpha\|$ for all $z \in F$, and $\alpha \in V$.
      (d) $\|\alpha + \beta\| \leqslant \|\alpha\| + \|\beta\|$ for all $\alpha, \beta \in V$.

A vector space V together with a function $\|*\| : V \mapsto \mathbb{R}$ satisfying (a) through (d) of (3.7) is called a normed linear vector space. Thus, any real or complex inner product space $(V, \langle *, * \rangle)$ is a normed linear vector space with norm given by $\|\alpha\| = \sqrt{\langle \alpha, \alpha \rangle}$.

We can use the length function to compute distances between vectors in an inner product space.

Definition 3.8   Let $(V, \langle *, * \rangle)$ be an inner product space over F. The distance, $d(\alpha, \beta)$, between two vectors $\alpha, \beta \in V$ is the nonnegative real number $d(\alpha, \beta) = \|\alpha - \beta\|$.

The distance between two vectors in V obviously depends on the particular inner product being used.

Example 3.9   Let $V = \mathbb{R}^2$. Let $\langle \alpha, \beta \rangle = \alpha^t \beta$ denote the standard inner product on $\mathbb{R}^2$, and let $\langle *, * \rangle_1$ denote the inner product given in Example 1.4. Suppose $\{\varepsilon_1, \varepsilon_2\}$ is the canonical basis of $\mathbb{R}^2$.

In $(\mathbb{R}^2, \langle *, * \rangle)$

$$d(\varepsilon_1, \varepsilon_2) = \|\varepsilon_1 - \varepsilon_2\| = \|[1, -1]^t\|$$

$$= (\langle [1, -1]^t, [1, -1]^t \rangle)^{1/2} = \sqrt{2}$$

In $(\mathbb{R}^2, \langle *, * \rangle_1)$,

$$d(\varepsilon_1, \varepsilon_2) = \|\varepsilon_1 - \varepsilon_2\| = \|[1, -1]^t\|$$

$$= (\langle [1, -1]^t, [1, -1]^t \rangle_1)^{1/2} = 1$$

Thus, before computing distances, we must specify what inner product is being used on V.

Suppose that $(V, \langle *, * \rangle)$ is a fixed inner product space over F. The distance function satisfies the following formulas.

3.10 (a) $d(\alpha, \beta) \geqslant 0$ for all $\alpha, \beta \in V$.

(b) $d(\alpha, \beta) = 0$ if and only if $\alpha = \beta$.

(c) $d(\alpha, \beta) = d(\beta, \alpha)$ for all $\alpha, \beta \in V$.

(d) $d(\alpha, \beta) \leqslant d(\alpha, \gamma) + d(\gamma, \beta)$ for all $\alpha, \beta, \gamma \in V$.

The proofs of these assertions are all straightforward. See Exercise 3.2.

In topology, a set S together with a function $d: S \times S \mapsto \mathbb{R}$ which satisfies (a) through (d) of (3.10) is called a metric space. Thus, every real or complex inner product space $(V, \langle *, * \rangle)$ is a metric space with distance function given by $d(\alpha, \beta) = \|\alpha - \beta\|$. We could now develop a calculus in V using the distance function d. I invite the reader to consult [2] or [6] to see how this is done.

We can now discuss the principal topic of this section, namely least squares problems. Suppose $(V, \langle *, * \rangle)$ is a fixed inner product space over F. Let W be a proper subspace of V. Thus, $(o) \neq W \neq V$. Suppose $\alpha \in V$. One of the most important problems in linear algebra is finding a vector $P(\alpha) \in W$ which is closest to α in the distance function $d(\alpha, \beta) = \|\alpha - \beta\|$. Thus, we seek a vector $P(\alpha) \in W$ such that $d(\alpha, P(\alpha)) \leqslant d(\alpha, \beta)$ for all $\beta \in W$. If $\alpha \in W$, then of course, α is the vector in W closest to α. But if $\alpha \notin W$, then it is not even clear that such a vector $P(\alpha)$ exists. However, if W is finite dimensional, we will show that there is a unique vector $P(\alpha) \in W$ such that $d(\alpha, P(\alpha)) \leqslant d(\alpha, \beta)$ for all $\beta \in W$. Furthermore, we will exhibit a simple formula for $P(\alpha)$.

If $V = \mathbb{R}^n$ and $\langle \alpha, \beta \rangle = \alpha^t \beta$ is the standard inner product on $\mathbb{R}^n$, then

$$\|\alpha - \beta\| = ((a_1 - x_1)^2 + \cdots + (a_n - x_n)^2)^{1/2}$$

Here $\alpha = [a_1 \cdots a_n]^t$, and $\beta = [x_1 \cdots x_n]^t$. Finding a vector $P(\alpha) \in W$ which is closest to α is equivalent to finding $[x_1 \cdots x_n]^t \in W$ such that $(a_1 - x_1)^2 + \cdots + (a_n - x_n)^2$ is as small as possible. Thus, we are trying to minimize a sum of squares. This is where the name "least squares problem" originated. Of course, for a general inner product space, e.g., Example 2.6, no sum of squares may be involved. At any rate, trying to find a vector in W which is closest to a given vector α is always called a least squares problem.

If $\dim W < \infty$, then a solution to the least squares problem always exists.

Theorem 3.11 Let $(V, \langle *, * \rangle)$ be an inner product space over F. Let W be a finite dimensional subspace of V. Let $\alpha \in V$. Then there exists a unique vector $P(\alpha) \in W$ such that $\|\alpha - P(\alpha)\| \leqslant \|\alpha - \beta\|$ for all $\beta \in W$.

Proof. Let us dispose of the trivial cases first. If $W = (o)$, then clearly $P(\alpha) = o$. If $W = V$, then $P(\alpha) = \alpha$. Hence, we may assume W is a proper subspace of V. Since W is finite dimensional, the Gram–Schmidt process guarantees that W has an orthogonal basis $\{\lambda_1, \ldots, \lambda_n\}$. Here $n = \dim W$. Define the vector $P(\alpha)$ as follows.

3.12
$$P(\alpha) = \sum_{k=1}^{n} (\langle \alpha, \lambda_k \rangle / \langle \lambda_k, \lambda_k \rangle) \lambda_k$$

Obviously, $P(\alpha) \in W$. Notice that $P(\alpha) = \alpha$ if and only if $\alpha \in W$. This follows from Theorem 1.13(b) in the real case and Theorem 2.10(b) in the complex case.

We first argue that $\alpha - P(\alpha)$ is orthogonal to W. Since any vector in W is a linear combination of $\lambda_1, \ldots, \lambda_n$, it suffices to show $\langle \alpha - P(\alpha), \lambda_j \rangle = 0$ for all $j = 1, \ldots, n$. Fix $j \in \{1, \ldots, n\}$. Then

$$\langle \alpha - P(\alpha), \lambda_j \rangle = \left\langle \alpha - \sum_{k=1}^{n} (\langle \alpha, \lambda_k \rangle / \langle \lambda_k, \lambda_k \rangle) \lambda_k, \lambda_j \right\rangle$$

$$= \langle \alpha, \lambda_j \rangle - \sum_{k=1}^{n} (\langle \alpha, \lambda_k \rangle / \langle \lambda_k, \lambda_k \rangle) \langle \lambda_k, \lambda_j \rangle$$

$$= \langle \alpha, \lambda_j \rangle - \langle \alpha, \lambda_j \rangle = 0$$

Therefore, $\alpha - P(\alpha)$ is orthogonal to every vector in W.

Now let β be an arbitrary vector in W. Then

3.13
$$\|\alpha - \beta\|^2 = \|(\alpha - P(\alpha)) + (P(\alpha) - \beta)\|^2$$

$$= \langle (\alpha - P(\alpha)) + (P(\alpha) - \beta), (\alpha - P(\alpha)) + (P(\alpha) - \beta) \rangle$$

$$= \langle \alpha - P(\alpha), \alpha - P(\alpha) \rangle + \langle \alpha - P(\alpha), P(\alpha) - \beta \rangle$$

$$+ \langle P(\alpha) - \beta, \alpha - P(\alpha) \rangle + \langle P(\alpha) - \beta, P(\alpha) - \beta \rangle$$

We have seen that $\alpha - P(\alpha)$ is orthogonal to W. Since $P(\alpha) - \beta \in W$,

$$\langle \alpha - P(\alpha), P(\alpha) - \beta \rangle = \langle P(\alpha) - \beta, \alpha - P(\alpha) \rangle = 0.$$

In particular, equation (3.13) can be rewritten as follows.

3.14
$$\|\alpha - \beta\|^2 = \|\alpha - P(\alpha)\|^2 + \|P(\alpha) - \beta\|^2$$

Now $\|P(\alpha) - \beta\| \geq 0$. Therefore, equation (3.14) implies

$$\|\alpha - P(\alpha)\| \leq \|\alpha - \beta\|$$

for any $\beta \in W$. Thus, $P(\alpha)$ is closest to α. Furthermore, if β is a vector in W such that $\beta \neq P(\alpha)$, then $\|P(\alpha) - \beta\| > 0$ and equation (3.14) implies $\|\alpha - P(\alpha)\| < \|\alpha - \beta\|$. Thus, β is not closest to α. We conclude that $P(\alpha)$ is the unique vector in W which is closest to α. □

The unique vector $P(\alpha)$ constructed in Theorem 3.11 is called the orthogonal projection of α onto W. Let us introduce this definition formally.

Definition 3.15 Let $(V, \langle *, * \rangle)$ be an inner product space over F, and suppose W is a finite dimensional subspace of V. Let $\alpha \in V$. The unique vector $P(\alpha) \in W$ such that $\|\alpha - P(\alpha)\| \leq \|\alpha - \beta\|$ for all $\beta \in W$ is called the orthogonal projection of α onto W.

Henceforth, we will denote the orthogonal projection of α onto W by $P_W(\alpha)$. Theorem 3.11 guarantees that $P_W(\alpha)$ exists for every vector $\alpha \in V$.

The reader might be wondering what happens if W is not finite dimensional. There are easy examples which show there may be no vector in W which is closest to α when W is infinite dimensional. Thus, $P_W(\alpha)$ may not exist if W is infinite dimensional. For a concrete example, see Exercise 3.18.

Suppose W is a finite dimensional subspace of $(V, \langle *, * \rangle)$. Then $P_W(\alpha)$ exists for every vector $\alpha \in V$. Thus, $P_W(*): V \mapsto W$ is a well-defined function from V to W. In our next theorem, we discuss some of the simple properties of this map $P_W(*)$.

Theorem 3.16 Let W be a finite dimensional subspace of some inner product space $(V, \langle *, * \rangle)$. Let $P_W(*): V \mapsto W$ denote the orthogonal projection of V onto W. Then

(a) $P_W(\alpha) = \sum_{k=1}^{n} (\langle \alpha, \lambda_k \rangle / \langle \lambda_k, \lambda_k \rangle) \lambda_k$ for any orthogonal basis $\{\lambda_1, \ldots, \lambda_n\}$ of W and any $\alpha \in V$.
(b) $\alpha - P_W(\alpha)$ is orthogonal to W for any $\alpha \in V$.
(c) $P_W(\alpha_1 + \alpha_2) = P_W(\alpha_1) + P_W(\alpha_2)$ for any $\alpha_1, \alpha_2 \in V$.
(d) $P_W(x\alpha) = xP_W(\alpha)$ for any $x \in F$ and $\alpha \in V$.
(e) $P_W(\alpha) = \alpha$ if and only if $\alpha \in W$.
(f) $P_W(P_W(\alpha)) = P_W(\alpha)$ for any $\alpha \in V$.

Proof. (a) Let $\{\lambda_1, \ldots, \lambda_n\}$ be an orthogonal basis of W. We have seen in the proof of Theorem 3.11, that $\sum_{k=1}^{n} (\langle \alpha, \lambda_k \rangle / \langle \lambda_k, \lambda_k \rangle) \lambda_k$ is the unique vector in W which is closest to α. Thus, $P_W(\alpha) = \sum_{k=1}^{n} (\langle \alpha, \lambda_k \rangle / \langle \lambda_k, \lambda_k \rangle) \lambda_k$. If $\{\lambda_1', \ldots, \lambda_n'\}$ is another orthogonal basis of W, then the same reasoning shows

$$P_W(\alpha) = \sum_{k=1}^{n} (\langle \alpha, \lambda_k' \rangle / \langle \lambda_k', \lambda_k' \rangle) \lambda_k'.$$

This proves (a).

(b) We showed in the proof of Theorem 3.11 that $\alpha - P_W(\alpha)$ is orthogonal to every vector in W.

(c) Let $\alpha_1, \alpha_2 \in V$. Let $\{\lambda_1, \ldots, \lambda_n\}$ be an orthogonal basis of W. Then from (a)

$$
\begin{aligned}
P_W(\alpha_1 + \alpha_2) &= \sum_{k=1}^{n} \frac{\langle \alpha_1 + \alpha_2, \lambda_k \rangle}{\langle \lambda_k, \lambda_k \rangle} \lambda_k \\
&= \sum_{k=1}^{n} \frac{\langle \alpha_1, \lambda_k \rangle}{\langle \lambda_k, \lambda_k \rangle} \lambda_k + \sum_{k=1}^{n} \frac{\langle \alpha_2, \lambda_k \rangle}{\langle \lambda_k, \lambda_k \rangle} \lambda_k \\
&= P_W(\alpha_1) + P_W(\alpha_2)
\end{aligned}
$$

(d) This is similar to (c). We leave it as an exercise.

(e) This statement is clear from the definition.

(f) Let $\alpha \in V$. Then $P_W(\alpha) \in W$. Therefore, $P_W(P_W(\alpha)) = P_W(\alpha)$ from (e). $\qquad\square$

Theorem 3.16(a) implies the orthogonal projection of α onto W does not depend on any specific orthogonal basis of W. Any orthogonal basis of W can be used to compute $P_W(\alpha)$. The name orthogonal projection comes from Theorem 3.16(b). For example, if W is a plane in $(\mathbb{R}^3, \langle \alpha, \beta \rangle = \alpha^t \beta)$, then $P_W(\alpha)$ is the vector in W from O to the foot of the perpendicular dropped from α to W. Thus, $P_W(\alpha)$ really is the orthogonal projection of α onto W.

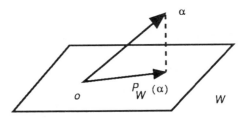

Theorems 3.16(c) and (d) imply $P_W(*)$ is a linear transformation from V to W. (e) says the restriction of $P_W(*)$ to W is the identity map on W. Since $W \subseteq V$, we can think of $P_W(*)$ as a linear transformation from V to V. Then (f) says $P_W^2 = P_W$ in $\text{Hom}(V, V)$. Such maps are called idempotent, linear transformations. Thus, the orthogonal projection of V onto W is an idempotent linear transformation in $\text{Hom}(V, V)$.

Suppose $(V, \langle *, * \rangle)$ is a finite dimensional, inner product space over F. Then we have seen in previous sections that any subspace W of V has an orthogonal complement $W^{\perp}$. Every vector $\alpha \in V$ can be written uniquely as $\alpha = \beta + \delta$ with $\beta \in W$ and $\delta \in W^{\perp}$. In terms of the orthogonal projection, $P_W(\alpha) = \beta$ and $\alpha - P_W(\alpha) = \delta$. Clearly, $P_W(\alpha) \in W$ and $\alpha - P_W(\alpha) \in W^{\perp}$ by Theorem 3.16(b). Since $\beta + \delta = \alpha = P_W(\alpha) + (\alpha - P_W(\alpha))$, we conclude that $\beta = P_W(\alpha)$ and $\delta = \alpha - P_W(\alpha)$.

We finish this section with an application of Theorem 3.11 in the theory of linear equations. Suppose $A \in M_{m \times n}(F)$ and let $B \in F^m$. Consider the linear equation $AX = B$. We have seen in Chapter II that $AX = B$ has a solution if and only if $\text{rk}(A) = \text{rk}(A \mid B)$. The equality of these ranks is in turn equivalent to $B \in CS(A)$. Suppose $B \notin CS(A)$. Then of course, there is no vector $\xi \in F^n$ such that $A\xi = B$. However, it makes sense to ask for a $\xi \in F^n$ such that $\|A\xi - B\|$ is as small as possible. Throughout this discussion, $\|\alpha\|$ is the length of α relative to the standard inner product on F^m.

Definition 3.17 A vector $\xi \in F^n$ is called a least squares solution to the equation $AX = B$ if $\|A\xi - B\| \leqslant \|A\gamma - B\|$ for all $\gamma \in F^n$.

Throughout this discussion, we will only consider the standard inner product, $\langle \alpha, \beta \rangle = \alpha^t \bar{\beta}$, on F^m. Thus, $\|A\gamma - B\| = ((A\gamma - B)^t \overline{(A\gamma - B)})^{1/2}$ for all $\gamma \in F^n$.

If $B \in CS(A)$, then there exists a vector $\gamma \in F^n$ such that $A\gamma = B$. In particular, if ξ is a least squares solution to $AX = B$, then $\|A\xi - B\| \leqslant \|A\gamma - B\| = 0$. Therefore, $A\xi = B$. Thus, if $AX = B$ has a solution, then any least squares solution to $AX = B$ is in fact a solution to $AX = B$. If $AX = B$ is inconsistent, i.e., has no solution, then a least squares solution to $AX = B$ is a vector $\xi \in F^n$ for which $A\xi$ is as close to B as possible.

Least squares solutions to $AX = B$ always exist. There is only one least squares solution to $AX = B$ if $\text{rk}(A) = n$.

Theorem 3.18 Let $A \in M_{m \times n}(F)$ and $B \in F^m$. Then the equation $AX = B$

has a least squares solution. If $\mathrm{rk}(A) = n$, then the equation $AX = B$ has a unique least squares solution.

Proof. $CS(A)$ is a subspace of F^m. By Theorem 3.11, there exists a unique vector $C \in CS(A)$ such that $\|B - C\| \leqslant \|B - Y\|$ for all $Y \in CS(A)$. The vector C is just the orthogonal projection, $P_{CS(A)}(B)$, of B onto $CS(A)$. Any vector $Y \in CS(A)$ has the form $A\gamma$ for some $\gamma \in F^n$. In particular, $C = A\xi$ for some $\xi \in F^n$. Then for any $\gamma \in F^n$,

$$\|A\xi - B\| = \|C - B\| = \|B - C\| \leqslant \|B - A\gamma\| = \|A\gamma - B\|$$

Thus, ξ is a least squares solution to $AX = B$.

If ξ is any least squares solution to $AX = B$, then $\|A\xi - B\| \leqslant \|A\gamma - B\|$ for all $\gamma \in F^n$. Since $P_{CS(A)}(B)$ is the unique vector in $CS(A)$ closest to B, we must have $A\xi = P_{CS(A)}(B)$. Thus, ξ is a least squares solution to $AX = B$ if and only if $A\xi = P_{CS(A)}(B)$.

Now suppose $\mathrm{rk}(A) = n$. Then the equation $AX = P_{CS(A)}(B)$ has only one solution since the columns of A are linearly independent. Thus, the least squares solution to $AX = B$ is unique in this case.   □

We can use the ideas in the proof of Theorem 3.18 to construct an equation least squares solutions must satisfy. We state the result in two parts depending on whether $F = \mathbb{R}$ or $\mathbb{C}$.

Theorem 3.19

(a) Let $A \in M_{m \times n}(\mathbb{R})$ and $B \in \mathbb{R}^m$. A vector $\xi \in \mathbb{R}^n$ is a least squares solution to $AX = B$ if and only if ξ is a solution to $(A^t A)X = A^t B$.

(b) Let $A \in M_{m \times n}(\mathbb{C})$ and $B \in \mathbb{C}^m$. A vector $\xi \in \mathbb{C}^n$ is a least squares solution to $AX = B$ if and only if $\bar{\xi}$ is a solution to $(A^t \bar{A})X = A^t \bar{B}$.

Proof. We will prove (b) and leave (a) as an exercise. We remind the reader that the conjugate $\bar{A}$ of A is obtained from A by conjugating the entries of A.

Suppose $\xi \in \mathbb{C}^n$ is a least squares solution to $AX = B$. Then we have seen in the proof of Theorem 3.18 that $A\xi = P_{CS(A)}(B)$. In particular, $A\xi - B$ is orthogonal to $CS(A)$ by Theorem 3.16(b). Write $CS(A) = L(A\varepsilon_1, \ldots, A\varepsilon_n)$ where $\{\varepsilon_1, \ldots, \varepsilon_n\}$ is the canonical basis of $\mathbb{C}^n$. Since $A\xi - B$ is orthogonal to $CS(A)$,

$$0 = \langle A\varepsilon_j, A\xi - B \rangle = (A\varepsilon_j)^t \overline{(A\xi - B)} = \varepsilon_j^t A^t (\bar{A}\bar{\xi} - \bar{B})$$

for all $j = 1, \ldots, n$. $\varepsilon_j^t Y = \text{Row}_j(Y)$ for any matrix Y. Therefore, $A^t(\overline{A\xi} - \overline{B}) = O$ and ξ is a solution to $(A^t\overline{A})X = A^t\overline{B}$.

Conversely, suppose $\overline{\xi}$ is a solution to $(A^t\overline{A})X = A^t\overline{B}$. Then the computations in the last paragraph show $A\xi - B$ is orthogonal to $CS(A)$. The proof of Theorem 3.11 then implies $A\xi = P_{CS(A)}(B)$. Thus, ξ is a least squares solution to $AX = B$. $\qquad\qquad\qquad\square$

The equation

3.20
$$(A^tA)X = A^tB, \quad \text{if } F = \mathbb{R}$$
$$(A^t\overline{A})X = A^t\overline{B}, \quad \text{if } F = \mathbb{C}$$

is called the normal equation of A. Theorem 3.19 implies that the least squares solutions to $AX = B$ are precisely the solutions to the normal equation when $F = \mathbb{R}$, or the conjugates of the solutions to the normal equation when $F = \mathbb{C}$. If $\text{rk}(A) = n$, then $A^t\overline{A}$ is nonsingular, and the normal equation has a unique solution.

Lemma 3.21 Let $A \in M_{m \times n}(\mathbb{C})$. If $\text{rk}(A) = n$, then $A^t\overline{A}$ is nonsingular.

Proof. $A^t\overline{A} \in M_{n \times n}(\mathbb{C})$. This matrix is nonsingular if $NS(A^t\overline{A}) = (O)$ (Theorem II.3.23). Let $\eta \in NS(A^t\overline{A})$. The map $\alpha \mapsto \bar{\alpha}$ is clearly 1-1 and onto. Therefore, $\eta = \bar{\alpha}$ for some $\alpha \in \mathbb{C}^n$. Then

$$\|A\alpha\|^2 = \langle A\alpha, A\alpha \rangle = (A\alpha)^t \overline{(A\alpha)} = \alpha^t A^t \overline{A}\bar{\alpha} = \alpha^t(A^t\overline{A})\eta = \alpha^t O = 0$$

Thus, $A\alpha = O$. Since the rank of A is n, $\alpha = O$. But then $\eta = \bar{\alpha} = O$. Therefore, $NS(A^t\overline{A}) = (O)$ and $A^t\overline{A}$ is nonsingular. $\qquad\square$

Example 3.22 Find the straight line in $\mathbb{R}^2$ which best fits the data (x_1, y_1), $\ldots, (x_n, y_n)$. Assume that $x_1, \ldots, x_n$ are all distinct.

A straight line in $\mathbb{R}^2$ has equation $y = mt + b$. We want to find real numbers m and b such that the following system of equations is satisfied.

3.23
$$\begin{bmatrix} x_1 & 1 \\ x_2 & 1 \\ \vdots & \vdots \\ x_n & 1 \end{bmatrix} \begin{bmatrix} m \\ b \end{bmatrix} = \begin{bmatrix} y_1 \\ y_2 \\ \vdots \\ y_n \end{bmatrix}$$

Of course, if $n \geqslant 3$, this linear system of equations will not have a solution in general. Set

$$A = \begin{bmatrix} x_1 & 1 \\ x_2 & 1 \\ \vdots & \vdots \\ x_n & 1 \end{bmatrix}, \quad X = \begin{bmatrix} m \\ b \end{bmatrix}, \quad B = \begin{bmatrix} y_1 \\ y_2 \\ \vdots \\ y_n \end{bmatrix}$$

The question: Find the straight line which best fits the data $(x_1, y_1), \ldots,$ (x_n, y_n), means find a least squares solution to the equation $AX = B$. If $n \geqslant 2$, then $\mathrm{rk}(A) = 2$ since $x_1, \ldots, x_n$ are all distinct. By Lemma 3.21, $A^t A$ is invertible. In particular, the normal equation $(A^t A)X = A^t B$ has the unique solution $\xi = (A^t A)^{-1} A^t B$. Theorem 3.19 implies the least squares solution $[m, b]^t$ to $AX = B$ is given by the following formula

$$\begin{bmatrix} m \\ b \end{bmatrix} = (A^t A)^{-1} A^t B = \begin{bmatrix} \sum\limits_{k=1}^{n} x_i^2 & \sum\limits_{k=1}^{n} x_i \\ \sum\limits_{k=1}^{n} x_i & n \end{bmatrix}^{-1} \begin{bmatrix} \sum\limits_{k=1}^{n} x_i y_i \\ \sum\limits_{k=1}^{n} y_i \end{bmatrix}$$

Example 3.24 Find the quadratic polynomial $q(t) = a + bt + ct^2 \in \mathbb{C}[t]$ which best fits the data $(-1, 0)$, $(0, 1)$, $(1, i)$, and $(i, 1 + i)$.

Consider the following equation:

3.25

$$\begin{bmatrix} 1 & -1 & 1 \\ 1 & 0 & 0 \\ 1 & 1 & 1 \\ 1 & i & -1 \end{bmatrix} \begin{bmatrix} a \\ b \\ c \end{bmatrix} = \begin{bmatrix} 0 \\ 1 \\ i \\ 1 + i \end{bmatrix}$$

If $q(t)$ passes through the points, then this equation is consistent. Again let

$$A = \begin{bmatrix} 1 & -1 & 1 \\ 1 & 0 & 0 \\ 1 & 1 & 1 \\ 1 & i & -1 \end{bmatrix}, \quad X = \begin{bmatrix} a \\ b \\ c \end{bmatrix}, \quad B = \begin{bmatrix} 0 \\ 1 \\ i \\ 1 + i \end{bmatrix}$$

It is easy to check $\mathrm{rk}(A \mid B) = 4$. Thus, the equation in (3.25) has no solution. The polynomial which best fits the data is the polynomial whose coefficients $[a, b, c]^t$ is a least squares solution to the equation $AX = B$. Since $\mathrm{rk}(A) = 3$, $A^t \bar{A}$ is invertible by Lemma 3.21. Thus, the equation

$AX = B$ has a unique least squares solution $\xi \in \mathbb{C}^3$ whose conjugate $\bar{\xi}$ satisfies the equation $(A^t\bar{A})X = A^t\bar{B}$. Therefore,

$$\bar{\xi} = (A^t\bar{A})^{-1}A^t\bar{B} = \begin{bmatrix} 4 & -i & 1 \\ i & 3 & -i \\ 1 & i & 3 \end{bmatrix}^{-1} \begin{bmatrix} 2 - 2i \\ 1 \\ -1 \end{bmatrix}$$

$$= \begin{bmatrix} 1/3 & i/6 & -1/6 \\ -i/6 & 11/24 & 5i/24 \\ -1/6 & -5i/24 & 11/24 \end{bmatrix} \begin{bmatrix} 2 - 2i \\ 1 \\ -1 \end{bmatrix}$$

$$= \begin{bmatrix} (5 - 3i)/6 \\ (3 - 13i)/24 \\ (-19 + 3i)/24 \end{bmatrix}$$

Thus,

$$\xi = \begin{bmatrix} (5 + 3i)/6 \\ (3 + 13i)/24 \\ (-19 - 3i)/24 \end{bmatrix}$$

and

$$q(t) = (5 + 3i)/6 + ((3 + 13i)/24)t + ((-19 - 3i)/24)t^2.$$

Exercises for Section 3

3.1   Let $(V, \langle *, * \rangle)$ be an inner product space over F. Show that for all α, $\beta \in V$, $|\,\|\alpha\| - \|\beta\|\,| \leqslant \|\alpha - \beta\|$.

3.2   Prove the assertions in (3.10).

3.3   Find the orthogonal projection, $P_W(\alpha)$, in the following problems:
   (a) $(V, \langle *, * \rangle) = (\mathbb{R}^4, \langle \alpha, \beta \rangle = \alpha^t\beta)$, $W = L([1\ 1\ 0\ 0]^t, [1\ 2\ 0\ 0]^t)$, $\alpha = [1\ 1\ 1\ 1]^t$.
   (b) $(V, \langle *, * \rangle) = (\mathbb{C}^3, \langle \alpha, \beta \rangle = \alpha^t\bar{\beta})$, $W = L([1\ i\ 0]^t, [1\ 1\ 0]^t)$, $\alpha = [1\ 1\ i]^t$.
   (c) $(V, \langle *, * \rangle) = (\mathscr{P}_3, \langle f, g \rangle = \int_0^1 f(t)g(t)\,dt)$, $W = L(1 + t, t^2)$, $\alpha = t^3$.

3.4   Give the proof of Theorem 3.6 when $F = \mathbb{R}$.

3.5   Find a matrix representation of each map $P_W(*)$ in Exercise 3.3.

3.6   Prove Theorem 3.16(d).

3.7   Let l^2 denote the set of sequences in $\mathbb{R}$ which are square summable.

Thus, $\iota^2 = \{\{a_n\}_{n=0}^\infty \mid a_n \in \mathbb{R}$ for all n and $\sum_{n=0}^\infty a_n^2$ converges$\}$.

(a) Use the Cauchy–Schwarz inequality to show the following: If $\{a_n\}_{n=0}^\infty$, $\{b_n\}_{n=0}^\infty \in \iota^2$, then $\{a_n + b_n\}_{n=0}^\infty \in \iota^2$.

(b) Show that ι^2 is a real vector space with addition and scalar multiplication defined as follows:

$$\{a_n\}_{n=0}^\infty + \{b_n\}_{n=0}^\infty = \{a_n + b_n\}_{n=0}^\infty$$

and

$$x\{a_n\}_{n=0}^\infty = \{xa_n\}_{n=0}^\infty$$

(c) Show that

$$\langle \{a_n\}_{n=0}^\infty, \{b_n\}_{n=0}^\infty \rangle = \sum_{n=0}^\infty a_n b_n$$

defines a real inner product on ι^2.

3.8 Let $(V, \langle *, * \rangle)$ be a real inner product space. Show that two vectors $\alpha, \beta \in V$ are orthogonal if and only if $\|\alpha + \beta\|^2 = \|\alpha\|^2 + \|\beta\|^2$.

3.9 Let $(V, \langle *, * \rangle)$ be a complex inner product space. If $\alpha, \beta \in V$ are orthogonal, show $\|\alpha + \beta\|^2 = \|\alpha\|^2 + \|\beta\|^2$. Give an example which shows the converse of this statement is false.

3.10 Let $(V, \langle *, * \rangle)$ be an inner product space over F. Suppose $\lambda_1, \ldots, \lambda_n$ are pairwise orthogonal, nonzero vectors in V. Prove Bessel's inequality:

$$\sum_{k=1}^n |\langle \beta, \lambda_k \rangle|^2 / \|\lambda_k\|^2 \le \|\beta\|^2 \quad \text{for all } \beta \in V.$$

3.11 In Exercise 3.10, show we have equality if and only if $\beta \in L(\lambda_1, \ldots, \lambda_n)$.

3.12 Find the straight line $y = mt + b$ which best fits the data $(-1, 1)$, $(0, 2)$, $(1, 3)$, and $(2, 1)$ in $\mathbb{R}^2$. Graph this line and the points in question.

3.13 Find the straight line $y = mt + b$ which best fits the data $(-1, 0)$, $(0, 1)$, $(1, -1)$, and $(i, 1 + i)$ in $\mathbb{C}^2$.

3.14 Find the least squares solutions to the following equations:

(a)
$$\begin{bmatrix} 1 & -1 & 1 \\ 1 & 0 & 1 \\ 0 & 0 & 2 \\ 1 & 1 & -1 \end{bmatrix} \begin{bmatrix} x_1 \\ x_2 \\ x_3 \end{bmatrix} = \begin{bmatrix} 2 \\ 3 \\ 4 \\ 1 \end{bmatrix}$$

(b) $\begin{bmatrix} 1 & 1 & -1 \\ 0 & 1 & -3 \\ 1 & 1 & -1 \end{bmatrix} \begin{bmatrix} x_1 \\ x_2 \\ x_3 \end{bmatrix} = \begin{bmatrix} 2 \\ -1 \\ 1 \end{bmatrix}$

(c) $\begin{bmatrix} 1 & 0 & i \\ 1 & 1+i & -i \\ 1 & 2 & 1 \\ i & 0 & 0 \end{bmatrix} \begin{bmatrix} x_1 \\ x_2 \\ x_3 \end{bmatrix} = \begin{bmatrix} 2 \\ 3 \\ -i \\ 1 \end{bmatrix}$

3.15 Find a quadratic polynomial $q(t) = a + bt + ct^2 \in \mathbb{R}[t]$ which best fits the data $(-2, 4), (-1, 4), (0, 2), (1, 3)$, and $(2, 4)$. Use a computer for this computation.

3.16 Find a quadratic polynomial $q(t) = a + bt + ct^2 \in \mathbb{C}[t]$ which best fits the data $(-i, 1), (1 + i, 2), (0, 1), (1, 1 - i)$, and $(2, i)$. Use a computer.

3.17 Work out a formula for the polynomial

$$p(t) = a_0 + a_1 t + \cdots + a_n t^n \in \mathbb{R}[t]$$

which best fits the data $(x_0, y_0), \ldots, (x_m, y_m) \in \mathbb{R}^2$. Assume the $m + 1$ real numbers $x_0, \ldots, x_m$ are all distinct.

3.18 Let W denote the set of all real valued sequences in $\mathbb{R}$ which are eventually zero. Thus, $W = \{\{a_n\}_{n=0}^\infty \mid a_n \in \mathbb{R}$ for all n and $a_n = 0$ for all n sufficiently large$\}$.
(a) Show that W is a subspace of ι^2 (Exercise 3.7).
(b) Let $\alpha = \{1/n\}$. Show that $\alpha \in \iota^2$.
(c) Show there is no vector in W which is closest to α.

4. NORMAL MATRICES

Let $A \in M_{n \times n}(\mathbb{C})$. The conjugate transpose (or Hermitian conjugate) A^* of A is the $n \times n$ matrix whose (i, j)th entry is given by the following equation: $[A^*]_{ij} = \overline{[A]}_{ji}$. Thus, A^* is formed from A by conjugating every entry of the transpose of A. In symbols, $A^* = (\bar{A})^t$. If the entries of A are all real, i.e., $A \in M_{n \times n}(\mathbb{R})$, then clearly $A^* = A^t$.

In this section, $\langle *, * \rangle$ will always denote the standard inner product on F^n. Thus, $\langle \alpha, \beta \rangle = \alpha^t \beta$ if $F = \mathbb{R}$ and $\langle \alpha, \beta \rangle = \alpha^t \bar{\beta}$ if $F = \mathbb{C}$. There is an important relationship between A, A^*, and $\langle *, * \rangle$.

Theorem 4.1 Let $A \in M_{n \times n}(F)$. Then $\langle A\alpha, \beta \rangle = \langle \alpha, A^*\beta \rangle$ for all $\alpha, \beta \in F^n$.

Proof. Suppose $F = \mathbb{C}$. Then

$$\langle A\alpha, \beta \rangle = (A\alpha)^t \overline{\beta} = \alpha^t A^t \overline{\beta}$$

On the other hand,

$$\langle \alpha, A^*\beta \rangle = \alpha^t \overline{(A^*\beta)} = \alpha^t \overline{A^*} \overline{\beta} = \alpha^t A^t \overline{\beta}$$

Thus, $\langle A\alpha, \beta \rangle = \langle \alpha, A^*\beta \rangle$. A similar proof works when $F = \mathbb{R}$. □

We have seen in Chapter II that any matrix $A \in M_{n \times n}(F)$ induces a linear transformation $\xi \mapsto A\xi$ from F^n to F^n. If $(V, \langle *, * \rangle)$ is an inner product space over F and $T \in \mathrm{Hom}(V, V)$, then a linear transformation $T^* \in \mathrm{Hom}(V, V)$ is called the adjoint of T if $\langle T(\alpha), \beta \rangle = \langle \alpha, T^*(\beta) \rangle$ for all α, $\beta \in V$. It is not hard to see that if such a map T^* exists, then T^* is unique. Theorem 4.1 implies the adjoint of the linear transformation $\xi \mapsto A\xi$ on $(F^n, \langle *, * \rangle)$ is the map $\xi \mapsto A^*\xi$. For this reason, A^* is sometimes called the adjoint of A. Thus, if $F = \mathbb{R}$, then A^t is the adjoint of A. If $F = \mathbb{C}$, then A^* is the adjoint of A. In this text, we will refer to A^* as the conjugate transpose of A and not use the name adjoint again.

Definition 4.2 Let $A \in M_{n \times n}(\mathbb{C})$. A is unitary if $A^*A = I_n$.

We have noted after the proof of Theorem I.5.13, that $A^*A = I_n$ implies $AA^* = I_n$. Thus, a matrix $A \in M_{n \times n}(\mathbb{C})$ is unitary if and only if the conjugate transpose A^* of A is the inverse of A.

This is a good place to introduce the Kronecker delta function. Let $\Delta(n) = \{1, \ldots, n\}$. Define a function $\delta: \Delta(n) \times \Delta(n) \mapsto \{0, 1\}$ as follows.

4.3
$$\delta(i, j) = \begin{cases} 1, & \text{if } i = j \\ 0, & \text{if } i \neq j \end{cases}$$

Here $1 \leq i, j \leq n$. It is customary in mathematics to write the functional values $\delta(i, j)$ of the Kronecker delta function as δ_{ij}. Therefore, $\delta_{ij} = 1$ if $i = j$ and $\delta_{ij} = 0$ if $i \neq j$. In our next theorem, we will use the Kronecker delta function in one method for deciding when a matrix is unitary.

Theorem 4.4 Let $A \in M_{n \times n}(\mathbb{C})$ and suppose $A = [\xi_1 | \xi_2 | \cdots | \xi_n]$ is the column partition of A. The following statements are equivalent:

(a) A is unitary.

(b) $\langle A\alpha, A\beta \rangle = \langle \alpha, \beta \rangle$ for all $\alpha, \beta \in \mathbb{C}^n$.
(c) $\langle \xi_i, \xi_j \rangle = \delta_{ij}$ for all $i, j = 1, \ldots, n$.
(d) $A = M(C, B)$ where C and B are two orthonormal bases of $\mathbb{C}^n$.

Proof. (a)$\Leftrightarrow$(b) Suppose A is unitary. Then $A^*A = I_n$. Let $\alpha, \beta \in \mathbb{C}^n$. Theorem 4.1 implies $\langle A\alpha, A\beta \rangle = \langle \alpha, A^*A\beta \rangle = \langle \alpha, I_n\beta \rangle = \langle \alpha, \beta \rangle$.

Conversely, suppose A satisfies (b). Then $\langle \alpha, \beta \rangle = \langle A\alpha, A\beta \rangle = \langle \alpha, A^*A\beta \rangle$ for all $\alpha, \beta \in \mathbb{C}^n$. Therefore, $\langle \alpha, \beta - A^*A\beta \rangle = 0$. Fix β, and vary α. $\langle \alpha, \beta - A^*A\beta \rangle = 0$ implies the vector $\beta - A^*A\beta$ is orthogonal to all of $\mathbb{C}^n$. Since O is the only vector orthogonal to itself, $\beta - A^*A\beta = O$. But β is arbitrary. Therefore, $A^*A\beta = \beta$ for all $\beta \in \mathbb{C}^n$. Thus, $A^*A = I_n$ and A is unitary.

(a)$\Leftrightarrow$(c) If $A = [\xi_1 | \cdots | \xi_n]$, then $A^* = (\bar{\xi}_1^t; \ldots; \bar{\xi}_n^t)$. In particular, if A is unitary, then

$$I_n = A^*A = (\bar{\xi}_1^t; \ldots; \bar{\xi}_n^t)[\xi_1 | \cdots | \xi_n].$$

Thus, $\bar{\xi}_i^t \xi_j = \delta_{ij}$ for all $i, j = 1, \ldots, n$. Since δ_{ij} is a real number,

$$\delta_{ij} = \bar{\delta}_{ij} = \overline{(\bar{\xi}_i^t \xi_j)} = \xi_i^t \bar{\xi}_j = \langle \xi_i, \xi_j \rangle$$

Thus, (a) implies (c). If $\langle \xi_i, \xi_j \rangle = \delta_{ij}$ for all $i, j = 1, \ldots, n$, then $\xi_i^t \bar{\xi}_j = \delta_{ij}$. Therefore,

$$\delta_{ij} = \bar{\delta}_{ij} = \overline{(\xi_i^t \bar{\xi}_j)} = \bar{\xi}_i^t \xi_j$$

Thus, $A^*A = I_n$ and A is unitary.

Before arguing that (d) is equivalent to (a), recall the definition of $M(C, B)$. If B and $C = \{\beta_1, \ldots, \beta_n\}$ are two bases of $\mathbb{C}^n$, then $M(C, B)$ is the change of basis matrix defined by

$$M(C, B) = [[\beta_1]_B | [\beta_2]_B | \cdots | [\beta_n]_B].$$

(a)$\Leftrightarrow$(d) Suppose $A = M(C, B)$ where $C = \{\beta_1, \ldots, \beta_n\}$ and $B = \{\alpha_1, \ldots, \alpha_n\}$ are two orthonormal bases of $\mathbb{C}^n$. Let $M(C, B) = (m_{ij})$. Then

$$\beta_i = m_{1i}\alpha_1 + m_{2i}\alpha_2 + \cdots + m_{ni}\alpha_n$$

for all $i = 1, \ldots, n$. Since C is an orthonormal basis of $\mathbb{C}^n$, $\langle \beta_i, \beta_j \rangle = \delta_{ij}$ for all $i, j = 1, \ldots, n$. Thus,

$$\delta_{ij} = \langle \beta_i, \beta_j \rangle = \langle m_{1i}\alpha_1 + \cdots + m_{ni}\alpha_n, m_{1j}\alpha_1 + \cdots + m_{nj}\alpha_n \rangle$$

$$= \sum_{p,q=1}^n m_{pi}\bar{m}_{qj} \langle \alpha_p, \alpha_q \rangle$$

Since B is an orthonormal basis of $\mathbb{C}^n$, $\langle \alpha_p, \alpha_q \rangle = \delta_{pq}$ for all $p, q = 1, \ldots, n$.

Thus,

$$\sum_{p,q=1}^{n} m_{pi}\bar{m}_{qj}\langle \alpha_p, \alpha_q \rangle = \sum_{k=1}^{n} m_{ki}\bar{m}_{kj}$$

We have now shown that $\delta_{ij} = \sum_{k=1}^{n} m_{ki}\bar{m}_{kj}$ for all $i, j = 1, \ldots, n$.

Now consider the inner product of any two columns of $M(C, B)$.

$$\langle [\beta_i]_B, [\beta_j]_B \rangle = \langle [m_{1i} \quad \cdots \quad m_{ni}]^t, [m_{1j} \quad \cdots \quad m_{nj}]^t \rangle$$

$$= \sum_{k=1}^{n} m_{ki}\bar{m}_{kj} = \delta_{ij} \quad \text{for all } i, j = 1, \ldots, n$$

Thus, $M(C, B)$ satisfies (c). Since (c) and (a) are equivalent, $M(C, B)$ is unitary.

Conversely, suppose A is unitary. Then the columns $\xi_1, \ldots, \xi_n$ of A satisfy (c). In particular, $C = \{\xi_1, \ldots, \xi_n\}$ is an orthonormal basis of $\mathbb{C}^n$. The canonical basis $\underline{\varepsilon}$ of $\mathbb{C}^n$ is an orthonormal basis of $\mathbb{C}^n$. Since $\xi_i = [\xi_i]_{\underline{\varepsilon}}$ for all $i = 1, \ldots, n$, $A = M(C, \underline{\varepsilon})$. Thus, A is the change of bases matrix between the two orthonormal bases C and $\underline{\varepsilon}$ of $\mathbb{C}^n$. $\qquad \square$

If A is unitary, then Theorem 4.4(b) implies

$$\|A\alpha\|^2 = \langle A\alpha, A\alpha \rangle = \langle \alpha, \alpha \rangle = \|\alpha\|^2$$

for all $\alpha \in \mathbb{C}^n$. Thus, $\|A\alpha\| = \|\alpha\|$ for all $\alpha \in \mathbb{C}^n$. In particular, a unitary matrix when viewed as a linear transformation on $\mathbb{C}^n$ preserves lengths. Theorem 4.4(b) also implies the following corollary concerning the spectrum of A.

Corollary 4.5 Let $A \in M_{n \times n}(\mathbb{C})$. If A is unitary, then

$$\mathscr{S}_{\mathbb{C}}(A) \subseteq \{z \in \mathbb{C} \,|\, |z| = 1\}$$

Proof. Let $z \in \mathscr{S}_{\mathbb{C}}(A)$. Let α be an eigenvector of A belonging to z. Then $\alpha \neq 0$ and $A\alpha = z\alpha$. Since A is unitary, Theorem 4.4(b) implies $\|\alpha\|^2 = \langle \alpha, \alpha \rangle = \langle A\alpha, A\alpha \rangle = \|A\alpha\|^2 = \|z\alpha\|^2 = |z|^2 \|\alpha\|^2$. Since $\alpha \neq 0$, $\|\alpha\| \neq 0$. Thus, $|z|^2 = 1$. Since $|z| \geq 0$, we have $|z| = 1$. $\qquad \square$

The set $\{z \in \mathbb{C} \,|\, |z| = 1\}$ is called the unit circle in the complex plane $\mathbb{C}$. The spectrum of a unitary matrix is always some finite (nonempty) subset of the unit circle in $\mathbb{C}$.

Theorem 4.4(c) says the columns of a unitary matrix A form an orthonormal basis of $\mathbb{C}^n$. Conversely, if $\{\xi_1, \ldots, \xi_n\}$ is an orthonormal basis of $\mathbb{C}^n$, then $\xi_1, \ldots, \xi_n$ obviously satisfy (c). Thus, $A = [\xi_1 | \cdots | \xi_n]$ is unitary. It follows that unitary matrices are precisely those matrices in

$M_{n \times n}(\mathbb{C})$ whose columns form an orthonormal basis of $\mathbb{C}^n$. A closely related statement is Theorem 4.4(d). Unitary matrices are precisely the change of bases matrices between orthonormal bases of $\mathbb{C}^n$.

A unitary matrix all of whose entries are real numbers is called an orthogonal matrix. Thus, a matrix $A \in M_{n \times n}(\mathbb{R})$ is orthogonal if and only if $A^t A = A A^t = I_n$. We have the following analog of Theorem 4.4 for orthogonal matrices.

Theorem 4.6 Let $A \in M_{n \times n}(\mathbb{R})$. Suppose $A = [\xi_1 | \cdots | \xi_n]$ is the column partition of A. Then the following statements are equivalent:

(a) A is orthogonal.
(b) $\langle A\alpha, A\beta \rangle = \langle \alpha, \beta \rangle$ for all $\alpha, \beta \in \mathbb{R}^n$,
(c) $\langle \xi_i, \xi_j \rangle = \delta_{ij}$ for all $i, j = 1, \ldots, n$.
(d) $A = M(C, B)$ where C and B are orthonormal bases of $\mathbb{R}^n$.

The proof of Theorem 4.6 is basically the same as the proof of Theorem 4.4 only easier. We need not worry about conjugation symbols for real matrices. Since $M_{n \times n}(\mathbb{R}) \subset M_{n \times n}(\mathbb{C})$, every orthogonal matrix is a unitary matrix. In particular, for any orthogonal matrix A, $\mathscr{S}_{\mathbb{R}}(A) \subseteq \mathscr{S}_{\mathbb{C}}(A) \subseteq \{z \in \mathbb{C} \,|\, |z| = 1\}$. Of course, $\mathscr{S}_{\mathbb{R}}(A)$ could be empty. For example

$$A = \begin{bmatrix} 0 & -1 \\ 1 & 0 \end{bmatrix}$$

is an orthogonal matrix for which $\mathscr{S}_{\mathbb{R}}(A) = \varnothing$.

Definition 4.7 Let $A \in M_{n \times n}(\mathbb{C})$. A is Hermitian if $A = A^*$.

If A has all real entries, then A is Hermitian if and only if $A = A^t$. Thus, a real Hermitian matrix is just a symmetric matrix. In particular, we already know lots of examples of Hermitian matrices.

$$A = \begin{bmatrix} 1 & i \\ -i & 1 \end{bmatrix}$$

is a complex Hermitian matrix. The only theorem about Hermitian matrices which we will need is the following result.

Theorem 4.8 Let $A \in M_{n \times n}(\mathbb{C})$.

(a) A is Hermitian if and only if $\langle A\alpha, \beta \rangle = \langle \alpha, A\beta \rangle$ for all $\alpha, \beta \in \mathbb{C}^n$.
(b) If A is Hermitian, then $\mathscr{S}_{\mathbb{C}}(A) \subseteq \mathbb{R}$.

Proof. (a) Suppose A is Hermitian. Then $\langle A\alpha, \beta \rangle = \langle \alpha, A\beta \rangle$ by Theorem 4.1. Conversely, suppose $\langle A\alpha, \beta \rangle = \langle \alpha, A\beta \rangle$ for all $\alpha, \beta \in \mathbb{C}^n$. Since $\langle A\alpha, \beta \rangle = \langle \alpha, A^*\beta \rangle$ by Theorem 4.1,

$$0 = \langle \alpha, A\beta \rangle - \langle \alpha, A^*\beta \rangle = \langle \alpha, (A - A^*)\beta \rangle$$

for all $\alpha, \beta \in \mathbb{C}^n$. In particular,

$$0 = \langle (A - A^*)\beta, (A - A^*)\beta \rangle = \|(A - A^*)\beta\|^2.$$

Thus, $(A - A^*)\beta = O$. Since β is arbitrary, $A = A^*$. Thus, A is Hermitian.

(b) We have seen in Chapter III that the Fundamental Theorem of Algebra guarantees that $\mathscr{S}_{\mathbb{C}}(A) \neq \varnothing$ for any matrix A. Suppose A is Hermitian, and let $d \in \mathscr{S}_{\mathbb{C}}(A)$. Let α be an eigenvector of A belonging to d. Then using (a), we have

$$d\|\alpha\|^2 = d\langle \alpha, \alpha \rangle = \langle d\alpha, \alpha \rangle = \langle A\alpha, \alpha \rangle = \langle \alpha, A\alpha \rangle = \langle \alpha, d\alpha \rangle$$
$$= \bar{d}\langle \alpha, \alpha \rangle = \bar{d}\|\alpha\|^2$$

Since $\alpha \neq O$, $d = \bar{d}$. A complex number d is equal to its conjugate if and only if d is real. Thus, $d \in \mathbb{R}$, and the proof of (b) is complete. $\square$

Notice that Theorem 4.8(b) implies that a symmetric matrix in $M_{n \times n}(\mathbb{R})$ has a real eigenvalue. This result is important enough to list as a corollary.

Corollary 4.9 Any symmetric matrix $A \in M_{n \times n}(\mathbb{R})$ has a real eigenvalue. In fact, all the eigenvalues of A in $\mathbb{C}$ are real numbers.

Proof. Since the entries in A are real numbers and A is symmetric, $A = A^*$. Thus, A is Hermitian. Theorem 4.8(b) implies $\mathscr{S}_{\mathbb{C}}(A) \subseteq \mathbb{R}$. Since $\mathscr{S}_{\mathbb{C}}(A) \neq \varnothing$, A has a real eigenvalue, and furthermore, any eigenvalue of A is real. $\square$

Thus, if $A \in M_{n \times n}(\mathbb{R})$ is symmetric, then $\mathscr{S}_{\mathbb{R}}(A) \neq \varnothing$.

Definition 4.10 Let $A \in M_{n \times n}(\mathbb{C})$. A is skew-Hermitian if $A^* = -A$.

If the entries in A are all real numbers, then A is skew-Hermitian if and only if $A^t = -A$. Real skew-Hermitian matrices are called skew-symmetric matrices. For example

$$\begin{bmatrix} 0 & -1 \\ 1 & 0 \end{bmatrix} \text{ is a skew-symmetric matrix}$$

$\begin{bmatrix} 0 & i \\ i & 0 \end{bmatrix}$ is a skew-Hermitian matrix

Notice that the diagonal entries of any skew-Hermitian matrix must be zero. Also notice that Corollary 4.9 is not true for skew-symmetric matrices

$$A = \begin{bmatrix} 0 & -1 \\ 1 & 0 \end{bmatrix}$$ is skew-symmetric and orthogonal

but $\mathscr{S}_{\mathbb{R}}(A) = \varnothing$.

We have introduced a lot of names in this section. It might be helpful to present a chart of the complex matrices discussed so far and their corresponding real counterparts.

4.11

$M_{n \times n}(\mathbb{C})$	$M_{n \times n}(\mathbb{R})$
A unitary: $A^*A = I_n$	A orthogonal: $A^tA = I_n$
A Hermitian: $A^* = A$	A symmetric: $A^t = A$
A skew-Hermitian: $A^* = -A$	A skew-symmetric: $A^t = -A$

We can now define normal matrices.

Definition 4.12 Let $A \in M_{n \times n}(\mathbb{C})$. A is normal if $A^*A = AA^*$.

Thus, a matrix is normal if it commutes with its conjugate transpose. Unitary, Hermitian, and skew-Hermitian matrices are all normal. In particular, if the entries of A are all real numbers, and A is orthogonal, symmetric or skew-symmetric, then A is a normal matrix in $M_{n \times n}(\mathbb{R})$. The matrices appearing in the chart in (4.11) are certainly the most important examples of normal matrices. However, there are normal matrices which are not unitary. Hermitian or skew-Hermitian. For example

$$A = \begin{bmatrix} 1 & i \\ i & 1 \end{bmatrix}$$

is easily seen to be normal, but not unitary, Hermitian or skew-Hermitian.

In the main theorem in this section, we will show that every normal matrix is similar via a unitary matrix to a diagonal matrix. This will follow easily from our next theorem which is known as Schur's Theorem.

Theorem 4.13 (Schur)

(a)   Let $A \in M_{n \times n}(\mathbb{R})$ such that $\mathscr{S}_{\mathbb{C}}(A) \subseteq \mathbb{R}$. Then there exists an orthogonal matrix P such that $P^t A P$ is an upper triangular matrix.

(b)   Let $A \in M_{n \times n}(\mathbb{C})$. Then there exists a unitary matrix P such that $P^* A P$ is an upper triangular matrix.

Before we prove Theorem 4.13, notice that an orthogonal matrix P has inverse P^t. Thus, Theorem 4.13(a) says A is similar to an upper triangular matrix if all the eigenvalues of A are real. Similarly, if P is unitary, then P^* is the inverse of P. Thus, Theorem 4.13(b) says every complex matrix is similar to an upper triangular matrix.

Proof of Theorem 4.13.   The proofs of these two theorems are virtually the same. We will prove (a) and leave (b) to the exercises. Hence, suppose $A \in M_{n \times n}(\mathbb{R})$ and $\mathscr{S}_{\mathbb{C}}(A) \subseteq \mathbb{R}$. Since $\mathscr{S}_{\mathbb{C}}(A) \neq \varnothing$, A has a real eigenvalue d_1. Let $\xi_1 \in \mathbb{R}^n$ be an eigenvector of A belonging to d_1. Then $\xi_1 \neq O$, and any nonzero, scalar multiple of ξ_1 is another eigenvector of A belonging to d_1. Replacing ξ_1 with $\xi_1 / \|\xi_1\|$, we can assume $\|\xi_1\| = 1$. By Theorem II.2.9, there exists a basis of $\mathbb{R}^n$ of the form $\{\xi_1, \xi_2', \ldots, \xi_n'\}$. Applying the Gram–Schmidt process to this basis produces an orthonormal basis of $\mathbb{R}^n$ of the form $\{\xi_1, \xi_2, \ldots, \xi_n\}$. In particular, the matrix $P_1 = [\xi_1 | \xi_2 | \cdots | \xi_n]$ is an orthogonal matrix by Theorem 4.6(c).

Now

$$P_1^t A P_1 = (\xi_1^t; \xi_2^t; \ldots; \xi_n^t) A [\xi_1 | \xi_2 | \cdots | \xi_n]$$

$$= (\xi_1^t; \ldots; \xi_n^t)[A\xi_1 | \cdots | A\xi_n]$$

$$= \begin{bmatrix} \xi_1^t \\ \xi_2^t \\ \vdots \\ \xi_n^t \end{bmatrix} [d_1\xi_1 | A\xi_2 | \cdots | A\xi_n]$$

$$= \begin{bmatrix} \xi_1^t d_1 \xi_1 & \xi_1^t A\xi_2 & \cdots & \xi_1^t A\xi_n \\ \xi_2^t d_1 \xi_1 & \xi_2^t A\xi_2 & \cdots & \xi_2^t A\xi_n \\ \vdots & \vdots & & \vdots \\ \xi_n^t d_1 \xi_1 & \xi_n^t A\xi_2 & \cdots & \xi_n^t A\xi_n \end{bmatrix} = \begin{bmatrix} d_1 & * & \cdots & * \\ 0 & & & \\ & & A_1 & \\ 0 & & & \end{bmatrix}$$

As usual, the $*$s here denote entries whose exact values are not relevant to the proof. The zeros appear below d_1 in the first column of $P_1^t A P_1$

since $\{\xi_1, \ldots, \xi_n\}$ is an orthonormal basis of $\mathbb{R}^n$. For $i > 1$, $\xi_i^t d_1 \xi_1 = d_1 \langle \xi_i, \xi_1 \rangle = 0$.

If $n = 2$, then $P_1^t A P_1$ is upper triangular, and the proof is complete. If $n > 2$, we proceed by induction. Since P_1 is orthogonal, $P_1^t = P_1^{-1}$. Thus, A is similar to $P_1^t A P_1$. It follows from the proof of Corollary III.4.12 that A and $P_1^t A P_1$ have the same characteristic polynomial. The characteristic polynomial of $P_1^t A P_1$ is given by $C_{(P_1^t A P_1)}(t) = (t - d_1) C_{A_1}(t)$. This equation follows from Laplace's expansion down the first column of $tI_n - P_1^t A P_1$. In particular, every root of $C_{A_1}(t)$ is a root of $C_{(P_1^t A P_1)}(t) = C_A(t)$. Theorem III.4.9 implies $\mathscr{S}_{\mathbb{C}}(A_1) \subseteq \mathscr{S}_{\mathbb{C}}(A) \subseteq \mathbb{R}$. Thus, every eigenvalue of A_1 (in $\mathbb{C}$) is a real number.

Our induction assumption is that Theorem 4.13(a) holds for every matrix in $M_{(n-1) \times (n-1)}(\mathbb{R})$ which has all real eigenvalues. Hence, we can apply our induction assumption to A_1. There exists an orthogonal matrix $Q_1 \in M_{(n-1) \times (n-1)}(\mathbb{R})$ such that $Q_1^t A_1 Q_1 = U$, an upper triangular matrix.

Now consider the $n \times n$ matrix

$$
Q = \begin{bmatrix} 1 & 0 & \cdots & 0 \\ 0 & & & \\ \vdots & & Q_1 & \\ 0 & & & \end{bmatrix}
$$

Since Q_1 is an orthogonal matrix, the columns of Q_1 form an orthonormal basis of $\mathbb{R}^{n-1}$. Clearly then, the columns of Q form an orthonormal basis of $\mathbb{R}^n$. Thus, Q is an orthogonal matrix in $M_{n \times n}(\mathbb{R})$. Using Theorem I.3.10, we have

$$
Q^t (P_1^t A P_1) Q = \begin{bmatrix} 1 & O \\ O & Q_1^t \end{bmatrix} \begin{bmatrix} d_1 & * \\ O & A_1 \end{bmatrix} \begin{bmatrix} 1 & O \\ O & Q_1 \end{bmatrix}
$$

$$
= \begin{bmatrix} d_1 & *Q_1 \\ O & Q_1^t A_1 Q_1 \end{bmatrix} = \begin{bmatrix} d_1 & * \\ O & U \end{bmatrix}
$$

Since U is upper triangular, this last matrix is an $n \times n$ upper triangular matrix. The product of two orthogonal matrices is easily seen to be orthogonal. Thus, $P_1 Q$ is orthogonal and $(P_1 Q)^t A (P_1 Q)$ is upper triangular. This completes the proof. $\qquad \square$

Suppose $A \in M_{n \times n}(\mathbb{R})$ is symmetric. Then Corollary 4.9 implies $\mathscr{S}_{\mathbb{C}}(A) \subseteq \mathbb{R}$. Hence, we can apply Schur's Theorem to A. There exists an orthogonal matrix $P \in M_{n \times n}(\mathbb{R})$ such that $P^t A P = U$, an upper triangular matrix. The matrix U is symmetric since $U^t = (P^t A P)^t = P^t A^t P^{tt} = P^t A P = U$. An upper triangular matrix U is symmetric if and only if U is diagonal. Hence, we have the following corollary to Theorem 4.13(a).

Corollary 4.14 Let $A \in M_{n \times n}(\mathbb{R})$. If A is symmetric, then there exists an orthogonal matrix P such that $P^t A P$ is a diagonal matrix.

Corollary 4.14 says any symmetric matrix is similar (via an orthogonal matrix) to a diagonal matrix. In particular, the applications discussed at the end of Section 4 in Chapter III can always be carried out for symmetric matrices.

Theorem 4.13(b) says any matrix in $M_{n \times n}(\mathbb{C})$ is similar to an upper triangular matrix. Suppose A is Hermitian. Then there exists a unitary matrix P such that $P^* A P = U$, an upper triangular matrix. The matrix U is Hermitian since $U^* = (P^* A P)^* = P^* A^* P^{**} = P^* A P = U$. Clearly, an upper triangular matrix U is Hermitian if and only if U is diagonal with only real entries on the diagonal. Thus, we have the following corollary to Theorem 4.13(b).

Corollary 4.15 Let $A \in M_{n \times n}(\mathbb{C})$. If A is Hermitian, then there exists a unitary matrix P such that $P^* A P$ is a diagonal matrix with only real entries on the diagonal.

These last two corollaries are special cases of the following general result.

Theorem 4.16 Let $A \in M_{n \times n}(\mathbb{C})$. A is normal if and only if there exists a unitary matrix P such that $P^* A P$ is a diagonal matrix.

Proof. Suppose there exists a unitary matrix P such that $P^* A P = D$, a diagonal matrix. Then $A = P D P^*$. Since diagonal matrices are clearly normal,

$$A A^* = (P D P^*)(P D P^*)^* = P D P^* P^{**} D^* P^* = P D P^* P D^* P^* = P D D^* P^*$$

$$= P D^* D P^* = (P D^* P^*)(P D P^*) = A^* A$$

Thus, A is normal.

Conversely, suppose A is normal. By Schur's Theorem, there exists a unitary matrix P such that

4.17

$$P*AP = U = \begin{bmatrix} z_{11} & z_{12} & \cdots & z_{1n} \\ 0 & z_{22} & \cdots & z_{2n} \\ \vdots & \vdots & & \vdots \\ 0 & 0 & \cdots & z_{nn} \end{bmatrix} \quad \text{(upper triangular)}$$

Since A is normal, an argument similar to the one used in the first paragraph of this proof shows that $P*AP$ is normal. Thus, U is normal.

$$U* = \begin{bmatrix} \bar{z}_{11} & 0 & \cdots & 0 \\ \bar{z}_{12} & \bar{z}_{22} & & 0 \\ \vdots & \vdots & & \vdots \\ \bar{z}_{1n} & \bar{z}_{2n} & \cdots & \bar{z}_{nn} \end{bmatrix}$$

Since U is normal, $U*U = UU*$. In particular, we have the following equation.

4.18

$$\bar{z}_{11}z_{11} = [U*U]_{11} = [UU*]_{11} = z_{11}\bar{z}_{11} + z_{12}\bar{z}_{12} + \cdots + z_{1n}\bar{z}_{1n}$$

$$= \bar{z}_{11}z_{11} + \sum_{k=2}^{n} |z_{1k}|^2$$

Since each $|z_{1k}|^2$ is a nonnegative real number, equation (4.18) implies $z_{12} = z_{13} = \cdots = z_{1n} = 0$. Thus, U and $U*$ have the following form.

4.19

$$U = \begin{bmatrix} z_{11} & 0 & 0 & \cdots & 0 \\ 0 & z_{22} & z_{23} & \cdots & z_{2n} \\ \vdots & \vdots & \vdots & & \vdots \\ 0 & 0 & \cdots & \cdots & z_{nn} \end{bmatrix}, \quad U* = \begin{bmatrix} \bar{z}_{11} & 0 & \cdots & 0 \\ 0 & \bar{z}_{22} & \cdots & 0 \\ \vdots & \vdots & & \vdots \\ 0 & \bar{z}_{2n} & \cdots & \bar{z}_{nn} \end{bmatrix}$$

Again we have

$$|z_{22}|^2 = \bar{z}_{22}z_{22} = [U*U]_{22} = [UU*]_{22}$$

$$= z_{22}\bar{z}_{22} + \sum_{k=3}^{n} z_{2k}\bar{z}_{2k} = |z_{22}|^2 + \sum_{k=3}^{n} |z_{2k}|^2$$

Thus, $z_{23} = z_{24} = \cdots = z_{2n} = 0$. Now U and U^* have the form

4.20

$$
U = \begin{bmatrix}
z_{11} & 0 & 0 & \cdots & 0 \\
0 & z_{22} & 0 & \cdots & 0 \\
0 & 0 & z_{33} & \cdots & z_{3n} \\
\vdots & \vdots & \vdots & & \vdots \\
0 & 0 & 0 & \cdots & z_{nn}
\end{bmatrix},
$$

$$
U^* = \begin{bmatrix}
\bar{z}_{11} & 0 & 0 & \cdots & 0 \\
0 & \bar{z}_{22} & 0 & \cdots & 0 \\
0 & 0 & \bar{z}_{33} & \cdots & 0 \\
\vdots & \vdots & \vdots & & \vdots \\
0 & 0 & \bar{z}_{3n} & \cdots & \bar{z}_{nn}
\end{bmatrix}
$$

Repeating this argument $n - 1$ times, we have every off-diagonal entry of U is zero. Hence, U is a diagonal matrix, and the proof of Theorem 4.16 is complete. $\qquad\qquad\square$

We close this section with a corollary to Theorem 4.16 which enables us to determine which normal matrices are unitary and Hermitian by examining their eigenvalues.

Corollary 4.21 Let $A \in M_{n \times n}(\mathbb{C})$. Suppose A is normal.

(a) A is Hermitian if and only if $\mathscr{S}_{\mathbb{C}}(A) \subseteq \mathbb{R}$.
(b) A is unitary if and only if $\mathscr{S}_{\mathbb{C}}(A) \subseteq \{z \in \mathbb{C} \mid |z| = 1\}$.

Proof. (a) If A is Hermitian, then $\mathscr{S}_{\mathbb{C}}(A) \subseteq \mathbb{R}$ by Theorem 4.8(b). Conversely, suppose $\mathscr{S}_{\mathbb{C}}(A) \subseteq \mathbb{R}$. Thus, all eigenvalues of A are real. Since A is normal, Theorem 4.16 implies there exists a unitary matrix P such that $P^*AP = D = \text{Diag}(d_1, \ldots, d_n)$. Since $P^* = P^{-1}$, A is similar to D. In particular, Corollary III.4.12 implies $\mathscr{S}_{\mathbb{C}}(D) = \mathscr{S}_{\mathbb{C}}(A) \subseteq \mathbb{R}$. But we have seen in Example III.4.7, that $\mathscr{S}_{\mathbb{C}}(D)$ is precisely the distinct numbers in the list $d_1, \ldots, d_n$. In particular, every d_i is real, and consequently, $D = D^*$. Since $A = PDP^*$, we have

$$
A^* = (PDP^*)^* = PD^*P^* = PDP^* = A.
$$

Thus, A is Hermitian.

(b) If A is unitary, then $\mathscr{S}_{\mathbb{C}}(A) \subseteq \{z \in \mathbb{C} \,|\, |z| = 1\}$ by Corollary 4.5. Conversely, suppose $\mathscr{S}_{\mathbb{C}}(A) \subseteq \{z \in \mathbb{C} \,|\, |z| = 1\}$. Since A is normal, Theorem 4.16 implies there exists a unitary matrix P such that $P^*AP = D = \text{Diag}(d_1, \ldots, d_n)$. The same argument as in (a) shows the spectrum of A is the distinct numbers in the list $d_1, \ldots, d_n$. Thus, $d_i \in \{z \in \mathbb{C} \,|\, |z| = 1\}$ for all $i = 1, \ldots, n$. In particular,

$$DD^* = D^*D = \text{Diag}(d_1 \bar{d}_1, \ldots, d_n \bar{d}_n) = \text{Diag}(|d_1|^2, \ldots, |d_n|^2) = I_n$$

Since $A = PDP^*$, we have

$$A^*A = (PDP^*)^*(PDP^*) = PD^*P^*PDP^* = PD^*DP^* = PP^* = I_n.$$

Thus, A is unitary. $\qquad\qquad\qquad\qquad\qquad\qquad\qquad\qquad\qquad\square$

There is a similar result for skew-Hermitian matrices. We leave this to the exercises.

Exercises for Section 4

4.1 Let $A, B \in M_{n \times n}(\mathbb{C})$ be unitary (or orthogonal). Show AB and A^{-1} are unitary (or orthogonal). Is $A + B$ unitary?

4.2 Let $A \in M_{n \times n}(\mathbb{R})$. Suppose A has an inverse in $M_{n \times n}(\mathbb{C})$. Show A has an inverse in $M_{n \times n}(\mathbb{R})$.

4.3 Let $A \in M_{n \times n}(\mathbb{C})$ be unitary. Show the rows of A form an orthonormal basis of $V = M_{1 \times n}(\mathbb{C})$ with respect to a natural choice of inner product on V.

4.4 Prove Theorem 4.6.

4.5 Let $A \in M_{2 \times 2}(\mathbb{R})$ be orthogonal. Show A is similar to one of the following matrices:

$$\begin{bmatrix} \cos(\theta) & -\sin(\theta) \\ \sin(\theta) & \cos(\theta) \end{bmatrix} \quad \text{or} \quad \begin{bmatrix} \cos(\theta) & \sin(\theta) \\ \sin(\theta) & -\cos(\theta) \end{bmatrix}$$

Here $0 \leqslant \theta < \pi$.

4.6 Let $A \in M_{n \times n}(\mathbb{C})$ be skew-Hermitian. Show every eigenvalue of A is purely imaginary, i.e., $\mathscr{S}_{\mathbb{C}}(A) \subseteq \{z \in \mathbb{C} \,|\, \bar{z} = -z\}$.

4.7 Show any matrix in $M_{n \times n}(\mathbb{C})$ is a sum of a Hermitian and skew-Hermitian matrix.

4.8 Show unitary, Hermitian, and skew-Hermitian matrices are all normal.

4.9 Let $A \in M_{n \times n}(\mathbb{C})$. Show A is normal if and only if $\|A\alpha\| = \|A^*\alpha\|$ for all $\alpha \in \mathbb{C}^n$.

4.10 Let $A \in M_{n \times n}(\mathbb{C})$. Show A is normal if and only if $\text{Tr}(A^*A) = \sum_{k=1}^{n} |d_i|^2$ where $\{d_1, \ldots, d_n\} = \mathscr{S}_{\mathbb{C}}(A)$. The d_i are not necessarily distinct here.

4.11 Prove Theorem 4.13(b).

4.12 Let $A \in M_{n \times n}(\mathbb{C})$ be normal. Show that eigenvectors of A belonging to distinct eigenvalues of A are orthogonal.

4.13 Let $A \in M_{n \times n}(\mathbb{C})$ be normal. Show A is skew-Hermitian if and only if the eigenvalues of A are purely imaginary.

4.14 Is the product of two normal matrices normal?

4.15 Let $A, B \in M_{n \times n}(\mathbb{C})$ be normal matrices. Suppose AB is normal. Show BA is also normal.

4.16 For each matrix A listed below, find a unitary matrix P such that P^*AP is diagonal.

(a) $\begin{bmatrix} 1 & 2 \\ 2 & 3 \end{bmatrix}$

(b) $\begin{bmatrix} 1 & i \\ i & 1 \end{bmatrix}$

(c) $\begin{bmatrix} 0 & 1 & 1 \\ -1 & 0 & 2 \\ -1 & -2 & 0 \end{bmatrix}$

(d) $\begin{bmatrix} 4 & -1 & 1 \\ -1 & 4 & -1 \\ 1 & -1 & 4 \end{bmatrix}$

(e) $\begin{bmatrix} 1 & 0 & 0 & 0 \\ 0 & 0 & 1 & 0 \\ 0 & 1 & 0 & 0 \\ 0 & 0 & 0 & 1 \end{bmatrix}$

4.17 Let $A \in M_{n \times n}(\mathbb{C})$ be normal. Suppose $A^k = 0$ for some $k \geqslant 1$. Show $A = 0$.

4.18 Let $(V, \langle *, * \rangle)$ denote a finite dimensional, inner product space over F. Let $T \in \text{Hom}(V, V)$. Show that for every $\beta \in V$, there exists a unique vector $\gamma \in V$ such that $\langle T(\alpha), \beta \rangle = \langle \alpha, \gamma \rangle$ for all $\alpha \in V$.

4.19 Let $(V, \langle *, * \rangle)$ denote a finite dimensional, inner product space over F. Let $T \in \text{Hom}(V, V)$. Use Exercise 4.18 to show there exists a unique $T^* \in \text{Hom}(V, V)$ such that $\langle T(\alpha), \beta \rangle = \langle \alpha, T^*(\beta) \rangle$ for all α, $\beta \in V$.

5. REAL QUADRATIC FORMS

In this last section of the text, we will use Schur's Theorem to analyze real quadratic forms on $\mathbb{R}^n$. Throughout this discussion, $\langle *, * \rangle$ will denote the standard inner product on $\mathbb{R}^n$. Thus, $\langle \alpha, \beta \rangle = \alpha^t \beta$ for all $\alpha, \beta \in \mathbb{R}^n$. We begin with the definition of a real quadratic form.

Definition 5.1   Let $A \in M_{n \times n}(\mathbb{R})$ be a symmetric matrix. The function $q : \mathbb{R}^n \mapsto \mathbb{R}$ defined by $q(\alpha) = \langle A\alpha, \alpha \rangle$ is called the real quadratic form associated with A.

Definition 5.2   A function $q : \mathbb{R}^n \mapsto \mathbb{R}$ is called a real quadratic form if q is the real quadratic form associated with some symmetric matrix $A \in M_{n \times n}(\mathbb{R})$.

Suppose $q : \mathbb{R}^n \mapsto \mathbb{R}$ is a real quadratic form associated with the symmetric matrix $A \in M_{n \times n}(\mathbb{R})$. If $\alpha = [x_1 \ \cdots \ x_n]^t \in \mathbb{R}^n$, then we will often write $q(x_1, \ldots, x_n)$ for $q(\alpha)$. Thus, as α ranges over $\mathbb{R}^n$, $q(\alpha)$ is a function of n variables $x_1, \ldots, x_n$. If $A = (a_{ij})$, then

$$q(\alpha) = \langle A\alpha, \alpha \rangle = (A\alpha)^t \alpha = \alpha^t A^t \alpha = \alpha^t A \alpha$$

$$= [x_1 \ \cdots \ x_n] A [x_1 \ \cdots \ x_n]^t$$

$$= \sum_{i=1}^{n} \sum_{j=1}^{n} a_{ij} x_i x_j$$

Since A is symmetric, this sum can be rewritten as follows.

5.3
$$q(x_1, \ldots, x_n) = a_{11} x_1^2 + \cdots + a_{nn} x_n^2 + 2 \sum_{i<j} a_{ij} x_i x_j$$

In equation (5.3), $\sum_{i<j} a_{ij} x_i x_j$ means the sum of the real numbers $a_{ij} x_i x_j$ as (i, j) ranges over all ordered pairs for which $1 \leqslant i < j \leqslant n$. In many textbooks, equation (5.3) is taken to be the definition of a real quadratic form $q(x_1, \ldots, x_n)$ on $\mathbb{R}^n$.

We should point out here that the symmetric matrix defining the real quadratic form q is unique. By this we mean, if $q(\alpha) = \langle A\alpha, \alpha \rangle = \langle B\alpha, \alpha \rangle$ for all $\alpha \in \mathbb{R}^n$, and A, B are symmetric matrices, then $A = B$. To see this, first note that $A - B$ is symmetric and $\langle (A - B)\alpha, \alpha \rangle = 0$ for all $\alpha \in \mathbb{R}^n$. The eigenvalues of $A - B$ (in $\mathbb{C}$) are all real numbers by Corollary 4.9. Suppose d is an eigenvalue of $A - B$ and α is an eigenvector of $A - B$ belonging to d. Then we have

$$0 = \langle (A - B)\alpha, \alpha \rangle = \langle d\alpha, \alpha \rangle = d \|\alpha\|^2$$

Thus, $d = 0$. In particular, every eigenvalue of $A - B$ is zero. We have seen in Section 4 that the eigenvalues of $A - B$ fill out the diagonal of any diagonal matrix similar to $A - B$. Hence, $A - B = O$ by Corollary 4.14.

In particular, if $q(\alpha) = \langle A\alpha, \alpha \rangle$ for all $\alpha \in \mathbb{R}^n$ and A is symmetric, then we will call A *the* matrix of the quadratic form q. Equivalently, if q is written as in equation (5.3), then we will call the symmetric matrix $A = (a_{ij})$ the matrix of q.

Example 5.4 The quadratic from $q(x, y) = ax^2 + bxy + cy^2$ on $\mathbb{R}^2$ can be written as

$$q(x, y) = \begin{bmatrix} x & y \end{bmatrix} \begin{bmatrix} a & b/2 \\ b/2 & c \end{bmatrix} \begin{bmatrix} x \\ y \end{bmatrix}$$

$$A = \begin{bmatrix} a & b/2 \\ b/2 & c \end{bmatrix}$$

is the matrix of $q(x, y)$.

Equivalently, we could write

$$q(x, y) = ax^2 + cy^2 + 2(b/2)xy$$

Thus, from (5.3) we again get

$$A = \begin{bmatrix} a & b/2 \\ b/2 & c \end{bmatrix}$$

is the matrix of $q(x, y)$.

If $q : \mathbb{R}^n \mapsto \mathbb{R}$ is a real quadratic form on $\mathbb{R}^n$, then the equation

5.5 $$q(x_1, \ldots, x_n) = 1$$

is called the equation of q. In the calculus, students are interested in finding all solutions to this equation when $n = 2$ or 3. This is equivalent to drawing the graph of $q(x_1, \ldots, x_n) = 1$ in $\mathbb{R}^n$. The way this type of problem is solved is to make a linear substitution of variables,

$$\begin{bmatrix} x_1 & \cdots & x_n \end{bmatrix}^t = P \begin{bmatrix} x_1' & \cdots & x_n' \end{bmatrix}^t,$$

where P is an invertible matrix in $M_{n \times n}(\mathbb{R})$. This is done in a judicious manner so that the resulting quadratic form $q(P[x_1' \quad \cdots \quad x_n']^t)$ in $x_1', \ldots, x_n'$ is easy to identify. Consider the following familiar example from the calculus.

Example 5.6 Draw the graph of $q(x, y) = ax^2 + bxy + cy^2 = 1$ in $\mathbb{R}^2$.

There exists a suitable rotation of the coordinate axes such that $q(x, y)$ in the new coordinate system x', y' has the form $a_1(x')^2 + a_2(y')^2$. A rotation of the coordinate axes (in the counter-clockwise direction) through an angle θ is given by the following linear substitution.

5.7
$$\begin{bmatrix} x \\ y \end{bmatrix} = \begin{bmatrix} \cos(\theta) & -\sin(\theta) \\ \sin(\theta) & \cos(\theta) \end{bmatrix} \begin{bmatrix} x' \\ y' \end{bmatrix}$$

Notice that the matrix

$$P = \begin{bmatrix} \cos(\theta) & -\sin(\theta) \\ \sin(\theta) & \cos(\theta) \end{bmatrix}$$

is orthogonal.

Let

$$A = \begin{bmatrix} a & b/2 \\ b/2 & c \end{bmatrix}$$

be the matrix of q Substituting $[x, y]^t = P[x', y']^t$ into the equation $q(x, y) = 1$, gives

$$1 = q(x, y) = [x, y]A[x, y]^t = [x', y']P^t A P[x', y']^t$$
$$= a_1(x')^2 + a_2(y')^2$$

(if θ has been chosen correctly). Hence, the equation $ax^2 + bxy + cy^2 = 1$ has been replaced with the simpler equation $a_1(x')^2 + a_2(y')^2 = 1$. It is now a simple matter to identify this graph (ellipse, hyperbola, etc.) depending on what the signs of a_1 and a_2 are.

Notice in this substitution, that the matrix A of q is replaced by $P^t A P$.

We can carry out the computations in Example 5.6 for any quadratic form q on $\mathbb{R}^n$. Suppose $q(x_1, \ldots, x_n)$ is a real quadratic form on $\mathbb{R}^n$ with associated symmetric matrix A. Let $P \in M_{n \times n}(\mathbb{R})$ be invertible. For every $[x_1 \cdots x_n]^t \in \mathbb{R}^n$, there exists a unique $[x'_1 \cdots x'_n]^t \in \mathbb{R}^n$ such that

$$[x_1 \cdots x_n]^t = P[x'_1 \cdots x'_n]^t$$

since P is invertible. Let $\alpha = [x_1 \cdots x_n]^t$ and $\alpha' = [x'_1 \cdots x'_n]^t$. Then

$$q(x_1, \ldots, x_n) = q(\alpha) = q(P\alpha') = \langle AP\alpha', P\alpha' \rangle = (AP\alpha')^t P\alpha'$$
$$= (\alpha')^t (P^t A P)\alpha' = q'(x'_1, \ldots, x'_n).$$

Here q' is the real quadratic form in $x'_1, \ldots, x'_n$ given by the symmetric matrix $P^t A P$. The equation $q(x_1, \ldots, x_n) = 1$ becomes $q'(x'_1, \ldots, x'_n) = 1$. We have now proven the following lemma.

Lemma 5.8   Let $q(x_1, \ldots, x_n)$ be a real quadratic form on $\mathbb{R}^n$ with associated symmetric matrix A. Let $P \in M_{n \times n}(\mathbb{R})$ be invertible. Set

$$\begin{bmatrix} x_1 \\ x_2 \\ \vdots \\ x_n \end{bmatrix} = P \begin{bmatrix} x_1' \\ x_2' \\ \vdots \\ x_n' \end{bmatrix}$$

and

$$q'(x_1', \ldots, x_n') = [x_1', \ldots, x_n'](P^t A P) \begin{bmatrix} x_1' \\ x_2' \\ \vdots \\ x_n' \end{bmatrix}$$

Then $q(x_1, \ldots, x_n) = q'(x_1', \ldots, x_n')$ for all $x_1, \ldots, x_n \in \mathbb{R}$.

Thus, the real quadratic form $q(x_1, \ldots, x_n)$ with associated symmetric matrix A is changed by a linear substitution,

$$[x_1 \quad \cdots \quad x_n]^t = P[x_1' \quad \cdots \quad x_n']^t$$

into the real quadratic form $q'(x_1', \ldots, x_n')$ with associated, symmetric matrix $P^t A P$. The linear substitution

$$[x_1 \quad \cdots \quad x_n]^t = P[x_1' \quad \cdots \quad x_n']^t$$

is a change of coordinates in $\mathbb{R}^n$. Hence, we insist that P be invertible.
Consider the following simple example.

Example 5.9   Let $q(x, y) = xy$ on $\mathbb{R}^2$. The matrix of q is

$$A = \begin{bmatrix} 0 & 1/2 \\ 1/2 & 0 \end{bmatrix}$$

Let

$$P = \begin{bmatrix} 1 & -1 \\ 1 & 1 \end{bmatrix}$$

and set

$$\begin{bmatrix} x \\ y \end{bmatrix} = \begin{bmatrix} 1 & -1 \\ 1 & 1 \end{bmatrix} \begin{bmatrix} x' \\ y' \end{bmatrix}$$

Then

$$P^t AP = \begin{bmatrix} 1 & 0 \\ 0 & -1 \end{bmatrix}$$

and

$$q'(x', y') = [x', y'](P^t AP)[x', y']^t = (x')^2 - (y')^2$$

Thus

$$q(x, y) = xy = (x' - y')(x' + y') = (x')^2 + (y')^2 = q'(x', y')$$

The matrices A and $P^t AP$ in Lemma 5.8 come from quadratic forms which are describing the same thing in two different coordinate systems. This prompts us to introduce the following definition.

Definition 5.10 Two symmetric matrices A, $B \in M_{n \times n}(\mathbb{R})$ are said to be congruent if $P^t AP = B$ for some invertible matrix P.

Thus, if $q'(x'_1, \ldots, x'_n)$ is obtained from $q(x_1, \ldots, x_n)$ via the linear substitution

$$[x_1 \quad \cdots \quad x_n]^t = P[x'_1 \quad \cdots \quad x'_n]^t$$

(with P invertible), then the associated symmetric matrices of q and q' are congruent. If A and B are two symmetric matrices which are congruent, we will write $A \equiv B$. We will need the following lemma regarding the relation $\equiv$.

Lemma 5.11 Let A, B, and C be symmetric matrices in $M_{n \times n}(\mathbb{R})$. Then

(a) $A \equiv A$.
(b) $A \equiv B$ if and only if $B \equiv A$.
(c) If $A \equiv B$ and $B \equiv C$, then $A \equiv C$.

Proof. (a) $I_n^t A I_n = A$. Therefore, $A \equiv A$.

(b) Suppose $A \equiv B$. Then there exists an invertible matrix $P \in M_{n \times n}(\mathbb{R})$ such that $P^t AP = B$. We have seen in Theorem I.2.35(d) that $(P^{-1})^t = (P^t)^{-1}$. Therefore,

$$(P^{-1})^t B(P^{-1}) = (P^{-1})^t P^t AP(P^{-1}) = A$$

Thus, $B \equiv A$.

(c) If $A \equiv B$, and $B \equiv C$, then $P^t AP = B$ and $Q^t BQ = C$ for invertible matrices P and Q. Then

$$(PQ)^t A(PQ) = Q^t P^t APQ = Q^t BQ = C.$$

Thus, $A \equiv C$. □

Recall that a relation satisfying (a), (b), and (c) in Lemma 5.11 is called an equivalence relation. Thus, $\equiv$ is an equivalence relation on the subspace of all symmetric matrices in $M_{n \times n}(\mathbb{R})$. If A is a given symmetric matrix in $M_{n \times n}(\mathbb{R})$, then the congruence class of A is the set of all symmetric matrices in $M_{n \times n}(\mathbb{R})$ which are congruent to A. Obviously, the congruence class of A contains A itself. If $q(x_1, \ldots, x_n)$ is the real quadratic form defined by A, then Lemma 5.8 implies that the symmetric matrix of $q'(x'_1, \ldots, x'_n)$ (obtained from q via a linear substitution) is in the congruence class of A. Thus, the whole question of how simple can we make q via a linear substitution $\alpha = P\alpha'$ is equivalent to what is the simplest symmetric matrix in the congruence class of A?

We already have all the tools needed to give a good answer to this question. The result is called Sylvester's law of inertia. Before stating Sylvester's theorem, we need two preliminary results.

Suppose $E = E_{ij}$ is a transposition in $M_{n \times n}(\mathbb{R})$. We can assume $1 \leqslant i \leqslant j \leqslant n$. Then $E^2 = I_n$. The columns of E are the same as the columns of I_n except that columns i and j (in I_n) have been interchanged. In particular, the columns of E form an orthonormal basis of $\mathbb{R}^n$. Theorem 4.6(c) implies E is an orthogonal matrix. In particular, $E^t = E^{-1} = E$.

If $D = \text{Diag}(d_1, \ldots, d_n)$, then a simple computation shows

$$EDE = E^t DE$$

$$= \text{Diag}(d_1, \ldots, d_{i-1}, d_j, d_{i+1}, \ldots, d_{j-1}, d_i, d_{j+1}, \ldots, d_n)$$

Thus, multiplying D on the left and right with E interchanges the ith and jth entries of the diagonal of D.

Now suppose P is an arbitrary permutation matrix. Then P is a finite product, $P = E_1 E_2 \cdots E_r$, of transpositions E_i. Since each E_i is orthogonal, P is orthogonal. If D is a diagonal matrix, then

$$P^t DP = [E_1 \quad \cdots \quad E_r]^t D[E_1 \quad \cdots \quad E_r] = E_r(\cdots(E_2(E_1 DE_1)E_2)\cdots)E_r$$

The nature of this product makes the following statement clear: If $\sigma \in S_n$ and if $D = \text{Diag}(d_1, \ldots, d_n)$, then there exists a permutation matrix P such that $P^t DP = \text{Diag}(d_{\sigma(1)}, d_{\sigma(2)}, \ldots, d_{\sigma(n)})$.

Now suppose A is a symmetric matrix in $M_{n \times n}(\mathbb{R})$. By Corollary 4.14, there exists an orthogonal matrix P_1 such that $P_1^t A P_1 = \text{Diag}(d_1, \ldots, d_n)$. The real numbers $d_1, \ldots, d_n$ are just the eigenvalues of A. Each distinct d_i appears with a certain multiplicity in the list $d_1, \ldots, d_n$. Suppose r of the numbers $d_1, \ldots, d_n$ are positive, s are negative, and the rest are zero. Then r, $s \geqslant 0$, $r + s \leqslant n$, and $r + s < n$ if some $d_i = 0$. For instance, if $D = \text{Diag}(1, 1, 1, -2, 0, 0, -2, 3, 4)$, then $n = 9, r = 5$, and $s = 2$. There is a permutation $\sigma \in S_n$ such that $d_{\sigma(1)}, \ldots, d_{\sigma(r)} > 0, d_{\sigma(r+1)}, \ldots, d_{\sigma(r+s)} < 0$, and $d_{\sigma(i)} = 0$ for $i > r + s$. From our discussion in the last paragraph, we conclude there exists a permutation matrix P_2 such that

$$P_2^t(P_1^t A P_1)P_2 = \text{Diag}(d_{\sigma(1)}, \ldots, d_{\sigma(r)}, d_{\sigma(r+1)}, \ldots, d_{\sigma(r+s)}, 0, \ldots, 0)$$

In this discussion it is perfectly possible that r, or s, or both r and s might be zero. If A has no positive eigenvalues, then $r = 0$. If A has no negative eigenvalues, then $s = 0$. If $r = s = 0$, the A is clearly zero. Since a permutation matrix is orthogonal, P_2 is orthogonal. Hence, $P = P_1 P_2$ is orthogonal. We have now proven the following lemma.

Lemma 5.12 Let $A \in M_{n \times n}(\mathbb{R})$ be symmetric. Then there exists an orthogonal matrix P such that

$$P^t A P = \text{Diag}(d_1, \ldots, d_r, d_{r+1}, \ldots, d_{r+s}, 0 \cdots 0)$$

Here $d_i > 0$ for $i \leqslant r$, $d_i < 0$ for $r + 1 \leqslant i \leqslant r + s$ and $0 \leqslant r + s \leqslant n$.

The other result we will need for the proof of Sylvester's law is the following dimension formula.

Lemma 5.13 Let V be a finite dimensional vector space over F. For any subspaces W_1 and W_2 of V,

$$\dim W_1 + \dim W_2 = \dim(W_1 + W_2) + \dim(W_1 \cap W_2)$$

Proof. The proof of this theorem is easy. We will give a brief sketch of what to do and leave the details to the exercises. Let $B = \{\alpha_1, \ldots, \alpha_r\}$ be a basis of $W_1 \cap W_2$. Expand B to a basis $C_1 = \{\alpha_1, \ldots, \alpha_r, \beta_1, \ldots, \beta_s\}$ of W_1. Also expand B to a basis $C_2 = \{\alpha_1, \ldots, \alpha_r, \gamma_1, \ldots, \gamma_t\}$ of W_2. Thus, $\dim(W_1 \cap W_2) = r$, $\dim W_1 = r + s$, and $\dim W_2 = r + t$. The proof is completed by showing that

$\{\alpha_1, \ldots, \alpha_r, \beta_1, \ldots, \beta_s, \gamma_1, \ldots, \gamma_t\}$ is a basis of $W_1 + W_2$. $\qquad \square$

We can now present Sylvester's law of inertia.

Theorem 5.14 (Sylvester's Law of Inertia)   Let $A \in M_{n \times n}(\mathbb{R})$ be symmetric. Then there exists an invertible matrix P such that

$$P^t A P = \begin{bmatrix} I_r & 0 & 0 \\ \hline 0 & -I_s & 0 \\ \hline 0 & 0 & 0 \end{bmatrix}$$

Furthermore, the two integers $r, s \geqslant 0$ determine A up to congruence. That is, two symmetric matrices are congruent if and only if they have the same r and s.

Proof.   Since A is symmetric, Lemma 5.12 implies there exists an orthogonal matrix P_1 such that

$$P_1^t A P_1 = D = \text{Diag}(d_1, \ldots, d_r, d_{r+1}, \ldots, d_{r+s}, 0, \ldots, 0)$$

where $d_i > 0$ for $i \leqslant r$, $d_i < 0$ for $r + 1 \leqslant i \leqslant r + s$, and $r + s \leqslant n$. Set

$$P_2 = \text{Diag}(1/\sqrt{d_1}, \ldots, 1/\sqrt{d_r}, 1/\sqrt{-d_{r+1}}, \ldots, 1/\sqrt{-d_{r+s}}, 1, \ldots, 1)$$

P_2 is invertible since no diagonal entry of P_2 is zero.

$$P_2^t D P_2 = P_2 D P_2 = \text{Diag}(1, \ldots, 1, -1, \ldots, -1, 0, \ldots, 0)$$

Thus

$$(P_1 P_2)^t A (P_1 P_2) = \begin{bmatrix} I_r & 0 & 0 \\ \hline 0 & -I_s & 0 \\ \hline 0 & 0 & 0 \end{bmatrix}$$

Since $P = P_1 P_2$ is a product of invertible matrices, P is invertible. This completes the first part of the proof of Sylvester's law.

Now suppose two symmetric matrices A and B in $M_{n \times n}(\mathbb{R})$ have the same r and s. Thus, there exist invertible matrices P and Q such that

$$P^t A P = \begin{bmatrix} I_r & 0 & 0 \\ \hline 0 & -I_s & 0 \\ \hline 0 & 0 & 0 \end{bmatrix} = Q^t B Q$$

Then clearly, $A \equiv P^t A P \equiv Q^t B Q \equiv B$. Thus, two symmetric matrices are congruent if they have the same r and s.

Now suppose A and B are symmetric matrices in $M_{n \times n}(\mathbb{R})$ which are congruent. By the first part of this proof, there exist invertible matrices P

and Q such that

$$P^tAP = \left[\begin{array}{c|c|c} I_r & O & O \\ \hline O & -I_s & O \\ \hline O & O & O \end{array}\right] = L$$

and

$$Q^tBQ = \left[\begin{array}{c|c|c} I_{r'} & O & O \\ \hline O & -I_{s'} & O \\ \hline O & O & O \end{array}\right] = K.$$

We claim $r = r'$ and $s = s'$. Notice that

$$L = P^tAP \equiv (P^{-1})^t(P^tAP)(P^{-1}) = A \equiv B = (Q^{-1})^t(Q^tBQ)(Q^{-1}) \equiv K$$

Hence, $L \equiv K$. In particular, $K = T^tLT$ for some invertible matrix $T \in M_{n \times n}(\mathbb{R})$.

Let $p = n - (r + s)$ and $p' = n - (r' + s')$. Then $p \times p$ ($p' \times p'$) is the size of the zero matrix in the lower right-hand corner of L (K). Since T is invertible, $\text{rk}(L) = \text{rk}(K)$. Therefore, $r + s = \text{rk}(L) = \text{rk}(K) = r' + s'$. In particular, it suffices to show $r = r'$.

Suppose $r \neq r'$. We can assume $r < r'$. Since $r + s = r' + s'$, $s > s'$. Notice then that $s > 0$ and $r' > 0$. Let U be the subset of $\mathbb{R}^n$ consisting of all column vectors whose first r and last p coordinates are zero. Thus,

$$U = \{[0 \quad \cdots \quad 0 \quad x_{r+1} \quad \cdots \quad x_{r+s} \quad 0 \quad \cdots \quad 0]^t \in \mathbb{R}^n \,|\, x_{r+j} \in \mathbb{R},$$
$$j = 1, \ldots, s\}.$$

Clearly, U is a subspace of $\mathbb{R}^n$, and $\dim U = s > 0$. If

$$\gamma = [0 \quad \cdots \quad 0 \quad x_{r+1} \quad \cdots \quad x_{r+s} \quad 0 \quad \cdots \quad 0]^t$$

is a nonzero vector in U, then

$$\langle L\gamma, \gamma \rangle = \gamma^t L^t \gamma = \gamma^t L \gamma = -\sum_{i=1}^{s} x_{r+i}^2 < 0$$

Thus, we have proven the following assertion.

5.15 $\langle L\gamma, \gamma \rangle < 0$ for any nonzero vector $\gamma \in U$.

Let W be the subset of $\mathbb{R}^n$ consisting of all column vectors whose $r' + 1$, $\ldots, r' + s'$ entries are zero. Thus,

$$W = \{[x_1 \quad \cdots \quad x_{r'} \quad 0 \quad \cdots \quad 0 \quad x_{r'+s'+1} \quad \cdots \quad x_n]^t \in \mathbb{R}^n \,|\, x_j \in \mathbb{R}\}$$

for $j = 1, \ldots, r'$ and $j = r' + s' + 1, \ldots, n$. Clearly W is a subspace of $\mathbb{R}^n$, and $\dim W = n - s' = r' + p'$. Notice $\dim W > 0$ since $r' > 0$. For any vector

$$\beta = [x_1 \quad \cdots \quad x_{r'} \quad 0 \quad \cdots \quad 0 \quad x_{r'+s'+1} \quad \cdots \quad x_n]^t \in W$$

$$\langle K\beta, \beta \rangle = (K\beta)^t\beta = \beta^t K^t \beta = \beta^t K\beta = \sum_{i=1}^{r'} x_i^2 \geq 0$$

Let $TW = \{T\beta \mid \beta \in W\}$. Then TW is a subspace of $\mathbb{R}^n$ and $\dim TW = \dim W = n - s'$ since T is invertible. Using Theorem 4.1, we have for any $\beta \in W$,

$$0 \leq \langle K\beta, \beta \rangle = \langle T^t LT\beta, \beta \rangle = \langle LT\beta, T\beta \rangle$$

We have now proven the following assertion.

5.16 $\qquad\qquad\qquad \langle L\gamma, \gamma \rangle \geq 0 \quad$ for all $\gamma \in TW$

We finish this proof by showing that the assertions in (5.15) and (5.16) lead to a contradiction. Since $s > s'$, $\dim TW + \dim U = (n - s') + s > n$ On the other hand, Lemma 5.13 implies

$$n \geq \dim(TW + U) = \dim TW + \dim U - \dim(TW \cap U)$$

Since $\dim TW + \dim U > n$, we conclude that $\dim(TW \cap U)$ is positive. In particular, $TW \cap U \neq (0)$. But, if γ is a nonzero vector in $TW \cap U$, then (5.15) implies $\langle L\gamma, \gamma \rangle < 0$ and (5.16) implies $\langle L\gamma, \gamma \rangle \geq 0$. This is clearly impossible. We conclude that $r = r'$ and, consequently, $s = s'$. $\qquad\square$

We had noted in the proof of Theorem 5.14 that if

$$P^t AP = \left[\begin{array}{c|c|c} I_r & O & O \\ \hline O & -I_s & O \\ \hline O & O & O \end{array}\right]$$

then $r + s = \mathrm{rk}(A)$. The integer $r - s$ is called the signature of A. We will let $\mathrm{sig}(A)$ denote the signature of A. Theorem 5.14 implies that the two integers $\mathrm{rk}(A)$ and $\mathrm{sig}(A)$ depend only on the congruence class of A. In particular, if

$$Q^t AQ = \left[\begin{array}{c|c|c} I_{r'} & O & O \\ \hline O & -I_{s'} & O \\ \hline O & O & O \end{array}\right]$$

for some invertible matrix Q, then $P^t A P \equiv A \equiv Q^t A Q$ implies $r = r'$ and $s = s'$. Thus, the integers $rk(A)$ and $sig(A)$ do not depend on any particular P or Q.

Since $r = (1/2)(rk(A) + sig(A))$ and $s = (1/2)(rk(A) - sig(A))$, Theorem 5.14 implies the two integers $rk(A)$ and $sig(A)$ completely determine the congruence class of A. Thus, two $n \times n$, symmetric matrices are congruent if and only if they have the same rank and signature.

Suppose $A \in M_{n \times n}(\mathbb{R})$ is symmetric and $q(x_1, \ldots, x_n)$ is the quadratic form associated with A. Using the same notation as in Lemma 5.8, we have the following corollary to Theorem 5.14.

Corollary 5.17 Let $q(x_1, \ldots, x_n)$ be a real quadratic form on $\mathbb{R}^n$ with associated symmetric matrix A. Then there exists a linear substitution,

$$[x_1 \quad \cdots \quad x_n]^t = P[x_1' \quad \cdots \quad x_n']^t$$

(P invertible), such that

$$q'(x_1', \ldots, x_n') = \sum_{i=1}^{r} (x_i')^2 - \sum_{i=r+1}^{r+s} (x_i')^2.$$

Here $r = (1/2)(rk(A) + sig(A))$ and $s = (1/2)(rk(A) - sig(A))$.

Proof. By Theorem 5.14, there exists an invertible matrix $P \in M_{n \times n}(\mathbb{R})$ such that

$$P^t A P = \begin{bmatrix} I_r & O & O \\ O & -I_s & O \\ O & O & O \end{bmatrix}$$

$r + s = rk(A)$ and $r - s = sig(A)$. Thus,

$$r = (1/2)(rk(A) + sig(A))$$

and

$$s = (1/2)(rk(A) - sig(A)).$$

The linear substitution

$$[x_1 \quad \cdots \quad x_n]^t = P[x_1' \quad \cdots \quad x_n']^t,$$

with $\alpha = [x_1 \quad \cdots \quad x_n]^t$ and $\alpha' = [x'_1 \quad \cdots \quad x'_n]^t$, has the following effect on q

$$q(x_1, \ldots, x_n) = q(\alpha) = q(P\alpha') = \langle A(P\alpha'), P\alpha' \rangle$$

$$= (\alpha')^t(P^tAP)(\alpha') = [x'_1 \quad \cdots \quad x'_n] \begin{bmatrix} I_r & 0 & 0 \\ \hline 0 & -I_s & 0 \\ \hline 0 & 0 & 0 \end{bmatrix} \begin{bmatrix} x'_1 \\ x'_2 \\ \vdots \\ x'_n \end{bmatrix}$$

$$= \sum_{i=1}^{r} (x'_i)^2 - \sum_{i=r+1}^{r+s} (x'_i)^2 = q'(x'_1, \ldots, x'_n)$$

$\square$

Corollary 5.17 gives us a method for studying the equation $q(x_1, \ldots, x_n) = 1$ in $\mathbb{R}^n$. We find a suitable linear substitution

$$[x_1 \quad \cdots \quad x_n]^t = P[x'_1 \quad \cdots \quad x'_n]^t$$

such that P^tAP has the form given in Corollary 5.17. Then the equation $q(x_1, \ldots, x_n) = 1$ is equivalent to the equation

$$\sum_{i=1}^{r} (x'_i)^2 - \sum_{i=r+1}^{r+s} (x'_i)^2 = 1$$

This last equation is obviously easier to analyze than the general equation in (5.3). For example, if $n = 2$ and $r = 2$, then $(x'_1)^2 + (x'_2)^2 = 1$ is clearly the equation of an ellipse in the prime coordinate system. If $n = 2$ and $r = s = 1$, then $(x'_1)^2 - (x'_2)^2 = 1$ is the equation of a hyperbola in the prime coordinate system. Let us return to Example 5.9.

Example 5.18   Let $q(x, y) = xy$ on $\mathbb{R}^2$. We have seen that the matrix of q is

$$A = \begin{bmatrix} 0 & 1/2 \\ 1/2 & 0 \end{bmatrix}$$

$$\mathscr{S}_{\mathbb{R}}(A) = \{1/2, -1/2\}$$

$$\alpha_1 = \begin{bmatrix} 1/\sqrt{2} \\ 1/\sqrt{2} \end{bmatrix}$$

is a unit eigenvector of A belonging to $1/2$.

$$\alpha_2 = \begin{bmatrix} -1/\sqrt{2} \\ 1/\sqrt{2} \end{bmatrix}$$

is a unit eigenvector of A belonging to $-1/2$. Thus,

$$P_1 = \begin{bmatrix} 1/\sqrt{2} & -1/\sqrt{2} \\ 1/\sqrt{2} & 1/\sqrt{2} \end{bmatrix}$$

is an orthogonal matrix for which $P_1^t A P_1 = \mathrm{Diag}(1/2, -1/2)$. Set $P_2 = \mathrm{Diag}(\sqrt{2}, \sqrt{2})$. Then $P = P_1 P_2$ is an invertible matrix for which

$$P^t A P = \begin{bmatrix} 1 & 0 \\ 0 & -1 \end{bmatrix}$$

Thus, $r = 1$, and $s = 1$ for the matrix A. In particular, $\mathrm{rk}(A) = 2$ and $\mathrm{sig}(A) = 0$.

Substituting

$$\begin{bmatrix} x \\ y \end{bmatrix} = P \begin{bmatrix} x' \\ y' \end{bmatrix} = \begin{bmatrix} 1 & -1 \\ 1 & 1 \end{bmatrix} \begin{bmatrix} x' \\ y' \end{bmatrix}$$

in $xy = q(x, y)$ gives $q'(x', y') = (x')^2 - (y')^2$. Thus, $q(x, y) = 1$ is a hyperbola.

We have been careful in this section to discuss only quadratic forms

$$a_{11}x_1^2 + \cdots + a_{nn}x_n^2 + \sum_{i<j} a_{ij}x_i x_j$$

on $\mathbb{R}^n$. A quadratic function $g(x_1, \ldots, x_n)$ on $\mathbb{R}^n$ is any function of the following form.

5.19
$$g(x_1, \ldots, x_n) = \sum_{i=1}^{n} \sum_{j=1}^{n} a_{ij}x_i x_j + \sum_{k=1}^{n} b_k x_k + c$$

Thus, a general quadratic function $g(x_1, \ldots, x_n)$ is a sum of a quadratic form $\sum_{i=1}^{n} \sum_{j=1}^{n} a_{ij}x_i x_j$, a linear function $\sum_{k=1}^{n} b_k x_k$ and a constant c. There is a corresponding result to Corollary 5.17 for quadratic functions. Namely, there exists an affine transformation

$$X = PX' + K \ (P \in M_{n \times n}(\mathbb{R}), K \in \mathbb{R}^n)$$

which reduces $g(x_1, \ldots, x_n)$ to one of the following two forms:

$$(x_1')^2 + \cdots + (x_r')^2 - (x_{r+1}')^2 - \cdots - (x_{r+s}')^2 + x_{r+s+1}'$$

or

$$(x_1')^2 + \cdots + (x_r')^2 - (x_{r+1}')^2 - \cdots -(x_{r+s}')^2 + c$$

The main work in proving this assertion is reducing the quadratic form part of $g(x_1, \ldots, x_n)$ to the difference of two sums of squares. We have already done this in Theorem 5.14. The rest of the argument is fairly easy. A complete proof can be found in [1].

Exercises for Section 5

5.1   Verify that $q(x_1, \ldots, x_n)$ is the sum given in equation (5.3).

5.2   Derive a trigonometric formula for the θ in equation (5.7) which eliminates the cross product term bxy from $q(x, y)$ in Example 5.6.

5.3   Let $S = \{A \in M_{n \times n}(\mathbb{R}) \mid A = A^t\}$. For any $A \in S$, let $[A]$ denote the congruence class containing A. Prove the following assertions:
   (a) $A \in [A]$.
   (b) If $A, B \in S$, then $[A] = [B]$ if and only if $A \equiv B$.
   (c) If $A, B \in S$, then $[A] = [B]$ or $[A] \cap [B] = \varnothing$.

5.4   In Exercise 5.3, show that S is a disjoint union of finitely many congruence classes. (Hint: Use Theorem 5.14.)

5.5   Let $A \in M_{n \times n}(\mathbb{R})$ and suppose A is similar to a diagonal matrix $D = \mathrm{Diag}(d_1, \ldots, d_n)$ with $d_1, \ldots, d_n \in \mathbb{R}$.
   (a) Give an example where $d_1, \ldots, d_n$ are not distinct.
   (b) Show that $\mathscr{S}_{\mathbb{R}}(A) \subseteq \{d_1, \ldots, d_n\}$.
   (c) Show that $C_A(t) = \prod_{j=1}^{r}(t - e_j)s_j$ in $\mathbb{R}[t]$ where $e_1, \ldots, e_r$ are the distinct real numbers in the list $d_1, \ldots, d_n$.
   (d) Show that if $d_i = e_j$, then d_i appears s_j times on the diagonal of D.

5.6   Complete the details of the proof of Lemma 5.13.

5.7   For the following quadratic forms on $\mathbb{R}^2$, find a linear substitution $[x \;\; y]^t = P[x' \;\; y']^t$ for which $q'(x', y')$ has the form in Corollary 5.17.
   (a) $q(x, y) = -2x^2 + 2xy - 2y^2$.
   (b) $q(x, y) = x^2 + 4xy + 3y^2$.

5.8   For the following quadratic forms on $\mathbb{R}^3$, find a linear substituting $[x \;\; y \;\; z]^t = P[x' \;\; y' \;\; z']^t$ for which $q'(x', y', z')$ has the form in Corollary 5.17.
   (a) $q(x, y, z) = 4x^2 + 4y^2 + 4z^2 - 2xy + 2xz - 2yz$.
   (b) $q(x, y, z) = x^2 + y^2 + 2z^2 + 2xz + 2yz$.

5.9   Draw the graph of $q = 1$, and find the rank and signature of each of the quadratic forms given in Exercises 5.7 and 5.8.

5.10   Show that $z = x^2 + 4xy + y^2$ has a saddle point at the origin by applying Corollary 5.17 to the quadratic form $x^2 + 4xy + y^2$.

5.11 Find all the linear substitutions

$$[x_1 \quad \cdots \quad x_n]^t = P[x_1' \quad \cdots \quad x_n']^t$$

which change $x_1^2 + \cdots + x_n^2$ to $(x_1')^2 + \cdots + (x_n')^2$.

5.12 A real quadratic form $q(x_1, \ldots, x_n)$ on $\mathbb{R}^n$ is said to be positive-definite if $q(x_1, \ldots, x_n) > 0$ for all $(x_1, \ldots, x_n) \neq (0, \ldots, 0)$. Show that q is positive-definite if and only if there exists a linear substitution

$$[x_1 \quad \cdots \quad x_n]^t = P[x_1' \quad \cdots \quad x_n']^t$$

such that $q'(x_1', \ldots, x_n') = (x_1')^2 + \cdots + (x_n')^2$.

5.13 A symmetric matrix A is positive-definite if $\langle A\alpha, \alpha \rangle > 0$ for all $\alpha \in \mathbb{R}^n - (O)$. Show a symmetric matrix A is positive-definite if and only if $A = P^tP$ for some nonsingular matrix P.

5.14 Show $q(x, y) = ax^2 + bxy + cy^2$ is positive-definite on $\mathbb{R}^2$ if and only if $a > 0$ and $4ac - b^2 > 0$.

5.15 Return to Exercise 5.12. A real quadratic form $q(x_1, \ldots, x_n)$ on $\mathbb{R}^n$ is said to be negative-definite if $q(x_1, \ldots, x_n) < 0$ for all $(x_1, \ldots, x_n) \neq (0, \ldots, 0)$. Let A be the associated symmetric matrix of q. Show that

(a) q is positive-definite if and only if all eigenvalues of A are positive.

(b) q is negative-definite if and only if all eigenvalues of A are negative.

Notation

I_V	the identity map on V	126
$\text{Hom}(V, W)$	the set of all linear transformations from V to W	128
$\text{Ker}(T)$	the kernel of T	128
$\text{Im}(T)$	the image of T	128
$V \cong W$	V is isomorphic to W	129
$\Gamma(B, C)(T)$	the matrix representation of T with respect to B and C	138
$\Delta(n)$	$\{1, 2, \ldots, n\}$	155
S_n	the set of all permutations on $\Delta(n)$	156
$\sigma = \begin{bmatrix} 1 & 2 & \cdots & n \\ j_1 & j_2 & \cdots & j_n \end{bmatrix}$	the representation of the permutation σ	156
$(i_1, \ldots, i_r)$	an r-cycle	160
$\text{sgn}(\sigma)$	the sign of the permutation σ	163
$\det(A)$	the determinant of the matrix A	163
$\text{Diag}(d_1, \ldots, d_n)$	an $n \times n$ diagonal matrix with $d_1, \ldots, d_n$ as diagonal entries	167
$(R_1; \ldots; R_n)$	the row partition $\begin{bmatrix} R_1 \\ \vdots \\ \hline R_n \end{bmatrix}$	168
$M_{ij}(A)$	the (i, j)th minor of A	182
$\text{cof}_{ij}(A)$	the (i, j)th cofactor of A	182
$\text{adj}(A)$	the adjoint of A	187
$\mathscr{S}_F(A)$	the spectrum of A	195
$C_A(t)$	the characteristic polynomial of A	197
$\langle *, * \rangle$	an inner product	217
$(V, \langle *, * \rangle)$	an inner product space	218
$\|\alpha\|$	the length of the vector α	220
$W^\perp$	the orthogonal complement of W	228
$\text{Tr}(A)$	the trace of A	230
$\text{Re}(z)$	the real part of z	232
$\text{Im}(z)$	the imaginary part of z	232
$|z|$	the modulus of z	232
$\bar{\beta}$	the conjugate of β	233
$d(\alpha, \beta)$	the distance between α and β	242
$P_W(\alpha)$	the orthogonal projection of α onto W	245
δ_{ij}	the Kronecker delta function	254
$A \equiv B$	A is congruent to B	271
$\text{sig}(A)$	the signature of A	276

References

1. G. Birkhoff and S. MacLane, *A Survey Of Modern Algebra*, 3rd ed., MacMillan, New York, 1965.

2. W. C. Brown, *A Second Course In Linear Algebra*, Wiley, New York, 1988.

3. I. N. Herstein, *Topics In Algebra*, 2nd ed., Wiley, New York, 1975.

4. M. Hirsch and S. Smale, *Differential Equations, Dynamical Systems, and Linear Algebra*, Academic Press, New York, 1974.

5. K. Hoffman and R. Kunze, *Linear Algebra*, 2nd edn., Prentice Hall, Englewood Cliffs, NJ, 1971.

6. L. Loomis and S. Sternberg, *Advanced Calculus*, Addison Wesley, Reading, MA, 1980.

7. D. G. Northcott, *Multilinear Algebra*, Cambridge University Press, New York, 1984.

Answers and Suggestions for Selected Exercises

Chapter I, Section 1

1.3 $\begin{bmatrix} 1 & 2 & 3 \\ 0 & 1 & 2 \end{bmatrix}, \left[\begin{array}{cc|c} 1 & 2 & 3 \\ 0 & 1 & 2 \end{array}\right], \left[\begin{array}{cc|c} 1 & 2 & 3 \\ 0 & 1 & 2 \end{array}\right], \left[\begin{array}{cc|c} 1 & 2 & 3 \\ 0 & 1 & 2 \end{array}\right],$

$\begin{bmatrix} 1 & 2 & 3 \\ 0 & 1 & 2 \end{bmatrix}, \left[\begin{array}{c|c|c} 1 & 2 & 3 \\ 0 & 1 & 2 \end{array}\right], \left[\begin{array}{cc|c} 1 & 2 & 3 \\ 0 & 1 & 2 \end{array}\right], \left[\begin{array}{c|c|c} 1 & 2 & 3 \\ 0 & 1 & 2 \end{array}\right]$

1.4 $\begin{bmatrix} 1/2 & -1/3 & 1/4 & -1/5 \\ -1/3 & 1/4 & -1/5 & 1/6 \\ 1/4 & -1/5 & 1/6 & -1/7 \end{bmatrix}$

1.6 If the lines are parallel and not equal, then the system has no solution. If the lines are the same, then the system has infinitely many solutions. If the lines are distinct and not parallel, then the system has precisely one solution.

1.7 A circle and a parabola can have 0, 1, 2, 3, or 4 intersection points. Write out specific equations for these five cases.

1.8 $x + 2y = 1$

$4y + 3z = 1$

If z is chosen arbitrarily, then $x = (1 + 3z)/2$, $y = (1 - 3z)/4$, and z are the only solutions.

1.11 $x + y = 0$
$x + y = 1$

1.12 $x + y = 1$
$2x + 2y = 2$

1.13 (a) $x_1 = 1$, $x_2 = 2$
(d) $x = 2$, $y = 3$, $z = -1$
(f) $x = 1$, $y = 2$, $z = 1/2$, $w = 3$, $r = 1/2$

1.16 $x = (c_1 d - c_2 b)/\Delta$, $y = (ac_2 - cc_1)/\Delta$

1.17 $A = \begin{bmatrix} 1 & 0 \\ 0 & 1 \end{bmatrix}$, $B = \begin{bmatrix} 1 & 0 \\ 0 & 0 \end{bmatrix}$

Chapter I, Section 2

2.1 (a) $\begin{bmatrix} 4 & -14 \\ -28 & 6 \end{bmatrix}$

(c) $\begin{bmatrix} -4 & 1 \\ -18 & -1 \end{bmatrix}$

(f) $\begin{bmatrix} 1 & 7 \\ -3 & 14 \end{bmatrix}$

2.3 If $AB = BA$, then $A^2 - B^2 = (A - B)(A + B)$. Is this true if $AB \neq BA$?

2.5 Show the system of equations

$$\begin{bmatrix} 1 & 1 & 1 \\ 2 & 0 & -1 \\ 4 & 1 & 1 \end{bmatrix} \begin{bmatrix} x \\ y \\ z \end{bmatrix} = \begin{bmatrix} 0 \\ 0 \\ 0 \end{bmatrix}$$

has only the trivial solution: $x = 0$, $y = 0$, $z = 0$.

2.8 $X = \begin{bmatrix} 0 & x \\ 0 & y \end{bmatrix}$, x, y arbitrary.

2.10 (a) $\begin{bmatrix} x \\ y \end{bmatrix} = \begin{bmatrix} 1/5 & 3/5 \\ -1/5 & 2/5 \end{bmatrix} \begin{bmatrix} 1 \\ 2 \end{bmatrix}$

(b) $\begin{bmatrix} x \\ y \\ z \end{bmatrix} = \begin{bmatrix} 1 & -1 & 2 \\ -1/2 & 1 & -3/2 \\ -1/2 & 0 & 1/2 \end{bmatrix} \begin{bmatrix} 2 \\ 3 \\ 4 \end{bmatrix}$

2.11 A diagonal matrix $D \in M_{n \times n}$ is invertible, if and only if $[D]_{ii} \neq 0$ for all $i = 1, \ldots, n$.

2.17 (a) E_{11} and E_{22} are idempotent.

2.18 Suppose $A = (a_{ij}) \in M_{n \times n}$ is lower triangular. Thus, $a_{ij} = 0$ whenever $j > i$. First show A is nilpotent if $[A]_{ii} = 0$ for all $i = 1, \ldots, n$. Then worry about what happens if some diagonal entry of A is nonzero.

Chapter I, Section 3

3.1 Let $A_1 = \begin{bmatrix} 1 & 2 \\ 3 & 1 \end{bmatrix}$, $A_2 = \begin{bmatrix} 4 \\ 0 \end{bmatrix}$, $A_3 = \begin{bmatrix} 1 & 0 \\ 0 & 1 \end{bmatrix}$

$A_4 = \begin{bmatrix} 0 & 1 \end{bmatrix}$, $A_5 = \begin{bmatrix} 1 \end{bmatrix}$, $A_6 = \begin{bmatrix} -1 & 2 \end{bmatrix}$

$A_7 = \begin{bmatrix} 1 & 0 \\ -1 & 1 \end{bmatrix}$, $A_8 = \begin{bmatrix} 2 \\ -1 \end{bmatrix}$, $A_9 = \begin{bmatrix} 3 & 1 \\ 2 & 4 \end{bmatrix}$

$B_1 = \begin{bmatrix} 1 \\ 2 \end{bmatrix}$, $B_2 = \begin{bmatrix} 1 & 2 \\ -1 & 0 \end{bmatrix}$, $B_3 = \begin{bmatrix} 4 \end{bmatrix}$, $B_4 = \begin{bmatrix} 2 & 2 \end{bmatrix}$, $B_5 = \begin{bmatrix} 1 \\ 3 \end{bmatrix}$

$B_6 = \begin{bmatrix} 0 & 1 \\ 0 & 1 \end{bmatrix}$

Then $AB = \begin{bmatrix} \begin{array}{c|c|c} A_1 & A_2 & A_3 \\ \hline A_4 & A_5 & A_6 \\ \hline A_7 & A_8 & A_9 \end{array} \end{bmatrix} \begin{bmatrix} \begin{array}{c|c} B_1 & B_2 \\ \hline B_3 & B_4 \\ \hline B_5 & B_6 \end{array} \end{bmatrix}$

$= \begin{bmatrix} \begin{array}{c|c} A_1 B_1 + A_2 B_3 + A_3 B_5 & A_1 B_2 + A_2 B_4 + A_3 B_6 \\ \hline A_4 B_1 + A_5 B_3 + A_6 B_5 & A_4 B_2 + A_5 B_4 + A_6 B_6 \\ \hline A_7 B_1 + A_8 B_3 + A_9 B_5 & A_7 B_2 + A_8 B_4 + A_9 B_6 \end{array} \end{bmatrix}$

$= \begin{bmatrix} \begin{array}{c|cc} 22 & 7 & 11 \\ \hline 8 & 2 & 7 \\ \hline 11 & 1 & 3 \\ 15 & 5 & 10 \\ 11 & -4 & 2 \end{array} \end{bmatrix}$

3.5 $\begin{bmatrix} \begin{array}{c|c|c|c} A_{11} & 0 & \cdots & 0 \\ \hline 0 & A_{22} & \cdots & 0 \\ \hline \vdots & \vdots & & \vdots \\ \hline 0 & 0 & \cdots & A_{rr} \end{array} \end{bmatrix}^k = \begin{bmatrix} \begin{array}{c|c|c|c} A_{11}^k & 0 & \cdots & 0 \\ \hline 0 & A_{22}^k & \cdots & 0 \\ \hline \vdots & \vdots & & \vdots \\ \hline 0 & 0 & \cdots & A_{rr}^k \end{array} \end{bmatrix}$

3.6 (a) $\begin{bmatrix} 13 & 14 & 0 & 0 \\ 21 & 6 & 0 & 0 \\ 0 & 0 & 1 & 6 \\ 0 & 0 & 0 & 1 \end{bmatrix}$

(b) $\begin{bmatrix} 37 & 54 & 0 & 0 & 0 \\ 81 & 118 & 0 & 0 & 0 \\ 0 & 0 & 7 & 0 & 5 \\ 0 & 0 & 27 & 1 & 19 \\ 0 & 0 & 10 & 0 & 7 \end{bmatrix}$

3.8 $\begin{bmatrix} A & C \\ O & B \end{bmatrix}^{-1} = \begin{bmatrix} A^{-1} & -A^{-1}CB^{-1} \\ O & B^{-1} \end{bmatrix}$

3.11 $\begin{bmatrix} 1 & 1 & 1 \\ 1 & 1 & 1 \\ 1 & 1 & 1 \end{bmatrix} \begin{bmatrix} x \\ y \\ z \end{bmatrix} = \begin{bmatrix} 1 \\ 0 \\ 1 \end{bmatrix}$

3.12 Show $CS(A) = \mathbb{R}^3$ and use (3.21).

Chapter I, Section 4

4.1 (a) $\begin{bmatrix} 2 & 4 & -2 \\ 1 & -1 & 5 \end{bmatrix} \underset{r}{\sim} \begin{bmatrix} 1 & 0 & 3 \\ 0 & 1 & -2 \end{bmatrix}.$

Therefore, $x = 3$ and $y = -2$.

(c) $\begin{bmatrix} 1 & 1 & -1 & 1 \\ 2 & 1 & 2 & 2 \\ 8 & 5 & 4 & 7 \end{bmatrix} \underset{r}{\sim} \begin{bmatrix} 1 & 0 & 3 & 0 \\ 0 & 1 & -4 & 0 \\ 0 & 0 & 0 & 1 \end{bmatrix}$ No solutions.

(f) $\begin{bmatrix} 1 & 1 & 1 & 0 \\ 1 & 2 & -1 & 1 \\ 4 & 6 & 0 & 2 \\ 2 & 3 & 0 & 1 \end{bmatrix} \underset{r}{\sim} \begin{bmatrix} 1 & 0 & 3 & -1 \\ 0 & 1 & -2 & 1 \\ 0 & 0 & 0 & 0 \\ 0 & 0 & 0 & 0 \end{bmatrix}$

Let z be arbitrary. Then $y = 1 + 2z$, $x = -1 - 3z$.

4.4 A and C are in echelon form.

4.5 $A \underset{r}{\sim} \begin{bmatrix} 1 & 0 & -3 & 0 \\ 0 & 1 & 2 & 0 \\ 0 & 0 & 0 & 1 \end{bmatrix}$,

$C \underset{r}{\sim} \begin{bmatrix} 1 & 0 & -1 & 0 & -3 \\ 0 & 1 & 1/2 & 0 & -1/4 \\ 0 & 0 & 0 & 1 & 1/2 \end{bmatrix}$

4.9   What if size $(A) \neq$ size (B)?

4.10 (a) $\begin{bmatrix} 2 & 4 & 2 & 2 \\ 0 & 1 & 2 & 2 \\ 0 & 0 & 1 & 0 \end{bmatrix} \underset{r}{\sim} \begin{bmatrix} 1 & 0 & 0 & -3 \\ 0 & 1 & 0 & 2 \\ 0 & 0 & 1 & 0 \end{bmatrix}$

Therefore, $3w, -2w, 0, w$ is a solution for any w.

(c) $\begin{bmatrix} 1 & 0 & 0 & 0 & 0 \\ 0 & 0 & 1 & 2 & 0 \\ 0 & 0 & 1 & 2 & 3 \end{bmatrix} \underset{r}{\sim} \begin{bmatrix} 1 & 0 & 0 & 0 & 0 \\ 0 & 0 & 1 & 2 & 0 \\ 0 & 0 & 0 & 0 & 1 \end{bmatrix}$

Then $0, y, -2u, u, 0$ is a solution for any y and u.

4.11 $\begin{bmatrix} 1 & 0 \\ 0 & 1 \end{bmatrix}$ is not equivalent to $\begin{bmatrix} 1 & 0 \\ 0 & 0 \end{bmatrix}$

4.15 No. $\begin{bmatrix} 1 & 0 \\ 0 & 1 \end{bmatrix}\begin{bmatrix} 2 \\ 4 \end{bmatrix} = \begin{bmatrix} 2 \\ 4 \end{bmatrix}$ and $\begin{bmatrix} 1 & 0 & 2 \\ 0 & 1 & 4 \end{bmatrix} \underset{c}{\sim} \begin{bmatrix} 2 & 0 & 2 \\ 0 & 1 & 4 \end{bmatrix}$

but $\begin{bmatrix} 2 & 0 \\ 0 & 1 \end{bmatrix}\begin{bmatrix} 2 \\ 4 \end{bmatrix} \neq \begin{bmatrix} 2 \\ 4 \end{bmatrix}$

4.17 If $A \neq O$, then $A \approx \left[\begin{array}{c|c} I_k & O \\ \hline O & O \end{array}\right]$

Chapter I, Section 5

5.2   (a) $E_{12}(-3)E_2(-1/5)E_{21}(-2)E_{12}E_{21}(-1)A = \begin{bmatrix} 1 & 0 \\ 0 & 1 \end{bmatrix}$

5.5   (a) $\begin{bmatrix} 1 & 0 & 0 \\ 2 & 1 & 0 \\ 1 & -1 & 3 \end{bmatrix}^{-1} = \begin{bmatrix} 1 & 0 & 0 \\ -2 & 1 & 0 \\ -1 & 1/3 & 1/3 \end{bmatrix}$

(d) $\begin{bmatrix} 1 & 0 & -1 & 1 \\ 0 & 1 & 2 & 1 \\ -1 & 1 & 1 & 0 \\ 1 & 2 & 1 & -2 \end{bmatrix}^{-1}$

$= \begin{bmatrix} 1/10 & 3/10 & -7/10 & 2/10 \\ 6/10 & -2/10 & 8/10 & 2/10 \\ -5/10 & 5/10 & -5/10 & 0 \\ 4/10 & 2/10 & 2/10 & -2/10 \end{bmatrix}$

5.8   (a) $p(t) = 1 + (1/2)t + (3/2)t^2$

5.9   No. Given two distinct points in $\mathbb{R}^2$, there are at least two distinct quadratic polynomials whose graphs pass through these two points. Find a concrete example.

5.13
$$\begin{bmatrix} 5 & 1 & 6 & 13 \\ 1 & 0 & 1 & 2 \\ 1 & 1 & 2 & 5 \end{bmatrix} \approx \begin{bmatrix} 1 & 0 & 0 & 0 \\ 0 & 1 & 0 & 0 \\ 0 & 0 & 0 & 0 \end{bmatrix}$$

and
$$\begin{bmatrix} 6 & 0 & -1 & 4 \\ 1 & 1 & 2 & 3 \\ 2 & 3 & 1 & 2 \end{bmatrix} \approx \begin{bmatrix} 1 & 0 & 0 & 0 \\ 0 & 1 & 0 & 0 \\ 0 & 0 & 1 & 0 \end{bmatrix}$$

Show that the second and fourth matrix here cannot be equivalent.

5.14   Use Theorem 2.35(d).

5.17
$$\begin{bmatrix} 3 & 5 \\ 1 & 2 \end{bmatrix}\begin{bmatrix} 1 & 0 \\ 0 & 2 \end{bmatrix}\begin{bmatrix} 2 & -5 \\ -1 & 3 \end{bmatrix} = \begin{bmatrix} -4 & 15 \\ -2 & 7 \end{bmatrix}$$

5.18 $c_1 = 10.9497$, $c_2 = 1$, $c_3 = -17.8995$, $c_4 = 8.9497$.

Chapter I, Section 6

6.1   (b)
$$\begin{bmatrix} 1 & 0 & 1 & 1 & 1 \\ 2 & 1 & 1 & 4 & 6 \\ 3 & 1 & 2 & 0 & 1 \end{bmatrix}$$
$$= \begin{bmatrix} 1 & 0 & 0 \\ 2 & 1 & 0 \\ 3 & 1 & 1 \end{bmatrix}\begin{bmatrix} 1 & 0 & 1 & 1 & 1 \\ 0 & 1 & -1 & 2 & 4 \\ 0 & 0 & 0 & -5 & -6 \end{bmatrix}$$

(d)
$$\begin{bmatrix} 0 & 1 & 0 \\ 0 & 0 & 1 \\ 1 & 0 & 0 \end{bmatrix}\begin{bmatrix} 1 & 3 & 2 \\ 2 & 6 & 8 \\ 2 & 8 & 1 \end{bmatrix} = \begin{bmatrix} 1 & 0 & 0 \\ 1 & 1 & 0 \\ 1/2 & 0 & 1 \end{bmatrix}\begin{bmatrix} 2 & 6 & 8 \\ 0 & 2 & -7 \\ 0 & 0 & -2 \end{bmatrix}$$

6.2   (d)
$$\begin{bmatrix} 1 & 3 & 2 \\ 2 & 6 & 8 \\ 2 & 8 & 1 \end{bmatrix}\begin{bmatrix} x \\ y \\ z \end{bmatrix}$$
$$= \begin{bmatrix} 1 \\ 0 \\ 1 \end{bmatrix} \Leftrightarrow \begin{bmatrix} 1 & 0 & 0 \\ 1 & 1 & 0 \\ 1/2 & 0 & 1 \end{bmatrix}\begin{bmatrix} 2 & 6 & 8 \\ 0 & 2 & -7 \\ 0 & 0 & -2 \end{bmatrix}\begin{bmatrix} x \\ y \\ z \end{bmatrix} = \begin{bmatrix} 0 \\ 1 \\ 1 \end{bmatrix}$$

$$\begin{bmatrix} 1 & 0 & 0 \\ 1 & 1 & 0 \\ 1/2 & 0 & 1 \end{bmatrix}\begin{bmatrix} y_1 \\ y_2 \\ y_3 \end{bmatrix} = \begin{bmatrix} 0 \\ 1 \\ 1 \end{bmatrix} \Rightarrow \begin{matrix} y_1 = 0 \\ y_2 = 1 \\ y_3 = 1 \end{matrix}$$

$$\begin{bmatrix} 2 & 6 & 8 \\ 0 & 2 & -7 \\ 0 & 0 & -2 \end{bmatrix}\begin{bmatrix} x \\ y \\ z \end{bmatrix} = \begin{bmatrix} 0 \\ 1 \\ 1 \end{bmatrix} \Rightarrow \begin{matrix} x = 46/8 \\ y = -5/4 \\ z = -1/2 \end{matrix}$$

6.6 $\begin{bmatrix} 0 & 0 \\ 0 & 0 \end{bmatrix} = \begin{bmatrix} 1 & 0 \\ 0 & 1 \end{bmatrix}\begin{bmatrix} 0 & 0 \\ 0 & 0 \end{bmatrix} = \begin{bmatrix} 1 & 0 \\ 1 & 1 \end{bmatrix}\begin{bmatrix} 0 & 0 \\ 0 & 0 \end{bmatrix}$

6.9 (a) $\begin{bmatrix} 0 & 1 & 0 \\ 1 & 0 & 0 \\ 0 & 0 & 1 \end{bmatrix}\begin{bmatrix} 0 & 2 & 1 \\ 1 & 1 & 2 \\ 3 & 4 & 1 \end{bmatrix}$

$$= \begin{bmatrix} 1 & 0 & 0 \\ 0 & 1 & 0 \\ 3 & 1/2 & 1 \end{bmatrix}\begin{bmatrix} 1 & 0 & 0 \\ 0 & 2 & 0 \\ 0 & 0 & -11/2 \end{bmatrix}\begin{bmatrix} 1 & 1 & 2 \\ 0 & 1 & 1/2 \\ 0 & 0 & 1 \end{bmatrix}$$

(d) $\begin{bmatrix} 0 & 1 & 0 \\ 0 & 0 & 1 \\ 1 & 0 & 0 \end{bmatrix}\begin{bmatrix} 1 & 3 & 2 \\ 2 & 6 & 8 \\ 2 & 8 & 1 \end{bmatrix}$

$$= \begin{bmatrix} 1 & 0 & 0 \\ 1 & 1 & 0 \\ 1/2 & 0 & 1 \end{bmatrix}\begin{bmatrix} 2 & 0 & 0 \\ 0 & 2 & 0 \\ 0 & 0 & -2 \end{bmatrix}\begin{bmatrix} 1 & 3 & 4 \\ 0 & 1 & -7/2 \\ 0 & 0 & 1 \end{bmatrix}$$

6.11   If A is symmetric, then $A = LDL^t$.

6.16   First show that any transposition E is symmetric. Therefore, $E^2 = E^t E = I$. Then use Theorem 2.35 for the general case.

Chapter II, Section 1

1.1 $\mathbb{C} = L(1, i)$. Here $i = \sqrt{-1}$.
1.2   (a), (c), and (d) are subspaces.
1.3   (a) and (b) are subspaces.
1.5   For each $i, j = 1, \ldots, n$, let E_{ij} be the $n \times n$ matrix having a 1 in its (i,j)th entry and zeros everywhere else. Then

$$W = L(E_{11}, \ldots, E_{nn}, E_{ij} + E_{ji} \quad \text{where } 1 \leqslant i < j \leqslant n).$$

1.7   Make suitable adjustments in the answer to Exercise 1.5.

1.11   If $x_1 a_1 + \cdots + x_n a_n = 0$ for all $x_1, \ldots, x_n \in \mathbb{R}$, then $a_1 = \cdots = a_n = 0$. Thus, $\mathbf{W} \neq \mathbb{R}^n$. Since $n > 1$, there exist scalars $x_1, \ldots, x_n \in \mathbb{R}$, not all zero, such that $x_1 a_1 + \cdots + x_n a_n = 0$. Thus, $\mathbf{W} \neq (O)$. Is this exercise true if $n = 1$?

1.14   Let W_i be a straight line through the origin.

1.17   Show that $\begin{bmatrix} 1 & 0 & 1 \\ 1 & 1 & 0 \\ 1 & -1 & 2 \end{bmatrix} \begin{bmatrix} x \\ y \\ z \end{bmatrix} = \begin{bmatrix} 1 \\ 0 \\ 0 \end{bmatrix}$ has no solution.

1.21 W is not in general a subspace of F^n. If $B \neq O$, and $A\xi_1 = A\xi_2 = B$, then $A(\xi_1 + \xi_2) = 2B$.

Chapter II, Section 2

2.1 $\{\alpha, \beta\}$ and $\{\alpha, \beta, \delta\}$ are linearly independent.

2.2 $\{1, i\}$ is a basis of $\mathbb{C}$ as a vector space over $\mathbb{R}$.

2.3 $x = 2$.

2.5   If α_3 is any polynomial of degree three, then $\{1, \alpha_1, \alpha_2, \alpha_3\}$ is a basis of $\mathscr{P}_3$.

2.8 $B = \{\alpha, \beta, \gamma\}$ is a basis of $W = L(\alpha, \beta, \gamma) \subseteq V$.

$$[\alpha + \beta]_B = \begin{bmatrix} 1 \\ 1 \\ 0 \end{bmatrix}, \ [\beta + \gamma]_B = \begin{bmatrix} 0 \\ 1 \\ 1 \end{bmatrix}$$

and $[\gamma + \alpha]_B = \begin{bmatrix} 1 \\ 0 \\ 1 \end{bmatrix}$ in F^3.

The vectors

$$\begin{bmatrix} 1 \\ 1 \\ 0 \end{bmatrix}, \begin{bmatrix} 0 \\ 1 \\ 1 \end{bmatrix}, \begin{bmatrix} 1 \\ 0 \\ 1 \end{bmatrix}$$

are linearly independent in F_3. Therefore, $\alpha + \beta, \beta + \gamma, \gamma + \alpha$ are linearly independent in W.

2.11   When $m \leqslant n$, a basis of V is $\{E_{ij} \mid 1 \leqslant i \leqslant j \leqslant n\}$. Here the E_{ij} are the same as in Exercise 1.5. Thus, $\dim(V) = mn - m(m-1)/2$.

2.12 $\dim V = n + n(n-1)/2$.

2.17   If $B = \{E_{11}, E_{12}, E_{13}, E_{21}, E_{22}, E_{23}\}$ (notation as in Exercise 1.5), then

$$[A_1]_B = [1 \quad 0 \quad 0 \quad 1 \quad 0 \quad 1]^t, \quad [A_2]_B = [2 \quad -1 \quad 1 \quad 1 \quad 2 \quad 3]^t,$$
$$[A_3]_B = [-1 \quad -1 \quad 1 \quad 2 \quad 0 \quad 1]^t$$

and $[A_4]_B = [7 \quad -2 \quad 2 \quad 3 \quad 6 \quad 10]^t$. Show these four vectors are linearly independent in F^6.

2.21   Let $X = M(B, C)M(C, B) \in M_{n \times n}$. By Theorem 2.28, $X[\gamma]_C = [\gamma]_C$ for all $\gamma \in V$. In particular, $X\xi = \xi$ for all $\xi \in F^n$. This easily implies $X = I_n$.

Chapter II, Section 3

3.1   (a)   Suppose $A = [x_1 \quad \cdots \quad x_n]$ with some x_i, say x_r, not equal to zero. Then $CS(A) = F$, $RS(A) = L([x_1 \quad \cdots \quad x_n])$, and

$$NS(A) = \left\{ [b_1 \quad \cdots \quad b_n]^t \in F^n \mid \sum_{i=1}^{n} b_i x_i = 0 \right\}$$

$\{1\}$ is a basis of $CS(A)$. $\{[x_1 \cdots x_n]\}$ is a basis of $RS(A)$. Let

$$\delta_1 = [-x_r \quad 0 \quad \cdots \quad 0 \quad x_1 \quad 0 \quad \cdots \quad 0]^t$$
$$\vdots$$
$$\delta_{r-1} = [0 \quad \cdots \quad 0 \quad -x_r \quad x_{r-1}0 \quad \cdots \quad 0]^t$$
$$\delta_{r+1} = [0 \quad \cdots \quad 0 \quad x_{r+1} \quad -x_r \quad 0 \quad \cdots \quad 0]^t$$
$$\vdots$$
$$\delta_n = [0 \quad \cdots \quad 0 \quad x_n \quad 0 \quad \cdots \quad 0 \quad -x_r]^t$$

The numbers $x_1, \ldots, x_{r-1}, x_{r+1}, \ldots, x_n$ are placed in the rth entry of each of these vectors.

$\{\delta_1, \ldots, \delta_{r-1}, \delta_{r+1}, \ldots, \delta_n\}$ is a basis of $NS(A)$. What happens if $A = O$?

(d) $CS(A) = \mathbb{R}^2$, basis $\{\varepsilon_1, \varepsilon_2\}$.
$RS(A) = L(\text{Row}_1(A), \text{Row}_2(A))$, basis $\{\text{Row}_1(A), \text{Row}_2(A)\}$.
$NS(A) = L(\alpha = [-2 \quad 1 \quad 1 \quad 0 \quad 0]^t, \beta = [-5 \quad 2 \quad 0 \quad 1 \quad 0]^t,$
$\gamma = [2 \quad -3 \quad 0 \quad 0 \quad 1]^t)$ basis $\{\alpha, \beta, \gamma\}$.

(e) $CS(A) = L(\alpha = [1 \quad -1 \quad 5]^t, \beta = [0 \quad 1 \quad -3]^t)$, basis $\{\alpha, \beta\}$.
$RS(A) = L(\xi = [1 \quad 0 \quad 2 \quad 1], v = [0 \quad 1 \quad 1 \quad 2])$, basis $\{\xi, v\}$.
$NS(A) = L(\gamma = [-2 \quad -1 \quad 1 \quad 0]^t, \quad \delta = [-1 \quad -2 \quad 0 \quad 1]^t)$, basis $\{\gamma, \delta\}$.

3.3   Since $rk(A) = 3$, $AX = B$ has a solution for any $B \in F^3$.

3.7 $rk([1 \quad 2]) = rk([1 \quad 3])$, but $[1 \quad 2]$ is not row equivalent to $[1 \quad 3]$.

3.10   (a) $rk(A) = 3$, $v(A) = 1$.

        (b) $rk(A) = 3$, $v(A) = 4$.

3.15   First show $v(AB) \leqslant v(A) + v(B)$, and then use Theorem 3.22.

3.16   Let $\alpha_1 = A^{k-1}(\xi)$, $\alpha_2 = A^{k-2}(\xi)$, ..., $\alpha_k = \xi$. Show $\alpha_1, \ldots, \alpha_j$ are linearly independent for each $j = 1, \ldots, k$ (use induction).

3.17   If $rk(A) = 1$, then $A = [x_1\xi \,|\, \cdots \,|\, x_n\xi]$ for some nonzero $\xi \in F^n$ and scalars $x_1, \ldots, x_n$ (not all zero) in F. Now compute A^2.

Chapter II, Section 4

4.1   (a), (b), and (e) are linear transformations.

4.2   (a) and (b) are linear transformations.

4.5   The map $A \mapsto A^*$ does not preserve scalar multiplication.

4.7 $T(a + bt + ct^2) = (a + b - c) - ct + (c - b)t^3 + (2c - b)t^4$.

4.10   Hint: The linear transformations from $\mathbb{R}[t]$ to $\mathbb{R}[t]$ are in a one to one correspondence with the set mappings from B to $\mathbb{R}[t]$.

4.13   For the surjective part of this question, suppose $T = 0$?

4.16 $Ker(T) = L([0 \quad -1/2 \quad 2 \quad 1]^t)$. $Im(T) = \mathbb{R}^3$.

4.20   Hint: Choose a basis $\{\alpha_1, \ldots, \alpha_r, \alpha_{r+1}, \ldots, \alpha_n\}$ of V such that $\{\alpha_1, \ldots, \alpha_r\}$ is a basis of $Ker(T)$. Then $r > 0$. Now use Theorem 4.22(b).

4.21   Use the same idea for the solution to Exercise 20.

Chapter II, Section 5

5.1   If $B = \{1, i\}$, then $\Gamma(B, B)(z \mapsto \bar{z}) = \begin{bmatrix} 1 & 0 \\ 0 & -1 \end{bmatrix}$.

5.2   (i) $\Gamma(\underline{\varepsilon}, \underline{\varepsilon})(TS) = \begin{bmatrix} 0 & 0 & 1 \\ 1 & 0 & 0 \\ 0 & 1 & 0 \end{bmatrix}$

        (iii) $\Gamma(\underline{\varepsilon}, \underline{\varepsilon})(TS^2 - ST + T) = \begin{bmatrix} 0 & 1 & 0 \\ 2 & 0 & 1 \\ -1 & 0 & 0 \end{bmatrix}$

5.3   Let $B = \{1, t, t^2, \ldots, t^n\}$ and $C = \{1, t, t^2, \ldots, t^{n+1}\}$. Then

$$\Gamma(B, C)(\mathscr{S}) = \begin{bmatrix} 0 & 0 & 0 & \cdots & 0 \\ 1 & 0 & 0 & \cdots & 0 \\ 0 & 1/2 & 0 & \cdots & 0 \\ 0 & 0 & 1/3 & \cdots & 0 \\ \vdots & \vdots & \vdots & & \vdots \\ 0 & 0 & 0 & \cdots & 1/n+1 \end{bmatrix}$$

5.5 $\begin{bmatrix} 1 & 0 & 0 & 0 & 0 & 0 \\ 0 & 1 & 0 & 0 & 0 & 0 \\ 0 & 0 & 0 & 1 & 0 & 0 \\ 0 & 0 & 0 & 0 & 1 & 0 \end{bmatrix}$

5.7 $[1 \quad 2 \quad 4 \quad \cdots \quad 2^n]$

5.9   (a) $\Gamma(B, B)(T) = \begin{bmatrix} 0 & 0 & 0 \\ 5/2 & 3 & -5 \\ -1/2 & 0 & 2 \end{bmatrix}$

5.10 $P = \begin{bmatrix} 2 & -1 & -1 \\ -1/2 & 1 & 1/2 \\ 1/2 & 0 & -1/2 \end{bmatrix}$

5.13   (d) $\begin{bmatrix} 1 & 0 \\ 1 & -1 \\ 2 & 3 \\ 1 & 1 \end{bmatrix}$

5.16   (b) $\begin{bmatrix} 1 & 0 & 0 & 0 \\ 2 & 1 & 0 & 0 \\ 0 & -1 & 0 & -3 \end{bmatrix}$

5.17   Let $B = \left\{ \begin{bmatrix} 1 \\ 2 \end{bmatrix}, \begin{bmatrix} 2 \\ 3 \end{bmatrix} \right\}$. Then $\Gamma(B, B)(T) = \begin{bmatrix} 1 & 0 \\ 0 & 2 \end{bmatrix}$

Chapter III, Section 1

1.8 $(1, 2) = (1, 2)(2, 1)(1, 2)$.
1.9 $(1, 4, 5, 6, 7, 8, 9, 2, 3)$

1.10 Show the product of these two permutations is the identity.

1.11 Hint: Show the map $\text{sgn}(*): S_n \mapsto \{-1, 1\}$ preserves multiplication.

1.14 (b) and (c) are even.

1.16 No. Give examples A, $B \in M_{2 \times 2}$ such that $\det(A + B) \neq \det(A) + \det(B)$, and $\det(xA) \neq x \det(A)$ for any non-zero scalar $x \in F$.

1.18 (a) 7   (b) 36

Chapter III, Section 2

2.1 (a) $\begin{bmatrix} 0 & 1 \\ 1 & 0 \end{bmatrix}\begin{bmatrix} 2 & 1 \\ 1 & 2 \end{bmatrix} = \begin{bmatrix} 1 & 0 \\ 2 & 1 \end{bmatrix}\begin{bmatrix} 1 & 2 \\ 0 & -3 \end{bmatrix} \cdot \det\begin{bmatrix} 2 & 1 \\ 1 & 2 \end{bmatrix}$

$$= (-1)(1)(-3) = 3$$

(d) $\begin{bmatrix} 1 & 0 & 0 \\ 0 & 0 & 1 \\ 0 & 1 & 0 \end{bmatrix}\begin{bmatrix} 1 & 0 & 1 \\ 2 & 0 & 1 \\ -1 & 1 & 2 \end{bmatrix} = \begin{bmatrix} 1 & 0 & 0 \\ -1 & 1 & 0 \\ 2 & 0 & 1 \end{bmatrix}\begin{bmatrix} 1 & 0 & 1 \\ 0 & 1 & 3 \\ 0 & 0 & -1 \end{bmatrix}$

Therefore, $\det\begin{bmatrix} 1 & 0 & 1 \\ 2 & 0 & 1 \\ -1 & 1 & 2 \end{bmatrix} = (-1)(1)(-1) = 1.$

2.4 Let $A = (a_{ij})$. Then $\det(xA) = \sum_{\sigma \in S_n} \text{sgn}(\sigma)(xa_{1\sigma(1)} \cdots xa_{n\sigma(n)}) = x^n(\sum_{\sigma \in S_n} \text{sgn}(\sigma)a_{1\sigma(1)} \cdots a_{n\sigma(n)}) = x^n \det(A)$.

2.9 Hint: First prove the identity is true when $\text{size}(A) = 1$. This will follow directly from the definition. Then proceed by induction.

2.17 $0 = \det\begin{bmatrix} x & y & 1 \\ a & b & 1 \\ c & d & 1 \end{bmatrix} = (b - d)x + (c - a)y + (ad - bc)$

Show this equation vanishes at (a, b) and (c, d).

Chapter III, Section 3

3.1 (a) 3

(b) $2\det\begin{bmatrix} 2 & 5 \\ -1 & -1 \end{bmatrix} - 1\det\begin{bmatrix} 2 & 5 \\ 0 & -1 \end{bmatrix} + 3\det\begin{bmatrix} 2 & 2 \\ 0 & -1 \end{bmatrix} = 2$

(e) $2\det\begin{bmatrix} 1 & 2 & 1 \\ 3 & 6 & 1 \\ 1 & 0 & 4 \end{bmatrix} - 1\det\begin{bmatrix} 0 & 2 & 1 \\ 3 & 6 & 1 \\ 2 & 0 & 4 \end{bmatrix} + \det\begin{bmatrix} 0 & 1 & 1 \\ 3 & 3 & 1 \\ 2 & 1 & 4 \end{bmatrix}$

$$-2 \det \begin{bmatrix} 0 & 1 & 2 \\ 3 & 3 & 6 \\ 2 & 1 & 0 \end{bmatrix} = -1$$

3.3 (a) $\text{adj} \begin{bmatrix} 1 & 0 & 0 \\ 2 & 1 & 0 \\ 1 & -1 & 3 \end{bmatrix} = \begin{bmatrix} 3 & 0 & 0 \\ -6 & 3 & 0 \\ -3 & 1 & 1 \end{bmatrix}$

(d) $\text{adj} \begin{bmatrix} 1 & 0 & -1 & 1 \\ 0 & 1 & 2 & 1 \\ -1 & 1 & 1 & 0 \\ 1 & 2 & 1 & -1 \end{bmatrix} = \begin{bmatrix} 0 & 2 & -6 & 2 \\ 4 & -2 & 6 & 2 \\ -4 & 4 & -4 & 0 \\ 4 & 2 & 2 & -2 \end{bmatrix}$

3.4 (a) $\begin{bmatrix} 1 & 0 & 0 \\ 2 & 1 & 0 \\ 1 & -1 & 3 \end{bmatrix}^{-1} = (1/3) \begin{bmatrix} 3 & 0 & 0 \\ -6 & 3 & 0 \\ -3 & 1 & 1 \end{bmatrix}$

(d) $\begin{bmatrix} 1 & 0 & -1 & 1 \\ 0 & 1 & 2 & 1 \\ -1 & 1 & 1 & 0 \\ 1 & 2 & 1 & -1 \end{bmatrix}^{-1} = (1/8) \begin{bmatrix} 0 & 2 & -6 & 2 \\ 4 & -2 & 6 & 2 \\ -4 & 4 & -4 & 0 \\ 4 & 2 & 2 & -2 \end{bmatrix}$

3.11 (a) $x = \det \begin{bmatrix} 6 & -2 \\ 5 & 3 \end{bmatrix} \bigg/ \det \begin{bmatrix} 1 & -2 \\ 2 & 3 \end{bmatrix} = 4$

$y = \det \begin{bmatrix} 1 & 6 \\ 2 & 5 \end{bmatrix} \bigg/ \det \begin{bmatrix} 1 & -2 \\ 2 & 3 \end{bmatrix} = -1$

(c) $x = 2$, $y = 0$, $z = 0$, $w = 1$

3.12 When n is odd, use Exercise 2.4 together with Theorem 2.20. When n is even, $\det(A)$ need not be zero. Give a concrete example.

3.13 $\begin{bmatrix} 0 & 1 \\ 1 & 0 \end{bmatrix}$

3.15 (a) All 2×2 minors vanish, rank $= 1$.
(b) All 3×3 minors vanish, rank $= 2$.

3.16 Hint: Any two matrix representations of T are similar, and similar matrices have the same determinant.

3.18 Hint: Rearrange the matrix units $\{E_{ij} | 1 \leqslant i, j \leqslant n\}$ of $M_{n \times n}$ such that the matrix representation of T consists of n diagonal blocks (size $= n \times n$) each being A. Then use Exercise 2.10.

3.20 Hint: $\eta = [1 \quad 1 \quad \cdots \quad 1]^t \in NS(A)$. Therefore $\mathrm{rk}(A) \leqslant n - 1$. We can assume $\mathrm{rk}(A) = n - 1$. Thus, $NS(A) = L(\eta)$. Show the columns of $\mathrm{adj}(A)$ lie in $NS(A)$. Repeat this argument for A^t.

Example:

$$\begin{bmatrix} 1 & 0 & -1 \\ 1 & 1 & -2 \\ -2 & -1 & 3 \end{bmatrix}$$

Chapter III, Section 4

4.1 (a) $\mathscr{S}_{\mathbb{R}}\left(\begin{bmatrix} -38 & 20 \\ -63 & 33 \end{bmatrix}\right) = \{-3, -2\}$

$NS\left(\begin{bmatrix} 35 & -20 \\ 63 & -36 \end{bmatrix}\right) = L([4/7 \quad 1]^t)$

$NS\left(\begin{bmatrix} 36 & -20 \\ 63 & -35 \end{bmatrix}\right) = L([5/9 \quad 1]^t)$

(c) $\mathscr{S}_{\mathbb{R}}\left(\begin{bmatrix} 11 & -2 & -6 \\ 8 & 0 & -5 \\ 17 & -4 & -9 \end{bmatrix}\right) = \{-1, 1, 2\}$

$NS\left(\begin{bmatrix} -12 & 2 & 6 \\ -8 & -1 & 5 \\ -17 & 4 & 8 \end{bmatrix}\right) = L([4/7 \quad 3/7 \quad 1]^t)$

$NS\left(\begin{bmatrix} -10 & 2 & 6 \\ -8 & 1 & 5 \\ -17 & 4 & 10 \end{bmatrix}\right) = L([2 \quad 1 \quad 3]^t)$

$NS\left(\begin{bmatrix} -9 & 2 & 6 \\ -8 & 2 & 5 \\ -17 & 4 & 11 \end{bmatrix}\right) = L([1 \quad 3/2 \quad 1]^t)$

4.3 $c_A(t) = t^n$, $\mathscr{S}_F(A) = \{0\}$.

4.6 $\mathscr{S}_{\mathbb{R}}(D) = \{0\}$. $\mathbb{R} - (0)$ are the eigenvectors of D.

4.9 I_n

4.10 See Exercise 3.

4.14 $A^{25} = \begin{bmatrix} 1 & -1 \\ -i & -i \end{bmatrix}\begin{bmatrix} i^{25} & 0 \\ 0 & (-i)^{25} \end{bmatrix}\begin{bmatrix} 1/2 & i/2 \\ -1/2 & i/2 \end{bmatrix} = \begin{bmatrix} 0 & -1 \\ 1 & 0 \end{bmatrix} = A$

4.15 (a) $X_k = \begin{bmatrix} 1 & 3 \\ 3 & 5 \end{bmatrix} \begin{bmatrix} (-1)^k & 0 \\ 0 & 3^k \end{bmatrix} \begin{bmatrix} -5/4 & 3/4 \\ 3/4 & -1/4 \end{bmatrix} \begin{bmatrix} 1 \\ 1 \end{bmatrix}$

Thus, $x_{1k} = (1/2)[(-1)^{k+1} + 3^{k+1}]$
$x_{2k} = (1/2)[3(-1)^{k+1} + 5(3^k)]$

4.16 (b) $\begin{bmatrix} 1/3 & -1/3 & -2/3 \\ 1/3 & 2/3 & 1/3 \\ 1/3 & -1/3 & 1/3 \end{bmatrix} \begin{bmatrix} 4 & -1 & 1 \\ -1 & 4 & -1 \\ 1 & -1 & 4 \end{bmatrix} \begin{bmatrix} 1 & 1 & 1 \\ 0 & 1 & -1 \\ -1 & 0 & 1 \end{bmatrix}$

$= \begin{bmatrix} 3 & 0 & 0 \\ 0 & 3 & 0 \\ 0 & 0 & 6 \end{bmatrix}$

$z_1 = -2/3$, $z_2 = 4/3$, $z_3 = 1/3$

$Y(x) = (-2/3)e^{3x} \begin{bmatrix} 1 \\ 0 \\ -1 \end{bmatrix} + (4/3)e^{3x} \begin{bmatrix} 1 \\ 1 \\ 0 \end{bmatrix} + (1/3)e^{6x} \begin{bmatrix} 1 \\ -1 \\ 1 \end{bmatrix}$

Therefore, $u(x) = (2/3)e^{3x} + (1/3)e^{6x}$
$v(x) = (4/3)e^{3x} - (1/3)e^{6x}$
$w(x) = (2/3)e^{3x} + (1/3)e^{6x}$

4.17 (e) Let $A = \begin{bmatrix} 1 & 0 \\ 0 & t \end{bmatrix}$

Then $\det(A) = t \neq 0$. Show A is not invertible.

Chapter IV, Section 1

1.4   Hint: Let B be an (ordered) basis of $\mathbb{R}^n$. Show $\langle \alpha, \beta \rangle = [\alpha]_B^t [\beta]_B$ determines an inner product on $\mathbb{R}^n$. Are these inner products on $\mathbb{R}^n$ all different?

1.10 Hint: Expand $\|\alpha + \beta\|^2 = \langle \alpha + \beta, \ \alpha + \beta \rangle$ and $\|\alpha - \beta\|^2 = \langle \alpha - \beta, \alpha - \beta \rangle$, using Equation (1.2).

1.12 (a) $\lambda_1 = [1 \ \ 1 \ \ 1 \ \ 1]^t$
$\lambda_2 = [3/4 \ \ -1/4 \ \ -5/4 \ \ 3/4]^t$
$\lambda_3 = [-48/44 \ \ 16/44 \ \ -8/44 \ \ 40/44]^t$

1.14 Apply the Gram–Schmidt process to the basis $1 + t$, 1, t^2
$\lambda_1 = 1 + t$
$\lambda_2 = 5 - 9t$
$\lambda_3 = (1/6) - t + t^2$

1.15   No. Consider $W_1 = L(\varepsilon_1, \varepsilon_2)$ and $W_2 = L(\varepsilon_2, \varepsilon_3)$ in $\mathbb{R}^3$.

1.17   Consider the same two subspaces as in Exercise 1.15.

Chapter IV, Section 2

2.2   No. $\sqrt{2} = |1 + i| \neq |1| + |i| = 2$.

2.6   No. Show $\langle [z_1 \ \ z_2]^t, [w_1 \ \ w_2]^t \rangle = z_1\bar{w}_1 + iz_2\bar{w}_2$ does not satisfy Lemma 2.3(d).

2.9   Any complex inner product $\langle \alpha, \beta \rangle$ on $\mathbb{C}$ has the form $\langle \alpha, \beta \rangle = c\alpha\bar{\beta}$ for some positive real number c.

2.11   Consider Example 2.5.

2.12   (a) Use a computer for this exercise.

$$\lambda_1 = [i \ \ 1 + i \ \ 1 \ \ 0]^t$$
$$\lambda_2 = [(5/4) - (3/2)i \ \ (-1/4) + (1/4)i \ \ (3/2) + (3/4)i \ \ i]^t$$
$$\lambda_3 = \begin{bmatrix} -0.4839 - 0.2581i \\ 0.9355 - 0.3226i \\ -0.3548 + 0.7742i \\ 0.2258 + 0.4839i \end{bmatrix}$$

(c) $\lambda_1 = [1 \ \ 0]^t$, $\lambda_2 = [-1/2 \ \ 1]^t$

2.14   Hint: Expand all norms $\|\gamma\|^2$ as $\langle \gamma, \gamma \rangle$.

2.16   Hint: Show the matrix

$$\begin{bmatrix} \langle \alpha_1, \alpha_1 \rangle & \cdots & \langle \alpha_n, \alpha_1 \rangle \\ \vdots & & \vdots \\ \langle \alpha_1, \alpha_n \rangle & \cdots & \langle \alpha_n, \alpha_n \rangle \end{bmatrix}$$

is invertible.

2.17   Hint: Let $W^{\perp} = \{\alpha \in V \mid \langle \alpha, \beta \rangle = 0 \text{ for all } \beta \in W\}$.

Chapter IV, Section 3

3.3   (a) $W = L(\varepsilon_1, \varepsilon_2)$. Therefore, $P_W([1 \ \ 1 \ \ 1 \ \ 1]^t) = [1 \ \ 1 \ \ 0 \ \ 0]^t$.

(c) $W = L(1 + t, (-1/4) - t/4 + t^2)$. Therefore,
$P_W(t^3) = (27/140)(1 + t) - (13/37)(-1/4 - t/4 + t^2)$.

3.5   (a) $\Gamma(\underline{\varepsilon}, \underline{\varepsilon})(P) = \begin{bmatrix} 1 & 0 & 0 & 0 \\ 0 & 1 & 0 & 0 \\ 0 & 0 & 0 & 0 \\ 0 & 0 & 0 & 0 \end{bmatrix}$

3.9   Let $\alpha = [1 \quad 0]^t$ and $\beta = [i \quad 0]^t$ in $\mathbb{C}^2$.

3.13 $(-1.4 + 0.6i)t + (0.4 - 0.6i)$

3.14   (a) Use a computer for this exercise:

$$X = \begin{bmatrix} 1.3571 \\ 1.4286 \\ 1.9286 \end{bmatrix}$$

3.15   We want the least squares solution to the following system of equations:

$$\begin{bmatrix} 4 & -2 & 1 \\ 1 & -1 & 1 \\ 0 & 0 & 1 \\ 1 & 1 & 1 \\ 4 & 2 & 1 \end{bmatrix} \begin{bmatrix} c \\ b \\ a \end{bmatrix} = \begin{bmatrix} 4 \\ 4 \\ 2 \\ 3 \\ 4 \end{bmatrix}$$

$c = 0.3571$, $b = -0.1$, $a = 2.6857$.

Chapter IV, Section 4

4.1   If A is unitary, then $-A$ is also unitary, but $A + (-A) = O$ is not unitary.

4.2   Hint: If A is invertible, then $A^{-1} = \text{adj}(A)/\det(A)$.

4.6   Hint: Use Theorem 4.1.

4.9   Hint: Again use Theorem 4.1 in comparing $\langle A\alpha, A\alpha \rangle$ with $\langle A^*\alpha, A^*\alpha \rangle$.

4.14   No. Find a concrete example using 2×2 symmetric matrices.

4.16   (b) $P = (1/\sqrt{2}) \begin{bmatrix} 1 & 1 \\ -1 & 1 \end{bmatrix}$

(e) $P = \begin{bmatrix} 0 & 1 & 0 & 0 \\ 1/\sqrt{2} & 0 & 1/\sqrt{2} & 0 \\ -1/\sqrt{2} & 0 & 1/\sqrt{2} & 0 \\ 0 & 0 & 0 & 1 \end{bmatrix}$

4.17   Hint: First show if B is Hermitian and $B^k(\alpha) = 0$, then $B\alpha = 0$. Then apply this argument to A^*A.

4.18 Hint: First show if $S \in \operatorname{Hom}_F(V, F)$, then there exists a unique vector $\gamma \in V$ such that $S(\alpha) = \langle \alpha, \gamma \rangle$ for all $\alpha \in V$. Make use of an orthonormal basis of V for this argument. Then apply this argument to $S(*) = \langle T(*), \beta \rangle$.

Chapter IV, Section 5

5.2 The substitution given in Equation (5.7) will eliminate the xy term in $q(x, y)$ provided θ is chosen as the acute angle for which $\cot 2\theta = (a - c)/b$.

5.7 (a) $A = \begin{bmatrix} -2 & 1 \\ 1 & -2 \end{bmatrix}$

Let $P = \begin{bmatrix} 1 & 1 \\ -1 & 1 \end{bmatrix} \begin{bmatrix} 1/\sqrt{6} & 0 \\ 0 & 1/\sqrt{2} \end{bmatrix}$

5.8 (a) $A = \begin{bmatrix} 4 & -1 & 1 \\ -1 & 4 & -1 \\ 1 & -1 & 4 \end{bmatrix}$.

Let $P = \begin{bmatrix} 1 & -1 & 1 \\ 1 & 1 & -1 \\ 0 & 2 & 1 \end{bmatrix} \begin{bmatrix} 1/\sqrt{6} & 0 & 0 \\ 0 & 1/\sqrt{18} & 0 \\ 0 & 0 & 1/\sqrt{18} \end{bmatrix}$

5.9 $\operatorname{sig} \begin{bmatrix} -2 & 1 \\ 1 & -2 \end{bmatrix} = -2,\ \operatorname{sig} \begin{bmatrix} 4 & -1 & 1 \\ -1 & 4 & -1 \\ 1 & -1 & 4 \end{bmatrix} = 3$

5.10 Hint: $A = \begin{bmatrix} 1 & 2 \\ 2 & 1 \end{bmatrix}$. If $P = \begin{bmatrix} 1 & 1 \\ -1 & 1 \end{bmatrix}$, then

$P^t A P = \begin{bmatrix} -2 & 0 \\ 0 & 6 \end{bmatrix}$

Index